NUMERICAL APPROXIMATION OF PARTIAL DIFFERENTIAL EQUATIONS

NORTH-HOLLAND MATHEMATICS STUDIES 133

NORTH-HOLLAND – AMSTERDAM • NEW YORK • OXFORD • TOKYO

NUMERICAL APPROXIMATION OF PARTIAL DIFFERENTIAL EQUATIONS

Selection of Papers Presented at the International Symposium on Numerical Analysis held at the Polytechnic University of Madrid, September 17-19, 1985

Edited by

Eduardo L. ORTIZ
Department of Mathematics
Imperial College of Science & Technology
London, United Kingdom

1987

NORTH-HOLLAND – AMSTERDAM • NEW YORK • OXFORD • TOKYO

ISBN: 0 444 70140 0

Publishers:
ELSEVIER SCIENCE PUBLISHERS B.V.
P.O. BOX 1991
1000 BZ AMSTERDAM
THE NETHERLANDS

Sole distributors for the U.S.A and Canada:
ELSEVIER SCIENCE PUBLISHING COMPANY, INC.
52 VANDERBILT AVENUE
NEW YORK, N.Y. 10017
U.S.A.

Library of Congress Cataloging-in-Publication Data

International Symposium on Numerical Analysis (1985 :
Polytechnic University of Madrid)
Numerical approximation of partial differential
equations.

(North-Holland mathematics studies ; 133)
1. Differential equations--Numerical solutions--
Congresses. I. Ortiz, Eduardo L. II. Title.
III. Series.
QA370.I576 1985 515.3'5 86-24116
ISBN 0-444-70140-0 (U.S.)

PRINTED IN THE NETHERLANDS

PREFACE

The International Symposium of Numerical Analysis

This volume contains a selection of papers on problems arising in the numerical solution of differential equations. They were read at or contributed to ISNA, the International Symposium of Numerical Analysis, held in Madrid on September 17-19, 1985. Papers dealing with other topics and presented at that meeting will be published in a separate collection.

ISNA was the initiative of computer scientists Rafael Portaencasa, Rector of the Polytechnic University of Madrid and Carlos Vega, Vice-Dean of the Faculty of Informatics at the same university. Over a period of years this dynamic university has run a fruitful cooperation research agreement in the field of Numerical Analysis with Charles University, Prague. It seemed then natural that ISNA should be developed as a joint effort of the universities of Madrid and Prague.

After some preliminary discussions in 1983, a Scientific Committee met in Madrid at the beginning of 1984 under the Chairmanship of Rector Portaencasa, with Professor Vega acting as Secretary. His members were Professors Wolfgang Hackbusch, University of Kiel; Oliver Pironneau, INRIA, Paris, and a distinguished group of Spanish academics: Professors Enrique Alarcon, Alfredo Bermudez de Castro, Carlos Conde Lozano, Jose Manuel Corrales, Covadonga Fernandez Baizan, Luis Ferragut Canals, Manuel Lopez Quero, Francisco Michavila Pitarch, Jose Luis Morant Ramon, Pilar Perez Alonso, Libia Perez Jimenez, Arturo Ribagorda Garnacho and Antonio Valle Sanchez. Professors Ivo Marek, from Czechoslovakia, Yuri Kuznetsov, from the USSR and John Whiteman, from England, were members of this Committee, but could not be present at the meeting. I also had the privilege of being invited to participate.

The organizers of that meeting suggested the list of invited speakers, the topics to be emphasized and possible dates. A second and equally live meeting took place in Prague, where we were the guests of Professors Zdenek Ceska, Rector of Charles University and numerical analyst Professor Ivo Marek, Fellow of the Czechoslovak Academy of Sciences. A subgroup of that Committee discussed there the final details, and a call for papers for the first ISNA meeting was sent out.

The symposium took place in September, a most pleasant time of the year to visit Madrid. Our deliberations were held at the Campus of the University of Madrid, where the Polytechnic University has its headquarters. It attracted over 120 active participants from more than 20 different countries. Both the quality and quantity of papers contributed to the symposium exceeded the most optimistic expectations of the Scientific Committee.

At the end of the Madrid Symposium it was agreed that ISNA II will take place in Czechoslovakia in the Summer of 1987.

Papers submitted covered a broad spectrum of our discipline: Linear Algebra; Numerical Methods of Approximation Theory; Computational Statistics; Analysis and Complexity of Algorithms; Numerical Methods for Differential Equations; Optimization; Special Problems of Science and Engineering; Inverse and Ill Posed Problems and Topics on the Teaching of Numerical Mathematics.

We decided against a large volume of proceedings, which would have been over one thousand pages in length and agreed to subdivide the papers accepted into sections. I was given the task of editing and introducing those in the field of differential equations and related techniques, which constitute the present volume.

The first Part contains papers concerned with some of the techniques of Approximation Theory which are basic to the numerical treatment of Differential Equations. This last topic is specifically considered in Parts II-V. In the first of them, numerical techniques based on discrete process such as Finite Differences, Finite Elements and the Method of Lines are considered. Methods based on polynomial or rational approximation, such as the Tau Method, Collocation, Pade and Spectral Techniques are discussed in Part III. The following section is devoted to Variational Inequalities, Conformal Transformation and asymptotic technqiues. Finally, Part V contains a number of papers dealing with concrete applications of differential equations to problems of Science and Engineering, where a variety of techniques are used in an innovative way to produce desired numerical results.

This meeting would not have been possible without the help and assistance of several leading scientific institutions and Government departments of Spain and Czechoslovakia. The former provided finance and a number of facilities; the latter made possible the participation of a large number of scientists from Czechoslovakia. Professor W. Hackbusch, who could not attend the meeting due to other commitments, gave valuable advice to our Committee in the preliminary meetings of Madrid and Prague. To all of them we wish to transmit the gratitude of the Scientific Committee.

During the days we spent in Madrid Professors Portaencasa, Vega and their Spanish colleagues did everything possible to make our stay pleasant. The program, concentrated into three days, was tight but broken by a delightful dinner at a medieval castle, just outside Madrid. The organizers of ISNA deserve our warm thanks for their fine achievement.

Finally, I would like to express the appreciation of the Scientific Committee of ISNA to its Patron, Juan Carlos I, King of Spain, for his firm support and consistent encouragement of scientific research.

Eduardo L. Ortiz
Imperial College
London, 1986

INTERNATIONAL SYMPOSIUM OF NUMERICAL ANALYSIS

under the Patronage of

His Majesty Juan Carlos I, King of Spain

and with the support of:

The Ministry of Education; Ministry of Culture; the Royal Academies of Arts and of Sciences; the National Institute for Scientific Research; the Secretary of State for Universities and for Scientific Research; the Boards of Scientific Policy, of University Education and of Technical and Scientific Cooperation, Spain and the Charles University, Prague, Czechoslovakia.

Chairmen

Rafael PORTAENCASA, Rector of the Polytechnic University of Madrid

and

Zdenek CESKA, Rector of Charles University, Prague

Advisors

Jacques-Louis LIONS, College de France

and

Yuri MARCHUK, Academy of Sciences of the USSR

Executive Sectetaries

Carlos VEGA, Polytechnic University, Madrid

and

Ivo MAREK, Charles University, Prague

Organized by:

POLYTECHNIC UNIVERSITY OF MADRID

CHARLES UNIVERSITY OF PRAGUE

CONTENTS

INVITED PAPERS

CONTRIBUTED PAPERS

PART I: RESULTS ON COMPUTATIONAL LINEAR ALGEBRA

Interpolation and Related Techniques

Computational Linear Algebra

PART II: DISCRETE VARIABLE METHODS

Finite Difference Methods, Finite Element Methods and Related Techniques

INVITED PAPERS

Numerical Approximation of Partial Differential Equations
E.L. Ortiz (Editor)

RECENT PROGRESS IN THE TWO-DIMENSIONAL APPROXIMATION OF THREE-DIMENSIONAL PLATE MODELS IN NONLINEAR ELASTICITY

Philippe G. CIARLET

Université Pierre et Marie Curie, and
Ecole Normale Supérieure, Paris, France.

The asymptotic expansion method, with the thickness as the parameter, is applied to the equilibrium and constitutive equations of nonlinear three-dimensional elasticity. Then the leading term of the expansion can be identified with the solution of well-known two dimensional nonlinear plate models, such as the von Kármán equations.

Recent progresses in the application of this method, such as the extension to more general constitutive equations and boundary conditions, the effect of the assumption of polyconvexity, the application to one-dimensional rod models, etc ..., are presented. Various open problems, regarding in particular the existence of corresponding three-dimensional solutions and the nature of admissible three-dimensional boundary conditions, are also discussed.

1. THE THREE-DIMENSIONAL CLAMPED PLATE MODEL. Latin indices : $i,j,p,\dots$, take their values in the set $\{1,2,3\}$; Greek indices : $\alpha,\beta,\mu,\dots$, take their values in the set $\{1,2\}$. The repeated index convention is systematically used in conjunction with the above rule. Let (e_i) be an orthonormal basis in $\mathbb{R}^3$, and let ω be a bounded open subset of the "horizontal" plane, spanned by (e_α), with a sufficiently smooth boundary γ. Given $\varepsilon>0$, we let

$$\Omega^\varepsilon = \omega \times]-\varepsilon,\varepsilon[\,,\ \Gamma_0^\varepsilon = \gamma \times [-\varepsilon,\varepsilon]\,,\ \Gamma_\pm^\varepsilon = \omega \times \{\pm\varepsilon\}.$$

Because the *thickness* 2ε is thought as being "small" compared to the dimensions of the set ω, the set $\overline{\Omega}^\varepsilon$ is called a *plate*, with *lateral surface* Γ_0^ε and *upper and lower faces* Γ_+^ε and Γ_-^ε. We are concerned with the problem of finding the *displacement vector field* $u^\varepsilon = (u_i^\varepsilon) : \overline{\Omega}^\varepsilon \to \mathbb{R}^3$ and the *second Piola-Kirchhoff stress tensor field* $\sigma^\varepsilon = (\sigma_{ij}^\varepsilon) : \overline{\Omega}^\varepsilon \to \mathbf{S}^3$ (we let $\mathbf{S}^3$ denote the space of symmetric matrices of order 3) of a three-dimensional body which occupies the set $\overline{\Omega^\varepsilon}$ in the absence of applied forces. The plate is subjected to *body forces*, of density $f^\varepsilon = (f_i^\varepsilon) : \overline{\Omega^\varepsilon} \to \mathbb{R}^3$, and to *surface forces*, of density $g^\varepsilon = (g_i^\varepsilon) : \Gamma_+^\varepsilon \cup \Gamma_-^\varepsilon \to \mathbb{R}^3$, on the upper and lower faces. For simplicity, we shall assume that both kind of applied forces are *dead loads*, i.e., both densities are independent of the displacement, and that the *horizontal components of the applied forces vanish*, i.e., $f_\alpha^\varepsilon = 0$ and $g_\alpha^\varepsilon = 0$. Finally, we assume that the plate is *clamped*, in the sense that it is subjected to the *boundary condition of place* $u_i^\varepsilon = 0$ on the lateral surface Γ_o^ε. Then the unknowns and the data are related by the following *equations of finite elastostatics*, which express the elastic equilibrium of the plate (we let $\partial_j^\varepsilon = \partial/\partial x_j^\varepsilon$, where $x^\varepsilon = (x_j^\varepsilon)$ denotes the generic point of the set $\overline{\Omega^\varepsilon}$) :

$$(1.1) \qquad -\partial_j^\varepsilon(\sigma_{ij}^\varepsilon + \sigma_{kj}^\varepsilon \partial_k^\varepsilon u_i^\varepsilon) = f_i^\varepsilon \text{ in } \Omega^\varepsilon,$$

$$(1.2) \qquad \sigma_{i3}^\varepsilon + \sigma_{k3}^\varepsilon \partial_k^\varepsilon u_i^\varepsilon = \pm g_i^\varepsilon \text{ on } \Gamma_\pm^\varepsilon,$$

$$(1.3) \qquad u_i^\varepsilon = 0 \text{ on } \Gamma_o^\varepsilon.$$

We assume that the material constituting the plate is *elastic*, *homogeneous*, and *isotropic*. Then its *constitutive equation* takes the form :

$$(1.4) \quad \sigma^\varepsilon = \hat{\sigma}^\varepsilon(I+\nabla u^\varepsilon) = \gamma_o^\varepsilon(C^\varepsilon)I + \gamma_1^\varepsilon(C^\varepsilon)C^\varepsilon + \gamma_2^\varepsilon(C^\varepsilon)(C^\varepsilon)^2, \text{ for } I + \nabla u^\varepsilon \in \mathbb{M}_+^3,$$

where $\mathbb{M}_+^3$ denotes the set of matrices F or order 3 with det $F > 0$,

$$(1.5) \qquad C^\varepsilon = (I+\nabla u^\varepsilon)^T(I+\nabla u^\varepsilon) = I + 2E^\varepsilon$$

denotes the *right Cauchy-Green deformation tensor*, and $\gamma_o^\varepsilon, \gamma_1^\varepsilon, \gamma_2^\varepsilon$ are real-valued functions of the three principal invariants of the tensor C^ε. If we assume that the set $\overline{\Omega^\varepsilon}$ is a *natural state*, i.e., that $\sigma^\varepsilon = 0$ if $u^\varepsilon = 0$ (note that $u^\varepsilon = 0$ implies $C^\varepsilon = I$), the Taylor expansion of (1.4) in terms of the *Green-St Venant strain tensor* E^ε defined in (1.5) takes the form

$$(1.6) \qquad \sigma^\varepsilon = \lambda^\varepsilon(\text{tr } E^\varepsilon)I + 2\mu^\varepsilon E^\varepsilon + o(|E^\varepsilon|),$$

where $\lambda^\varepsilon > 0$ and $\mu^\varepsilon > 0$ are the *Lamé constants* of the material constituting the plate.

We may assume that the material is *hyperelastic,* in which case there exists a *stored energy function* $\hat{W}^\varepsilon : \mathbb{M}^\varepsilon_+ \to \mathbb{R}$ such that

$$F\,\hat{\sigma}^\varepsilon(F) = \frac{\partial \hat{W}^\varepsilon}{\partial F}(F) \text{ for all } F \in \mathbb{M}^3_+. \tag{1.7}$$

Then finding solutions of (1.1)-(1.6) formally amounts to finding the stationary points of the *total energy*

$$J^\varepsilon(v^\varepsilon) = \int_{\Omega^\varepsilon} \hat{W}^\varepsilon(I+\nabla v^\varepsilon)\,dx^\varepsilon - \int_{\Omega^\varepsilon} f_3^\varepsilon v_3^\varepsilon\,dx^\varepsilon - \int_{\Gamma_+^\varepsilon \cup \Gamma_-^\varepsilon} g_3^\varepsilon v_3^\varepsilon\,da^\varepsilon, \tag{1.8}$$

when v^ε span a space of $\mathbb{R}^3$-valued functions satisfying the boundary condition $v^\varepsilon = 0$ on Γ_0^ε and the orientation - preserving condition $\det(I+\nabla v^\varepsilon) > 0$ in $\overline{\Omega^\varepsilon}$ (for details about the above model, see e.g. Ciarlet, 1985 ; Ciarlet, 1986 ; Germain, 1972 ; Gurtin, 1981 ; Hanyga, 1985 ; Marsden & Hughes, 1983 ; Truesdell & Noll, 1965 ; Wang & Truesdell, 1973).

The following existence result holds ($\mathbb{M}^3$ denotes the space of matrices of order 3 ; Cof $F \in \mathbb{M}^3$ is the cofactor matrix of $F \in \mathbb{M}^3$) :

THEOREM 1 (Ball, 1977). *Assume that the material is hyperelastic and that its stored energy function satisfies the following assumptions : There exists a convex function* $\hat{\mathbb{W}}^\varepsilon : \mathbb{M}^3 \times \mathbb{M}^3 \times]0,+\infty[$ *such that (polyconvexity)*

$$\hat{W}^\varepsilon(F) = \hat{\mathbb{W}}^\varepsilon(F, \mathrm{Cof}\,F, \det F) \textit{ for all } F \in \mathbb{M}^3_+ ; \tag{1.9}$$

$$\hat{W}^\varepsilon(F) \to +\infty \textit{ as } \det F \to 0^+; \tag{1.10}$$

there exist numbers $\alpha^\varepsilon > 0$, $\beta^\varepsilon \in \mathbb{R}$, $p^\varepsilon \geq 2$, $q^\varepsilon \geq p^\varepsilon/(p^\varepsilon - 1)$, $r^\varepsilon > 1$ *such that*

$$\hat{W}^\varepsilon(F) \geq \alpha^\varepsilon\{|F|^{p^\varepsilon} + |\mathrm{Cof}\,F|^{q^\varepsilon} + (\det F)^{r^\varepsilon}\} + \beta^\varepsilon \textit{ for all } F \in \mathbb{M}^3_+ ; \tag{1.11}$$

the linear form

$$v^\varepsilon \in W^{1,p}(\Omega^\varepsilon; \mathbb{R}^3) \to \int_{\Omega^\varepsilon} f_3^\varepsilon v_3^\varepsilon\,dx^\varepsilon + \int_{\Gamma_+^\varepsilon \cup \Gamma_-^\varepsilon} g_3^\varepsilon v_3^\varepsilon\,da^\varepsilon \tag{1.12}$$

is continuous. Then there exists at least one displacement u^ε *satisfying*

(1.13) $$J^\varepsilon(u^\varepsilon) = \inf\{J^\varepsilon(v^\varepsilon)\ ;\ v^\varepsilon \in U^\varepsilon\}$$

where the functional J^ε *is defined in* (1.8) *and the set* U^ε *of admissible displacements is given by*

(1.14) $$U^\varepsilon = \{v^\varepsilon \in W^{1,p}(\Omega^\varepsilon;\mathbb{R}^3)\ ;\ \mathrm{Cof}(I+\nabla v^\varepsilon) \in L^q(\Omega^\varepsilon;\mathbb{M}^3),\ \det(I+\nabla v^\varepsilon) \in L^r(\Omega^\varepsilon),$$
$$\det(I+\nabla v^\varepsilon) > 0 \text{ a.e. } in\ \Omega^\varepsilon,\ v^\varepsilon = 0\ on\ \Gamma_0^\varepsilon\}. \blacksquare$$

It is always possible (Ciarlet & Geymonat, 1982) to construct a class of simple stored energy functions satisfying assumptions (1.9)-(1.11), and whose expansion in terms of the Green-St Venant strain tensor E^ε agrees with the expansion (1.6) of an elastic material with arbitrary Lamé constants $\lambda^\varepsilon > 0$ and $\mu^\varepsilon > 0$. Unilateral boundary conditions (Ciarlet & Nečas, 1985a) and conditions guaranteeing almost everywhere injectivity (Ciarlet & Nečas, 1985b) can also be included in the definition of the set U^ε of (1.14).

2. APPLICATION OF THE ASYMPTOTIC EXPANSION METHOD.

For ease of exposition, we assume that the constitutive equation takes the simplified form

(2.1) $$\sigma^\varepsilon = \lambda^\varepsilon(\mathrm{tr}\ E^\varepsilon)I + 2\mu^\varepsilon E^\varepsilon.$$

Although the associated stored energy function is not polyconvex (Raoult, 1985b), we mention that the present approach can be extended, at the expense of various refined arguments however, to stored energy functions satisfying all the assumptions of Theorem 1 (cf. Sect. 4). A *variational formulation* of equations (1.1)-(1.3) and (2.1) then consists in expressing that the pair $(u^\varepsilon,\sigma^\varepsilon)$ satisfies

(2.2) $$(u^\varepsilon,\sigma^\varepsilon) \in V^\varepsilon \times \Sigma^\varepsilon,$$
where $V^\varepsilon = \{v^\varepsilon \in W^{1,4}(\Omega^\varepsilon;\mathbb{R}^3)\ ;\ v^\varepsilon = 0 \text{ on } \Gamma_0^\varepsilon\}$, $\Sigma^\varepsilon = L^2(\Omega^\varepsilon;\mathbf{S}^3)$,

(2.3) $$\int_{\Omega^\varepsilon} \left\{\left(\frac{1+\nu^\varepsilon}{E^\varepsilon}\right)\sigma_{ij}^\varepsilon - \frac{\nu^\varepsilon}{E^\varepsilon}\sigma_{pp}^\varepsilon\delta_{ij}\right\}\tau_{ij}^\varepsilon dx^\varepsilon - \int_{\Omega^\varepsilon}\tau_{ij}^\varepsilon E_{ij}^\varepsilon dx^\varepsilon = 0$$
$$\text{for all } \tau^\varepsilon = (\tau_{ij}^\varepsilon) \in \Sigma^\varepsilon,$$

(2.4) $$\int_{\Omega^\varepsilon}(\sigma^\varepsilon_{ij}+\sigma^\varepsilon_{kj}\partial^\varepsilon_k u^\varepsilon_i)\partial^\varepsilon_j v^\varepsilon_i dx^\varepsilon =$$
$$= \int_{\Omega^\varepsilon} f^\varepsilon_3 v^\varepsilon_3 dx^\varepsilon + \int_{\Gamma^\varepsilon_+ \cup \Gamma^\varepsilon_-} g^\varepsilon_3 v^\varepsilon_3 da^\varepsilon \text{ for all } v^\varepsilon = (v^\varepsilon_i) \in V^\varepsilon ,$$

where $E^\varepsilon = \dfrac{\mu^\varepsilon(3\lambda^\varepsilon+2\mu^\varepsilon)}{\lambda^\varepsilon+\mu^\varepsilon}$ and $\nu^\varepsilon = \dfrac{\lambda^\varepsilon}{2(\lambda^\varepsilon+\mu^\varepsilon)}$ are respectively the *Young modulus* and the *Poisson ratio* of the material.

Our first task consists in defining a problem equivalent to the variational problem (2.2)-(2.4), but now posed over a domain which *does not* depend on ε. We let

$$\Omega = \omega \times]-1,1[,\ \Gamma_0 = \gamma \times [-1,1],\ \Gamma_\pm = \omega \times \{\pm 1\}.$$

With each point $x^\varepsilon = (x^\varepsilon_i) \in \overline{\Omega}^\varepsilon$, we associate the point $x = (x_i) \in \overline{\Omega}$ by letting $x_\alpha = x^\varepsilon_\alpha$, $x_3 = x^\varepsilon_3/\varepsilon$, and with the spaces V^ε, Σ^ε of (2.2), we associate the spaces

(2.5) $$V = \{v \in W^{1,4}(\Omega;\mathbf{R}^3)\ ;\ v = 0 \text{ on } \Gamma_0\},\ \Sigma = L^2(\Omega;\mathbf{S}^3).$$

With the *unknown functions* $(u^\varepsilon,\sigma^\varepsilon) \in V^\varepsilon \times \Sigma^\varepsilon$, we associate functions $(u(\varepsilon),\sigma(\varepsilon)) \in V \times \Sigma$ defined by

(2.6) $$u^\varepsilon_\alpha = \varepsilon^2 u_\alpha(\varepsilon),\ u^\varepsilon_3 = \varepsilon u_3(\varepsilon),$$

(2.7) $$\sigma^\varepsilon_{\alpha\beta} = \varepsilon^2\sigma_{\alpha\beta}(\varepsilon),\ \sigma^\varepsilon_{\alpha 3} = \varepsilon^3\sigma_{\alpha 3}(\varepsilon),\ \sigma^\varepsilon_{33} = \varepsilon^4\sigma_{33}(\varepsilon),$$

and with the "*trial functions*" $(v^\varepsilon,\tau^\varepsilon) \in V^\varepsilon \times \Sigma^\varepsilon$ appearing in (2.3)-(2.4), we associate functions $(v,\tau) \in V \times \Sigma$ defined by

(2.8) $$v^\varepsilon_\alpha = \varepsilon^2 v_\alpha,\ v^\varepsilon_3 = \varepsilon v_3,$$

(2.9) $$\tau^\varepsilon_{\alpha\beta} = \varepsilon^2\tau_{\alpha\beta},\ \tau^\varepsilon_{\alpha 3} = \varepsilon^3\tau_{\alpha 3},\ \tau^\varepsilon_{33} = \varepsilon^4\tau_{33},$$

where $u^\varepsilon_\alpha = \varepsilon^2 u_\alpha(\varepsilon)$ means that $u^\varepsilon_\alpha(x^\varepsilon) = \varepsilon^2 u_\alpha(\varepsilon)(x)$ for all corresponding points $x^\varepsilon \in \overline{\Omega}^\varepsilon$ and $x \in \overline{\Omega}$, etc.... As regards the data, *we assume that there exist functions* $f_3 : \Omega \to \mathbf{R}$, $g_3 : \Gamma_+ \cup \Gamma_- \to \mathbf{R}$, *and constants* λ,μ, *independent of* ε, *such that*

(2.10) $$f^\varepsilon_3 = \varepsilon^3 f_3\ ,\ g^\varepsilon_3 = \varepsilon^4 g_3,\ \lambda^\varepsilon = \lambda,\ \mu^\varepsilon = \mu$$

(as we shall indicate in Sect. 3, these are not the only possible assumptions). The following result shows that the three-dimensional problem (2.2)-(2.4) posed over $\overline{\Omega}^{\varepsilon}$ is equivalent to a problem posed over the fixed domain $\overline{\Omega}$, which displays a very simple, more precisely, polynomial, by virtue of the simplified constitutive equation (2.1), dependence on ε.

THEOREM 2. *The pair* $(u(\varepsilon),\sigma(\varepsilon)) \in V \times \Sigma$ *satisfies :*

(2.11) $$A_0(\sigma(\varepsilon),\tau) + \varepsilon^2 A_2(\sigma(\varepsilon),\tau) + \varepsilon^4 A_4(\sigma(\varepsilon),\tau) + B(\tau,u(\varepsilon)) + C_0(\tau,u(\varepsilon),u(\varepsilon)) + \varepsilon^2 C_2(\tau,u(\varepsilon),u(\varepsilon)) = 0 \text{ for all } \tau \in \Sigma,$$

(2.12) $$B(\sigma(\varepsilon),v) + 2C_0(\sigma(\varepsilon),u(\varepsilon),v) + 2\varepsilon^2 C_2(\sigma(\varepsilon),u(\varepsilon),v) = F(v) \text{ for all } v \in V,$$

where the linear form F, *the bilinear forms* A_0, A_2, A_4, B, *and the trilinear forms* C_0, C_2 *are independent of* ε. *In particular*,

(2.13) $$A_0(\sigma,\tau) = \int_\Omega \left\{ \left(\frac{1+\nu}{E}\right) \sigma_{\alpha\beta} - \frac{\nu}{E} \sigma_{\mu\mu} \delta_{\alpha\beta} \right\} \tau_{\alpha\beta} dx, \quad B(\tau,v) = \int_\Omega \tau_{ij} \partial_j v_i dx,$$

(2.14) $$C_0(\sigma,u,v) = -\frac{1}{2} \int_\Omega \sigma_{ij} \partial_i u_3 \partial_j v_3 dx, \quad F(v) = -\int_\Omega f_3 v_3 dx - \int_{\Gamma_+ \cup \Gamma_-} g_3 v_3 da. \qquad \blacksquare$$

In view of this result, we are naturally led to formally define an *asymptotic expansion of* $(u(\varepsilon),\sigma(\varepsilon))$, *as*

(2.15) $$u(\varepsilon) = u^0 + \varepsilon^2 u^2 + \ldots, \quad \sigma(\varepsilon) = \sigma^0 + \varepsilon^2 \sigma^2 + \ldots .$$

Then, following the principle of the *asymptotic expansion method* (Lions, 1973), we equate to zero the factors of the successive powers ε^p, $p \geq 0$, found in (2.11)-(2.12) when $u(\varepsilon)$, $\sigma(\varepsilon)$ are replaced by their expansions (2.15). In this fashion, we find in particular that *the first term* (u^0,σ^0) *should be solution of :*

(2.16) $$A_0(\sigma^0,\tau) + B(\tau,u^0) + C_0(\tau,u^0,u^0) = 0 \text{ for all } \tau \in \Sigma,$$

(2.17) $$B(\sigma^0,v) + 2C_0(\sigma^0,u^0,v) = F(v) \text{ for all } v \in V.$$

We shall not consider here the computation of further terms, such as (σ^2,u^2) (see Destuynder 1980, 1981, in the linear case).

3. THE TWO-DIMENSIONAL CLAMPED PLATE MODEL. Our main results consist in establishing the *existence* of a least one solution to the "*limit problem*" (2.16)-(2.17), and in recognizing in (2.16)-(2.17) a *known two-dimensional plate model.*

THEOREM 3. (Ciarlet & Destuynder, 1979b). *Assume that* $f_3 \in L^2(\Omega)$, $g_3 \in L^2(\Gamma_+ \cup \Gamma_-)$. *Equations* (2.16)-(2.17) *have at least one solution* $(u^o, \sigma^o) = ((u_i^o), (\sigma_{ij}^o)) \in V \times \Sigma$, *which is obtained by first solving the two-dimensional problem : Find*

$$(3.1) \qquad (\zeta_\alpha) \in (H_0^1(\omega))^2 \text{ and } \zeta_3 \in H_0^2(\omega)$$

such that

$$(3.2) \qquad \frac{2E}{3(1-\nu^2)} \Delta^2 \zeta_3 - n_{\alpha\beta} \partial_{\alpha\beta} \zeta_3 = \int_{-1}^{1} f_3 dx_3 + \overset{+}{g}_3 + \overset{-}{g}_3 \text{ in } \omega,$$

$$(3.3) \qquad \partial_\alpha n_{\alpha\beta} = 0 \text{ in } \omega,$$

where $\overset{\pm}{g}_3(x_1,x_2) = g_3(x_1,x_2,\pm 1)$ *for* $(x_1,x_2) \in \omega$, *and*

$$(3.4) \qquad n_{\alpha\beta} = \frac{2E}{(1-\nu^2)} \left\{ (1-\nu) \frac{\partial_\alpha \zeta_\beta + \partial_\beta \zeta_\alpha}{2} + \nu\, \partial_\mu \zeta_\mu \delta_{\alpha\beta} \right\} + \frac{E}{(1-\nu^2)} \left\{ (1-\nu) \partial_\alpha \zeta_3 \partial_\beta \zeta_3 + \nu \partial_\mu \zeta_3 \partial_\mu \zeta_3 \delta_{\alpha\beta} \right\},$$

and secondly, by computing

$$(3.5) \qquad u_\alpha^o = \zeta_\alpha - x_3 \partial_\alpha \zeta_3, \quad u_3^o = \zeta_3,$$

$$(3.6) \qquad \sigma_{\alpha\beta}^o = \frac{1}{2} n_{\alpha\beta} - \frac{E}{(1-\nu^2)} x_3 \left\{ (1-\nu) \partial_{\alpha\beta} \zeta_3 + \nu \Delta \zeta_3 \delta_{\alpha\beta} \right\},$$

$$(3.7) \qquad \sigma_{\alpha 3}^o = -\frac{E}{2(1-\nu^2)} (1-x_3^2) \partial_\alpha \Delta \zeta_3 ,$$

$$(3.8) \qquad \sigma_{33}^o = \frac{1}{2}(1+x_3) \int_{-1}^{1} f_3 dx_3 - \int_{-1}^{x_3} f_3 dx_3 + \frac{1}{2}(\overset{+}{g}_3 - \overset{-}{g}_3) + \frac{1}{2} x_3 (\overset{+}{g}_3 + \overset{-}{g}_3) - \frac{E}{2(1-\nu^2)} (1-x_3^2) \{ (1-\nu) \partial_{\alpha\beta} \zeta_3 \partial_{\alpha\beta} \zeta_3 + \nu \partial_{\mu\mu} \zeta_3 \partial_{\mu\mu} \zeta_3 \} + \frac{E}{6(1-\nu^2)} x_3 (1-x_3^2) \Delta^2 \zeta_3 .$$

Conversely, and sufficiently smooth solution of (2.16)-(2.17) *is necessarily of the form* (3.5)-(3.8), *where* (ζ_i) *is solution of problem* (3.1)-(3.4). ■

In order to get more meaningful formulas, it would remain to go back to the set $\overline{\Omega}^\varepsilon$, simply by inverting correspondences (2.6)-(2.7). For lack of space, we shall not give the equations over $\overline{\Omega}^\varepsilon$ which correspond to equations (3.2)-(3.8) over $\overline{\Omega}$; it suffices to mention that the effect of formulas (2.6)-(2.7) is that appropriate powers of ε appear in equations (3.2)-(3.8), which are otherwise unaltered. The most important conclusion is that, up to this transformation, *the equations found in Theorem* 3 *coincide with the expressions found in the literature on nonlinear plate theory*, where they are usually derived from *a priori* assumptions of a geometrical or a mechanical nature, regarding notably the nature of the variation of the unknowns across the thickness of the plate. Without any such *a priori* assumption, we have obtained here in particular the standard *two-dimensional nonlinear clamped model* (3.1)-(3.4), together with a *displacement field of the Kirchhoff-Love type* (cf. (3.5)) and stress components which vary across the thickness of the plate as *linear, quadratic,* or *cubic* functions of the variable x_3 (cf. (3.6)-(3.8)), according to their nature.

Another important observation is that, instead of assumptions (2.10), *we could have as well assumed that*

$$f_3^\varepsilon = \varepsilon^{3+t} f_3,\ g_3^\varepsilon = \varepsilon^{4+t} g_3,\ \lambda^\varepsilon = \varepsilon^t \lambda,\ \mu^\varepsilon = \varepsilon^t \mu, \tag{3.9}$$

for any real number t. It is immediately realized, simply by inspecting equations (1.1)-(1.3) and (2.1), that equations (2.11)-(2.12) are left unaltered if we replace (2.10) by the more general assumptions (3.9). This observation then leads us to the following *definition :* The family of three-dimensional problems (1.1)-(1.3) and (2.1) is *asymptotically equivalent to the two-dimensional plate model* (3.1)-(3.4) if, when the unknown functions $(u^\varepsilon,\sigma^\varepsilon)$ are subjected to the changes (2.6)-(2.7), the leading terms (u^0,σ^0) in the formal expansions (2.15) is such that the functions u_i^0 are necessarily of the form (3.5) and they necessarily satisfy equations (3.1)-(3.4). It then follows from Theorems 2 and 3 that *assumptions* (3.9) *guarantee such an asymptotic equivalence.*

In essence, these say that *a two-dimensional plate model cannot be recognized as the leading term of a formal expansion of a three-dimensional solution, unless the data behave in an appropriate manner as* $\varepsilon \to 0$. For instance, a family of three-dimensional clamped plates made up of the *same* elastic material (which corresponds to $t = 0$ in (3.9), as in (2.10)) is asymptotically equivalent to (3.1)-(3.4) *only if* the applied force densities verify $f_3^\varepsilon = O(\varepsilon^3)$ and $g_3^\varepsilon = O(\varepsilon^4)$, which thus rules out the gravity (which corresponds to $f_3^\varepsilon = O(1)$) as a body force. In order to take gravity into account and still get the same two-dimensional plate model, it suffices to let $t = -3$ in (3.9), which means that $\lambda^\varepsilon = \varepsilon^{-3}\lambda$, $\mu^\varepsilon = \varepsilon^{-3}\mu$. In other words, the *rigidity* of the material constituting the plates must increase as ε^{-3} as $\varepsilon \to 0^+$.

To sum up if we are to find (3.1)-(3.4) as a limit problem, *the ratio between some appropriate measure of the body force and the Lamé constants must behave like* ε^3, *and the ratio between some appropriate measure of the surface force and the Lamé constants must behave like* ε^4 *as* $\varepsilon \to 0^+$. If these assumptions are not satisfied, the limit behavior may be that of a rigid body, or the limit problem may even "vanish", as in the linear case (Caillerie, 1980).

Another noteworthy observation is that, upon combining the changes of functions (2.6)-(2.7) with the asymptotic expansions (2.15), we obtain

$$u_\alpha^\varepsilon = O(\varepsilon^2),\ u_3^\varepsilon = O(\varepsilon), \tag{3.10}$$

$$\sigma_{\alpha\beta}^\varepsilon = O(\varepsilon^2),\ \sigma_{\alpha 3}^\varepsilon = O(\varepsilon^3),\ \sigma_{33}^\varepsilon = O(\varepsilon^4), \tag{3.11}$$

so that these asymptotic orders are mathematical consequences of assumptions (3.9). Hence they *do not* constitute an *a priori* assumption in the present theory : A straightforward computation indeed shows that, had we let e.g. $u_\alpha^\varepsilon = \varepsilon u_\alpha(\varepsilon)$ and $u_3^\varepsilon = u_3(\varepsilon)$ instead of $u_\alpha^\varepsilon = \varepsilon^2 u_\alpha(\varepsilon)$ and $u_3^\varepsilon = \varepsilon u_3(\varepsilon)$ in (2.6), then the first term of the asymptotic expansion of each function $u_i(\varepsilon)$ should then be of order $O(\varepsilon)$.

Notice that the orders found in (3.10) mathematically substantiate the intuitive idea that a two-dimensional plate model should remain valid as long as the vertical deflection remains of the order of the thickness (hence the horizontal displacements are of the order of the square of the thickness).

Oddly enough, *the orders* (3.10)-(3.11) *cannot be fully determined within the sole linearized theory,* i.e., when the three-dimensional problem (1.1)-(1.6) is replaced by the well-known system of three-dimensional "linear" elasticity, since in this case all orders (3.10)-(3.11) are only determined up to a "hanging" multiplicative factor ε^s, s arbitrary (this observation caused some confusion in the early attempts to justify the well-know biharmonic model in two-dimensional linear plate theory, as in Ciarlet & Destuynder, 1979a). The remedy is deceptively simply : it suffices to remember that the three-dimensional system of "linear" elasticity is *not* a model *per se ;* it is instead an "approximation" of the nonlinear problem (1.1)-(1.6), and nothing else (to make matters even worse, the justification of this approximation in the three-dimensional case is at present limited to some very special cases ; see Ciarlet, 1985, p. 113). Consequently, the two-dimensional linear clamped plate model can be expected to be a good model only within the range of validity of the nonlinear model (3.1)-(3.8) which it in effect approximates, i.e., when assumptions (3.9) hold. Since in the nonlinear case, no hanging factor appears, assumptions (3.9), together with their consequences (3.10)-(3.11), are thus also the only ones which make sense in the linear case.

To conclude this brief discussion, we also mention that a striking feature of the above asymptotic equivalence is that "*it partially linearizes the three-dimensional equations*", in that a system of *quasilinear* (i.e., with nonlinearities in the higher order terms) second-order equations reduces in the limit to a system of *semilinear* (i.e., with nonlinearities only in the lower order terms) fourth-order equations, whose mathematical properties are accordingly easier to study. In particular, the existence theory, as well as the bifurcation theory in the case of the von Kármán equations mentionned in the next section, for such two-dimensional nonlinear plate models are quite satisfactory, while they have at present no fully equivalent counterpart for the original three-dimensional problem.

4. EXTENSIONS AND OPEN PROBLEMS. We list here various possible extensions of the approach described here, as well as some related open problems.

(i) *The von Kármán and the Marguerre-von Kármán equations.* Assume for definiteness that the constitutive equation is of the simplified form (2.1). If the three-dimensional boundary condition of place $u_i^\varepsilon = 0$ on Γ_0^ε (cf. (1.3)) is replaced by

(4.1) $$\frac{1}{2\varepsilon}\int_{-\varepsilon}^{\varepsilon}(\sigma_{\alpha\beta}^\varepsilon + \sigma_{k\beta}^\varepsilon \partial_k^\varepsilon u_\alpha^\varepsilon)\nu_\beta dx_3^\varepsilon = h_\alpha^\varepsilon \text{ on } \gamma,$$

(4.2) $$u_1^\varepsilon \text{ and } u_2^\varepsilon \text{ do not depend on } x_3^\varepsilon \text{ on } \Gamma_0^\varepsilon,$$

(4.3) $$u_3^\varepsilon = 0 \text{ on } \Gamma_0^\varepsilon,$$

where (ν_α) denotes the unit outer normal along γ, the application of the asymptotic expansion method yields the well-known *von Kármán equations* (Ciarlet, 1980), provided the functions h_α^ε satisfy appropriate compatibility conditions, and ω is simply connected ; in particular, this approach clearly delineates the admissible boundary conditions for the Airy stress function. Various extensions are possible, such as the replacement of condition (4.1) by a more general condition of live loading (Blanchard & Ciarlet, 1983 ; incidentally, this shows that *different* three-dimensional problems may be asymptotically equivalent to the *same* two-dimensional problem), or such as the replacement of the set $\overline{\Omega}^\varepsilon$ by a more general "shallow shell", which then yields the Marguerre-von Kármán equations (Ciarlet & Paumier, 1985). It is also worth noticing that when the functions h_α^ε in (4.1) are of the form $h_\alpha^\varepsilon = -\lambda\nu_\alpha$, $\lambda \in \mathbb{R}$, one can show that the functions $-\lambda\zeta_3^\lambda$ (where ζ_3^λ now stands for the function so far denoted ζ_3, as in Theorem 3) converge as $\lambda \to +\infty$ in the space $H_0^1(\omega)$ to the solution ζ of the famed *membrane equation* $-\Delta\zeta = \int_{-1}^{1} f_3 dx_3 + g_3^+ + g_3^-$ (Ciarlet & Rabier, 1980).

(ii) *More general constitutive equations.* If the more general constitutive equation (1.4)-(1.6) is used instead of the simplified equation (2.1), a striking conclusion is that the application of the asymptotic expansion method still yields the same limit two-dimensional plate model (Davet, 1985) ! In this sense, *the von Kármán equations, or those of a clamped plate, have a generic character.* Another interesting result is that for such general constitutive equations, a *definition of two-dimensional polyconvexity* can be naturally deduced from that of three-dimensional polyconvexity as given by J. Ball (Quintela-Estevez, 1985). The advantages of such constitutive equations is that their associated stored energy function $\hat{W}^\varepsilon$ can be polyconvex and can reflect the fact that "infinite stress must accompany extreme strains" (Antman, 1976) in the form of the mathematical condition $\hat{W}^\varepsilon(F) \to +\infty$ as $\det F \to 0^+$.

(iii) *More general boundary conditions, loadings, and materials.* Consider for instance a rectangular plate, for which parallel edges in one direction are free (no applied surface force), while the others are either clamped as in (1.3), or subjected to boundary conditions of the form (4.1)-(4.3). Then, even in the linear case, a *boundary layer* already appears in the first term of the asymptotic expansion (2.15) (Blanchard, 1981 ; de Oliveira, 1981). Nevertheless this method still unambiguously yields a well-defined two-dimensional nonlinear plate model, which is not necessarily identical to those which are sometimes hastily chosen *a priori* ! Indeed, *it is one of the merits of the present method to clearly identify which two-dimensional plate model should correspond to a given set of three-dimensional boundary conditions.*

The extension to more general loadings, notably so as to include nonzero horizontal components f_α^ε and g_α^ε in (1.1) and (1.2), has been studied in the case of nonlinear clamped plates (Ciarlet & Destuynder, 1979b). It would be of interest to study their effect when they are combined with other types of boundary conditions, such as those corresponding to the von Kármán equations for instance.

Other useful extensions, notably in view of contemporary engineering applications, would consist in applying the asymptotic expansion method to nonlinear plates made of anisotropic materials (cf. Destuynder, 1980, and Gilbert, Hsiao & Schneider, 1983, in the linear case), and to nonlinear plates made of "periodic materials", such as composite materials, or materials with holes and/or stiffeners (cf. Caillerie, 1985, Davet & Destuynder, 1985, and Kohn & Vogelius, 1984 , 1985a , 1985b, in the linear case).

(iv) *Three-dimensional existence theory*. The only available existence theory for three-dimensional problems with "changing" boundary conditions of the kind considered here is that of J. Ball (cf. Theorem 1). This otherwise very powerful and elegant theory suffers from two drawbacks : one is the lack of regularity of the minimizers of the energy, which prevents the solutions to satisfy, even in a weak sense, the associated boundary value problem ; the other is that the solutions found as global minimizers of the energy may not be the "expected" ones, in the sense that the displacement may not be "small" if the forces are "small". We note in passing that these drawbacks are not encountered in the existence theory based on the implicit function theorem, but then this second approach is limited to *very special cases* of boundary conditions, such as a boundary condition of place everywhere on the boundary of the reference configuration (Bernadou, Ciarlet & Hu, 1984 ; Ciarlet & Destuynder, 1979b ; Marsden & Hughes, 1978 ; Valent, 1979).

By constrast, there exist satisfactory existence, regularity, uniqueness or multiplicity, and bifurcation results for two-dimensional nonlinear plate models associated with various kinds of "two-dimensional boundary conditions" (see e.g. Ciarlet & Rabier, 1980, and the references quoted therein). Therefore a natural idea would consist in constructing a three-dimensional solution $(u(\varepsilon),\sigma(\varepsilon))$ of (2.11)-(2.12) "close to" a two-dimensional solution (u^0,σ^0) of (3.1)-(3.8). For some specific classes of three-dimensional boundary conditions, recent progresses have been obtained in this direction by using the Nash-Moser implicit function theorem (Paumier, 1985).

One step further would consist in comparing the three-dimensional and the two-dimensional solutions, thus generalizing to the nonlinear case the known convergence theory of the linear case (Caillerie, 1980 ; Ciarlet & Kesavan, 1980 ; Destuynder, 1980 , 1981 ; Raoult, (1985a)). Strangely enough, the problem of *numerically* substantiating such theoretical convergence results is not even solved in the linear case, because the discrete problem becomes severely ill-conditioned when the thickness of the plate is of the order of the discretization parameter. A way out may consist in choosing a conventional method, such as the finite element method, for the variables x_α, and a spectral method for the variable x_3^ε.

(v) *One-dimensional theories*. By appropriately adapting the asymptotic expansion method, one can similarly justify known *one-dimensional nonlinear rod models* (Cimetière, Geymonat, Le Dret, Raoult & Tutek, 1985a , 1985b), thus generalizing to the nonlinear case earlier results in the linear case (Bermudez & Viaño, 1984 ; Tutek & Aganovic, 1985).

(vi) *Multi-dimensional structures*. A problem of paramount importance in practice, notably in aerospace engineering, consists in deriving a *global model* of a structure where e.g. a *three-dimensional* and a *one-dimensional substructures* co-exist, which should each be modeled as such. What is then a correct mathematical model at the *junction* between the two substructures ? Even in the linear case, this problem is completely open at present.

REFERENCES

Antman, S.S..- Ordinary differential equations of nonlinear elasticity, II : Existence and regularity theory for conservative boundary value problems, *Arch. Rat. Mech. Anal.* 61, 1976, 353-393.

Ball, J.M.- Convexity conditions and existence theorems in nonlinear elasticity, *Arch. Rational. Mech. Anal.* 63, 1977, 337-403.

Bermudez, A. ; Viaño, J.M.- Une justification des équations de la thermo-élasticité des poutres à section variable par des méthodes asymptotiques, *RAIRO Analyse Numérique* 18, 1984, 347-376.

Bernadou, M. ; Ciarlet, P.G. ; Hu, J.- On the convergence of the semi-discrete incremental method in nonlinear, three-dimensional, elasticity, *J. Elasticity* 14, 1984, 425-440.

Blanchard, D.- *Justification de Modèles de Plaques Correspondant à Différentes Conditions aux Limites,* Thesis, Université Pierre et Marie Curie, 1981.

Blanchard, D. ; Ciarlet, P.G.- A remark on the von Kármán equations, *Comput. Methods Appl. Mech. Engrg.* 37, 1983, 79-92.

Caillerie, D.- The effect of a thin inclusion of high rigidity in an elastic body, *Math. Meth. in the Appl. Sci.* 2, 1980, 251-270.

Caillerie, D.- *Homogenization in Elasticity,* in Lectures on Homogenization Techniques for Composite Media, CISM, Udine, 1985.

Ciarlet, P.G.- A justification on the von Kármán equations, *Arch. Rational Mech. Anal.* 73, 1980, 349-389.

Ciarlet, P.G.- *Elasticité Tridimensionnelle,* Masson, Paris, 1985.

Ciarlet, P.G.- *Mathematical Elasticity, vol.* I, North-Holland, Amsterdam, 1986.

Ciarlet, P.G. ; Destuynder, P.- A justification of the two-dimensional linear plate model, *J. Mécanique* 18, 1979a, 315-344.

Ciarlet, P.G. ; Destuynder, P.- A justification of a nonlinear model in plate theory, *Comp. Methods Appl. Mech. Engrg.* 17/18, 1979b, 227-258.

Ciarlet, P.G. ; Geymonat, G.- Sur les lois de comportement en élasticité non-linéaire compressible, *C.R. Acad. Sci. Paris Sér.* A, 295, 1982, 423-426.

Ciarlet, P.G. ; Kesavan, S.- Two-dimensional approximations of three-dimensional eigenvalues in plate theory, *Comput. Methods Appl. Mech. Engrg.* 26, 1980, 149-172.

Ciarlet, P.G. ; Nečas, J.- Unilateral problems in nonlinear, three dimensional elasticity, *Arch. Rational Mech. Anal.* 87, 1985a, 319-338.

Ciarlet, P.G. ; Nečas, J.- Injectivité presque partout, auto-contact, et non-interpénétrabilité en élasticité non linéaire tridimensionnelle, *C.R. Acad. Sci. Paris Sér.* A, 1985b, to appear.

Ciarlet, P.G. ; Paumier, J.C.- A justification of the Marguerre-von Kármán equations, 1985, to appear.

Ciarlet, P.G. ; Rabier, P.- *Les Equations de von Kármán,* Lectures Notes in Mathematics, vol. 826, Springer-Verlag, Berlin, 1980.

Cimetière, A. ; Geymonat, G. ; Le Dret, H. ; Raoult, A. ; Tutek, Z.- Une justification d'un modèle non linéaire de poutres à partir de l'élasticité tridimensionnelle, I, *C.R. Acad. Sci. Paris Sér.* A, 1985a, to appear.

Cimetière, A. ; Geymonat, G. ; Le Dret, H. ; Raoult, A. ; Tutek, Z.- Une justification d'un modèle non linéaire de poutres à partir de l'élasticité tridimensionnelle, II, *C.R. Acad. Sci. Paris Sér.* A, 1985b, to appear.

Davet, J.L.- Justification de modèles de plaques non linéaires pour des lois de comportement générales, *Modélisation Mathématique et Analyse Numérique*, 1985, à paraître.

Davet, J.L. ; Destuynder, P.- Singularités logarithmiques dans les effets de bord d'une plaque en matériaux composites, *J. Mécanique Théor. Appl.* 4 1985, 357-373.

Destuynder, P.- *Sur une Justification des Modèles de Plaques et de Coques par les Méthodes Asymptotiques*, Thèse, Université Pierre et Marie Curie, Paris, 1980.

Destuynder, P.- Comparaison entre les modèles tridimensionnels et bidimensionnels de plaques en élasticité, *RAIRO Analyse Numérique* 15, 1981, 331-369.

Germain, P.- *Mécanique des Milieux Continus*, Tome 1, Masson, Paris, 1972.

Gilbert, R.P. ; Hsiao, G.C. ; Schneider, M.- The two-dimensional linear orthotropic plate, *Applicable Analysis* 15, 1983, 147-169.

Gurtin, M.E.- *Introduction to Continuum Mechanics*, Academic Press, New York, 1981.

Hanyga, A.- *Mathematical Theory of Non-Linear Elasticity*, Polish Scientific Publishers, Warszawa, and Ellis Horwood, Chichester, 1985.

Kohn, R.V. ; Vogelius, M.- A new model for thin plates with rapidly varying thickness, *Int. J. Solids & Structures* 20, 1984, 333-350.

Kohn, R.V. ; Vogelius, M.- A new model for thin plates with rapidly varying thickness II : A convergence proof, *Quart. Appl. Math.* 1985a, to appear.

Kohn, R.V. ; Vogelius, M.- A new model for thin plates with rapidly varying thickness III : Comparison of different scalings, *Quart. Appl. Math.*, 1985b, to appear.

Lions, J.L.- *Perturbations Singulières dans les Problèmes aux Limites et en Contrôle Optimal*, Lecture Notes in Mathematics, Vol. 323, Springer-Verlag, Berlin, 1973.

Marsden, J.E. ; Hughes, T.J.R.- Topics in the mathematical foundations of elasticity, *Nonlinear Analysis and Mechanics : Heriot-Watt Symposium, Vol.* 2, pp. 30-285, 1978, Pitman, London.

Marsden, J.E. ; Hughes, T.J.R.- *Mathematical Foundations of Elasticity*, Prentice-Hall, Englewood Cliffs, 1983.

de Oliveira, M.P.- *Alguns Problemas em Optimização de Estruturas*, Universidade de Coimbra, 1981.

Paumier, J.C.- Thesis, Université Pierre et Marie Curie, Paris, 1985.

Quintela-Estevez, P.- Polyconvexité dans les équations de von Kármán, 1985, to appear.

Raoult, A.- Construction d'un modèle d'évolution de plaques avec terme d'inertie de rotation, 1985a, to appear.

Raoult, A.- Non-polyconvexity of the stored energy function of a Saint Venant-Kirchhoff material, *Aplikace Matematiky*, 1985b, à paraître.

Truesdell, C. ; Noll, W.- The non-linear field theories of Mechanics, *Handbuch der Physik, vol.* III/3, Springer, Berlin, 1965.

Tutek, Z. ; Aganovic, I.- A justification of the one-dimensional linear model of elastic beam, *Math. Mech. Appl. Sci.*, 1985, to appear.

Valent, T.- Teoremi di esistenza e unicità in elastostatica finita, *Rend. Sem. Mat. Univ. Padova* 60, 1979, 165-181.

Wang, C.-C. ; Truesdell, C.- *Introduction to Rational Elasticity*, Noordhoff, Groningen, 1973.

Numerical Approximation of Partial Differential Equations
E.L. Ortiz (Editor)

FORMULATION OF ALTERNATING-DIRECTION ITERATIVE METHODS FOR MIXED METHODS IN THREE SPACE

Jim Douglas,Jr., Ricardo Durán, and Paola Pietra
University of Chicago
Chicago, IL 60637, USA

Two alternating-direction iterative methods are formulated to treat the linear algebraic equations that arise from the mixed finite element approximation of a second order elliptic boundary value problem in three space variables. One is based on the Uzawa technique for iterating to a solution of a saddle point problem and the other on the Arrow-Hurwitz method for saddle point problems. The Uzawa version is analyzed in a special case. Some preliminary computational results are presented.

1. Introduction. We shall consider the solution of the algebraic equations arising in the approximate solution of the Dirichlet problem

$$-\mathrm{div}(\, a(x)\, \mathbf{grad} u\,) = f, \quad x \in \Omega, \tag{1.1a}$$
$$u = -g, \quad x \in \partial\Omega, \tag{1.1b}$$

by mixed finite element methods. Here, Ω is a bounded domain in $\mathbf{R}^3$. Let

$$\mathbf{q} = -a\, \mathbf{grad} u, \qquad c(x) = a(x)^{-1}. \tag{1.2}$$

Then (1.1a) can be factored into the form

$$c\mathbf{q} + \mathbf{grad} u = \mathbf{0}, \tag{1.3a}$$
$$\mathrm{div}\mathbf{q} = f. \tag{1.3b}$$

Let $\mathbf{V} = H(\mathrm{div},\Omega) = \{\mathbf{v} \in L^2(\Omega)^3 : \mathrm{div}\mathbf{v} \in L^2(\Omega)\}$ and $W = L^2(\Omega)$. The weak form of (1.1) appropriate for mixed methods results from testing (1.3a) against $\mathbf{V}$ and (1.3b) against W and seeking a solution $\{\mathbf{q},u\}$ in $\mathbf{V}\times W$:

$$(c\mathbf{q},\mathbf{v}) - (\mathrm{div}\mathbf{v},u) = \langle g,\mathbf{v}.\mathbf{n}\rangle, \quad \mathbf{v} \in \mathbf{V}, \tag{1.4a}$$
$$(\mathrm{div}\mathbf{q},w) = (f,w), \quad w \in W, \tag{1.4b}$$

where (,) indicates the inner product in either $L^2(\Omega)$ or $L^2(\Omega)^3$ and $\langle\, ,\, \rangle$ that in $L^2(\partial\Omega)$.

The mixed finite element approximation $\{\mathbf{q}_h,u_h\}$ of the solution of (1.4) is found by seeking $\{\mathbf{q}_h,u_h\}$ in a properly chosen subspace $\mathbf{V}_h\times W_h$ of $\mathbf{V}\times W$ such that

$$(c\mathbf{q}_h,\mathbf{v}) - (\mathrm{div}\mathbf{v},u_h) = \langle g,\mathbf{v}.\mathbf{n}\rangle, \quad \mathbf{v} \in \mathbf{V}_h, \tag{1.5a}$$

(1.5b) $(\operatorname{div}\mathbf{q}_h, w) = (f,w), \quad w \in W_h.$

The algebraic equations associated with (1.5) take the form

(1.6a) $Aq - Bu = g,$

(1.6b) $B^* q = f,$

where q is the vector of degrees of freedom of $\mathbf{q}_h$, A is the symmetric, positive-definite matrix generated by $c(x)$-weighted L^2 projection into $\mathbf{V}_h$, and the rectangular matrices $-B$ and B^* correspond to discrete gradient and divergence operators. The system (1.6) is positive-semidefinite and corresponds to the saddle point problem (1.4). For the three space variable problems being treated here, direct elimination solution of (1.6) rapidly becomes very expensive; moreover, conjugate gradient iterative procedures have also proved experimentally to be only marginally acceptable, though certainly less expensive than direct methods.

The object of this paper is to describe an alternating-direction iterative procedure for (1.6) when $\mathbf{V}_h \times W_h$ is a Raviart-Thomas space of index k over a decomposition of Ω consisting of rectangular parallelepipeds ("cubes") taken from a tensor-product grid. The Raviart-Thomas space can be constructed as follows. Let $\mathcal{T}_h = \{R\}$, where $\operatorname{diam}(R) \le h$ and the ratio of any two sides of R is bounded (the second condition is related to the control of the error in the approximate solution, rather than the algebraic problem of interest here). Denote by $Q_{i,j,k}$ the tensor product of polynomials of degree not greater than i, j, and k in x, y, and z, respectively. Let $R \in \mathcal{T}_h$ and define the restriction of the Raviart-Thomas space of index k to R to be

(1.7) $\mathcal{RT}(k,R) = \mathbf{V}(k,R) \times W(k,R) = [Q_{k+1,k,k} \times Q_{k,k+1,k} \times Q_{k,k,k+1}] \times Q_{k,k,k}.$

Then, let

(1.8a) $\mathbf{V}_h = \mathbf{V}(k, \mathcal{T}_h) = \{\mathbf{v} \in \mathbf{V} : \mathbf{v}|_R \in \mathbf{V}(k,R),\ R \in \mathcal{T}_h\},$

(1.8b) $W_h = W(k, \mathcal{T}_h) = \{w \in W : w|_R \in W(k,R),\ R \in \mathcal{T}_h\},$

(1.8c) $\mathcal{RT}_h = \mathcal{RT}(k, \mathcal{T}_h) = \mathbf{V}_h \times W_h.$

Two varieties of alternating-direction iterative techniques have been discussed for solving the algebraic equations for the mixed finite element method for problems in two space variables. Brown (Brown, 1982) introduced a method related to the Uzawa iterative procedure for saddle point problems; however, the alternating-direction method is based on an implicit discretization of the induced virtual parabolic problem in the time variable. Later, Douglas and Pietra (Douglas and Pietra, 1985) considered a different technique related to the Arrow-Hurwitz iteration for saddle point problems for both Raviart-Thomas spaces and some spaces due to Brezzi, Douglas, and Marini (Brezzi, Douglas, and Marini, 1985). We shall formulate analogues of both techniques for the three dimensional case; because of the imposed limitation of length of this article, we shall consider only Raviart-Thomas spaces here.

2. An Uzawa Alternating-Direction Method. Consider the parabolic problem that results from introducing the time derivative of u into equation (1.6b):

(2.1a) $$Aq - Bu = g,$$

(2.1b) $$D\frac{\partial u}{\partial t} + B^* q = f,$$

where D should be taken symmetric, positive-definite. Introduce also initial values for u :

(2.1c) $$u(0) = u^0 .$$

The system (2.1) represents the parabolic mixed method for the initial-boundary problem

(2.2a) $$d\frac{\partial u}{\partial t} - \mathrm{div}(a\,\mathbf{grad}u) = f, \quad x \in \Omega, \quad t > 0,$$

(2.2b) $$u = g, \quad x \in \partial\Omega, \quad t > 0,$$

(2.2c) $$u = u^0, \quad x \in \Omega, \quad t = 0,$$

where $d(x)$ is a conveniently chosen weight and u^0 is the initial guess at the solution of the stationary problem (1.1).

Since the natural basis for $\mathbf{V}_h$ consists of vector functions having only one nonzero component, the equations (1.6a) split into three sets of the form

(2.3a) $$A_i q_i - B_i u = g_i, \quad i = 1,2,3,$$

with A_i being block tridiagonal and B_i block bidiagonal when the parameters defining q_i are ordered in an x_i-orientation. The second equation, (1.6b), then has the form

(2.3b) $$B_1^* q_1 + B_2^* q_2 + B_3^* q_3 = f.$$

This splitting is necessary if an Uzawa-type method is to be computationally efficient; if no such splitting is available, Arrow-Hurwitz procedures can still be effective.

Our discretization in the direction of the virtual time variable will be based on an alternating-direction method (Douglas, 1962) first introduced in connection with finite difference methods for parabolic equations in three space variables. It produces a perturbation of the standard Crank-Nicolson time discretization and is formally second-order correct in the time increment. We also studied a perturbation of backward differencing in time corresponding to the Douglas-Rachford method (Douglas and Rachford, 1956); since our experimental results indicated a clear superiority for the Crank-Nicolson-like method, we present only it.

The alternating-direction algorithm is formulated as follows. First, let $u^0 \in W_h$ be chosen arbitrarily and determine $q^0 = (q_1^0, q_2^0, q_3^0)$ by (2.3a). Then, the general step of the iteration splits into three sweeps and a correction:

(2.4a) $A_1 q_1^* - B_1 u^* = g_1,$

(2.4b) $D(u^* - u^n)/\tau^n + B_1^*(q_1^* + q_1^n)/2 + B_2^* q_2^n + B_3^* q_3^n = f;$

(2.5a) $A_2 q_2^{**} - B_2 u^{**} = g_2,$

(2.5b) $D(u^{**} - u^n)/\tau^n + [B_1^*(q_1^* + q_1^n) + B_2(q_2^{**} + q_2^n)]/2 + B_3^* q_3^n = f;$

(2.6a) $A_3 q^{n+1} - B_3 u^{n+1} = g_3,$

(2.6b) $D(u^{n+1} - u^n)/\tau^n + [B_1(q_1^* + q_1^n) + B_2^*(q_2^{**} + q_2^n) + B_3^*(q_3^{n+1} + q_3^n)]/2 = f;$

(2.7a) $A_1 q_1^{n+1} - B_1 u^{n+1} = g_1,$

(2.7b) $A_2 q_2^{n+1} - B_2 u^{n+1} = g_2.$

The equations (2.7) can be interpreted as a corrector for the first two components of the vector field. Making this correction to have q_1 and q_2 consistent with u also has the effect of having the iteration restart cleanly on each virtual time step; thus, the choice of τ^n can be made independently of previous time steps.

Each x_i-sweep and each half of (2.7) can be solved as a locally one-dimensional system in the usual fashion for alternating-direction schemes for finite difference or finite element methods. Consequently, a full iteration can be done in a number of arithmetic operations that is a fixed multiple of the number of parameters that define the space $\mathcal{RT}_h$.

Note that the domain Ω has not been assumed to be a rectangular parallelepiped. The algorithm is computable on domains that are unions of cubic elements. More general domains can be treated if the boundary elements are handled with care.

3. Spectral Analysis for $\mathcal{RT}(0, \mathcal{T}_h)$. In the remainder of this paper the special case of the Laplace operator (i.e., $a(x) = 1$) on a cube will be analyzed in detail for the choice of a uniform mesh and the simplest Raviart-Thomas space $\mathcal{RT}(0, \mathcal{T}_h)$. All of the qualitative statements regarding the convergence of the iteration hold for higher index spaces.

Let $x_i = ih$, $y_j = jh$, $z_k = kh$, and $R_{ijk} = [x_i,x_{i+1}]\times[y_j,y_{j+1}]\times[z_k,z_{k+1}] = I_i\times J_j\times K_k$ for $i,j,k = 0,\ldots,N = h^{-1}$. Use the following symbols to denote the parameters defining $\mathcal{RT}_h$:

$$(3.1)\quad q_1(x_i,J_j,K_k)=\lambda_{ijk},\ q_2(I_i,y_j,K_k)=\mu_{ijk},\ q_3(I_i,J_j,z_k)=\eta_{ijk},\ u(R_{ijk})=u_{ijk},$$

and let $v_{ijk} = 6h^{-1}u_{ijk}$. Define the Simpson operators S_i (times six) by

$$(3.2a)\qquad S_1\lambda_{ijk} = \lambda_{i-1,j,k}+4\lambda_{ijk}+\lambda_{i+1,j,k},$$
$$(3.2b)\qquad S_2\mu_{ijk} = \mu_{i,j-1,k}+4\mu_{ijk}+\mu_{i,j+1,k},$$
$$(3.2c)\qquad S_3\eta_{ijk} = \eta_{i,j,k-1}+4\eta_{ijk}+\eta_{i,j,k+1}.$$

Interpret $\{\lambda^n,\mu^n,\eta^n;u^n\}$ as the error in the nth iterate in the Uzawa alternating-direction algorithm. Also, note that the solution of (2.3) can be extended to a uniform decomposition of all of $\mathbf{R}^3$ by periodicity, with u^n or v^n being odd and $\{\lambda^n,\mu^n,\eta^n\}$ being even. Let

$$(3.3)\qquad \delta_1 v_{ijk} = v_{ijk}-v_{i-1,j,k},\ \delta_2 v_{ijk} = v_{ijk}-v_{i,j-1,k},\ \delta_3 v_{ijk} = v_{ijk}-v_{i,j,k-1}.$$

Then the error satisfies the following relations:

$$(3.4a)\qquad v^0 \text{ arbitrary},$$
$$(3.4b)\qquad S_1\lambda^0 + \delta_1 v^0 = 0,$$
$$(3.4c)\qquad S_2\mu^0 + \delta_2 v^0 = 0,$$
$$(3.4d)\qquad S_3\eta^0 + \delta_2 v^0 = 0;$$

and, for $n = 0,1,\ldots$, and with d_i indicating the undivided forward difference with respect to the ith variable,

$$(3.5a)\qquad S_1\lambda^* + \delta_1 v^* = 0,$$
$$(3.5b)\qquad (v^* - v^n)/\tau^n + d_1(\lambda^*+\lambda^n)/2 + d_2\mu^n + d_3\eta^n = 0,$$

$$(3.6a)\qquad S_2\mu^{**} + \delta_2 v^{**} = 0,$$
$$(3.6b)\qquad (v^{**} - v^*)/\tau^n + d_2(\mu^{**} - \mu^n)/2 = 0,$$

$$(3.7a)\qquad S_3\eta^{n+1} + \delta_3 v^{n+1} = 0,$$
$$(3.7b)\qquad (v^{n+1} - v^{**})/\tau^n + d_3(\eta^{n+1} - \eta^n)/2 = 0,$$

$$(3.8a)\qquad S_1\lambda^{n+1} + \delta_1 v^{n+1} = 0,$$
$$(3.8b)\qquad S_2\mu^{n+1} + \delta_2 v^{n+1} = 0.$$

Equations (3.6b) and (3.7b) result from subtracting (2.5b) and (2.6b), respectively, from (2.4b) and (2.5b).

The system (3.4)-(3.8) can be analyzed by spectral arguments. Eigenfunctions can be taken in the form

(3.9a) $\lambda_{ijk}^n = L^n \cos\pi p x_i \sin\pi q(y_j+.5h) \sin\pi r(z_k+.5h),$

(3.9b) $\mu_{ijk}^n = M^n \sin\pi p(x_i+.5h) \cos\pi q y_j \sin\pi r(z_k+.5h),$

(3.9c) $\eta_{ijk}^n = H^n \sin\pi p(x_i+.5h) \sin\pi q(y_j+.5h) \cos\pi r z_k,$

(3.9d) $v_{ijk}^n = V^n \sin\pi p(x_i+.5h) \sin\pi q(y_j+.5h) \sin\pi r(z_k+.5h),$

where $L^n = L_{pqr}^n$, etc., and $p,q,r = 1,\ldots,N$. Let V_i be the ith component of $\mathbf{V}_h$. Although the dimension of V_i is larger than that of W_h, the eigenfunctions above for V_i span the subspace of V_i isomorphic to W_h given by $\mathrm{Im}A_i^{-1}B_i$, which contains the solution of the equations.

Let

(3.10) $\rho_p = 2\,(3 - 2\sin^2 .5\pi ph)^{-1} \sin^2 .5\pi ph\,.$

An elementary, but long, calculation shows that

(3.11) $$V^{n+1} = F(\tau^n,\rho_p,\rho_q,\rho_r)V^n = f(\tau^n\rho_p,\tau^n\rho_q,\tau^n\rho_r)V^n$$
$$= [1 - 2\tau^n(\rho_p+\rho_q+\rho_r)(1+\tau^n\rho_p)^{-1}(1+\tau^n\rho_q)^{-1}(1+\tau^n\rho_r)^{-1}]V^n.$$

By (3.7a) and (3.8),

(3.12) $L^{n+1} = \alpha_p V^{n+1},\quad M^{n+1} = \alpha_q V^{n+1},\quad H^{n+1} = \alpha_r V^{n+1},$

with $\alpha_p = (3 - 2\sin^2 .5\pi ph)\sin .5\pi ph$. Thus, the coefficient of the eigenfunction corresponding to the parameters $\{p,q,r\}$ for each component of the error in the vector field (and in its divergence) is reduced by the same factor as for the scalar variable.

The reduction factor F is less than one for any $\tau>0$; consequently, convergence of the iteration is assured for any sequence of virtual time steps bounded away from zero and infinity; however, as is typical for alternating-direction procedures, very rapid convergence requires proper selection of $\{\tau^n\}$. With a different definition of ρ_p, the function f of (3.11) occurs in the analysis of the alternating-direction procedure of Douglas (Douglas,1962). We shall adopt the algorithm of that paper to select a cycle of time steps of length $O(\log N)$ which will ensure an error reduction in L^2 for both the scalar and vector variables by a fixed factor less than one, e.g, by a half.

Let $\beta \le a \le \gamma$ and $0 \le b,c \le \gamma$. Then, if $\beta < 1 < \gamma$, $\max f(a,b,c) = \max(f(\beta,0,0), f(\gamma,\gamma,\gamma))$. If we ask that $f(\beta,0,0) = f(\gamma,\gamma,\gamma)$, β is determined in terms of γ. Define the time step cycle as follows:

(3.13a) $$\rho_m = \rho_1 \approx \pi^2 h^2/6, \qquad \rho_M = \rho_N \approx 2;$$

(3.13b) $$\tau^1 = 2\beta/\rho_m; \quad \tau^n = \beta\gamma^{-1}\tau^{n-1}, \ n=2,\dots,NC;$$

(3.13c) $$NC = [\log(\rho_M/\rho_m) \div \log(\beta\gamma^{-1})] + 1.$$

In each cycle given by (3.13) there occurs a reduction in the L^2 norm of the error by a factor

(3.14) $$\sigma = \sup\{ \prod_{n=1}^{NC} F(\tau^n,\rho_p,\rho_q,\rho_r) : \rho_m \le \rho_p,\rho_q,\rho_r \le \rho_M\}.$$

By sampling on γ, we minimized the rough estimate of the reduction per iteration given by $\sigma^{1/NC}$ in order to choose γ optimally. For $h = .1$, γ_{opt} was found to be about 1.71.

4. Arrow-Hurwitz Alternating-Direction Iteration. Arrow-Hurwitz iterative procedures for (1.6) are based on a transient problem coming from adding time derivatives of both q and u to the equations. It is convenient to introduce two virtual time variables:

(4.1a) $$D_1 \frac{\partial q}{\partial t_1} + Aq - Bu = g,$$

(4.1b) $$D_2 \frac{\partial u}{\partial t_2} + B^* q = f.$$

In order to initiate the evolution generated by (4.1), it is necessary to specify both q and u at $t_1=t_2=0$. In many applications of the mixed method to physical problems, a better guess is available for q than for u, so that having to specify q^0 is a strong advantage, provided that the iterative method makes good use of this additional information. Tests (Douglas and Pietra, 1985) on the two-dimensional analogue of the iteration to be defined below have shown a superiority of the Arrow-Hurwitz technique over the Uzawa version when the initial error in the vector field is small in comparison with that in the scalar variable, and preliminary testing of the three-dimensional versions have indicated confirmation of this superiority.

An Arrow-Hurwitz algorithm can be formulated as follows:

(4.2) $$q^0 \text{ and } u^0 \text{ arbitrary;}$$

$$E_1(q_1^* - q_1^n)/\xi + A_1 q_1^* - B_1 u^* = g_1, \tag{4.3a}$$
$$D(u^* - u^n)/\tau^n + B_1^*(q_1^* + q^n)/2 + B_2^* q_2^n + B_3^* q_3^n = f; \tag{4.3b}$$

$$E_2(q_2^{**} - q_2^n)/\xi + A_2 q_2^{**} - B_2 u^{**} = g_2, \tag{4.4a}$$
$$D(u^{**} - u^*)/\tau^n + B_2^*(q_2^{**} - q_2^n)/2 = 0, \tag{4.4b}$$

$$E_3(q_3^{n+1} - q_3^n)/\xi + A_3 q_3^{n+1} - B_3 u^{n+1} = g_3, \tag{4.5a}$$
$$D(u^{n+1} - u^{**})/\tau^n + B_3^*(q_3^{n+1} - q_3^n)/2 = 0, \tag{4.5b}$$

$$E_1(q_1^{n+1} - q_1^n)/\xi + A_1 q_1^{n+1} - B_1 u^{n+1} = g_1, \tag{4.6a}$$
$$E_2(q_2^{n+1} - q_2^n)/\xi + A_2 q_2^{n+1} - B_2 u^{n+1} = g_2. \tag{4.6b}$$

The Arrow-Hurwitz algorithm has not been analyzed in detail even for the simplest case of the example of the last section. We have chosen to select iteration parameters using a modification of the rule (3.13) above; this corresponds to the experience related to the two-dimensional Arrow-Hurwitz procedure (Douglas and Pietra, 1985). Change the definition of ρ_m and ρ_M to be

$$\rho_m = \xi(1+\xi)^{-1}\rho_1, \quad \rho_M = \xi(1+\xi)^{-1}\rho_N, \tag{4.7}$$

and then employ (3.13b) and (3.13c) to define a parameter cycle for $\{\tau^n\}$.

5. Experimental Results. Only preliminary computational experiments have been carried out so far. The test problem of section 3 ($a(x)=1$, $f=g=0$, Ω the unit cube, $\mathcal{T}_h$ a uniform partition) has been studied with four different initializations of u^0; for the Arrow-Hurwitz procedure q^0 was taken to be zero, since the object of this method is to take advantage of good initial values for the vector variable.

The examples are as follows:

(5.1) Example 1 $\quad u_h^0 = 1$, or $\quad v_{ijk}^0 = 6h^{-1}$

(5.2) Example 2 $\quad u_h^0 = x(1-x)y(1-y)z(1-z)$

(5.3) Example 3 $\quad u_{ijk}^0 = (-1)^{i+j+k}$

(5.4) Example 4 $\quad u_h^0 = \begin{cases} 1, & \text{if } \max(x,y,z) > .5, \\ -1, & \text{otherwise} \end{cases}$

We shall report the results of a few of the experiments that we have made. Let $h = .1$ and $\gamma = 1.71$. The cycle length for both the Uzawa procedure and the Arrow-Hurwitz one with $\xi = 10$ is four and the factor σ given by (3.14) for the Uzawa version is equal to .1114. The choice of ξ is related to the earlier two-dimensional work of Douglas and Pietra.

Consider some properties of the Uzawa tests first. Examples 2 and 3 correspond to taking as initial conditions on the scalar variable its (roughly) fundamental and highest frequency eigenfunctions, respectively. For these two tests the reduction in each cycle was constant, .085 for Example 2 and .066 for Example 3, while for Examples 1 and 4 the reduction was greater in the first cycle than in succeeding cycles, for which it was .085 for both examples. What was occurring was that the iteration parameters used worked better on the high frequency components than on the low frequency ones, and the low frequency components dominated the error after one cycle had been completed. An adjustment of the iteration parameters should bring the convergence rates together.

We also tested the backward difference version of the Uzawa procedure and the Arrow-Hurwitz iteration on the same examples as above. The backward difference Uzawa method took about three times as many iterations to achieve the same reduction in error as did the Crank-Nicolson Uzawa method; this experience is much the same as was observed years ago for the backward difference and Crank-Nicolson alternating-direction (i.e., Douglas-Rachford and Douglas) iterative methods for standard finite differences for the Dirichlet problem. The Arrow-Hurwitz procedure was more effective than the Uzawa one, as was to be expected both on the basis of the motivation for the introduction of the method and from experience with the two-dimensional methods of Douglas and Pietra. The following table lists the number of iterations required to reduce the norm given by

$$[\ \|u_h\|^2 + \|\mathbf{q}_h\|^2\ ]^{1/2}$$

for the four test problems, with $h = .1$ and $\gamma = 1.71$, to less than $.001\|u_h^0\|$:

	Example 1	Example 2	Example 3	Example 4	Total
Uzawa	13	13	19	13	58
Arrow-Hurwitz	10	10	1	10	31

Three of the four examples were run with $h = .05$ and $\gamma = 1.8$. The results were as shown below.

	Example 1	Example 2	Example 3	Example 4	Total
Uzawa	17	15	x	17	49
Arrow-Hurwitz	13	13	x	14	40

Note that only thirty per cent more Arrow-Hurwitz iterations were required for the 20×20×20 case as for the 10×10×10 case for the Raviart-Thomas space.

Acknowledgements. Douglas's research was supported in part by the National Science Foundation, that of Durán by the Consejo Nacional de Investigaciones Científicas y Técnicas de Argentina, and that of Pietra by the Consiglio Nazionale delle Ricerche.

REFERENCES

1. Brezzi,F., Douglas,J.,Jr., and Marini,L.D., Two families of mixed finite elements for second order elliptic problems, to appear in Numerische Mathematik, 1985.
2. Brown,D.C., Alternating-direction iterative schemes for mixed finite element methods for second order elliptic problems, Thesis, University of Chicago, 1982.
3. Douglas,J.,Jr., Alternating direction methods for three space variables, Numerische Mathematik 4(1962),pp. 41-63.
4. Douglas,J.,Jr., and Pietra,P., A description of some alternating-direction iterative techniques for mixed finite element methods, to appear in the proceedings of a SIAM/SEG/SPE conference held in Houston, January 1985.
5. Douglas,J.,Jr., and Rachford,H.H., On the numerical solution of heat conduction problems in two and three space variables, Transactions of the American Mathematical Society 82(1956),pp. 421-439.

Numerical Approximation of Partial Differential Equations
E.L. Ortiz (Editor)

ITERATIVE METHODS FOR SINGULAR SYSTEMS

Ivo Marek

Charles University of Prague

Prague, Czechoslovakia

A class of iterative methods of numerical solving linear systems $Ax = b$ with general rectangular matrix A is considered. The iterative methods are constructed by using subproper splittings of A. For systems with square singular matrix a use of incomplete Choleski decomposition is examined. Some particular effective methods are obtained for the case of A being a singular M-matrix.

1. Introduction

We are going to consider general linear systems

$$Ax = b, \tag{1.1}$$

where $b \in R^m = Y_2$ and A $B(Y_1, Y_2)$, $Y_1 = R^n$; here $B(Y_1, Y_2)$ denotes the space of bounded linear operators mapping Y_1 into Y_2. In particular $B(Y) = B(Y, Y)$.

It should be noted that no solubility hypotheses concerning (1.1) are made.

Let Y_1 and Y_2 are equipped by inner products denoted $(.,.)_j$, $j = 1,2$. As usual, we let A to be defined by

$$(Ax,y)_2 = (x, A^* y)_1$$

for all $x \in Y_1$ and $y \in Y_2$. Thus, $A \in B(Y_2, Y_1)$.

2. Subproper splittings and iterations

Let $A \in B(Y_1, Y_2)$ and

$$A = M-Q, \tag{2.1}$$

where $M, Q \in B(Y_1, Y_2)$. We call (2.1) a <u>splitting of A</u>. A splitting is said to be <u>subproper</u>, if simultaneously holds that

(2.2) $\quad R(A) \subset R(M) \quad$ and $\quad \ker A \supset \ker M$,

where $R(A) = \{ y \in Y_2 \ \exists \ x \in Y_1 : y = Ax \}$.

If equality takes place in both relations in (2.2) the splitting is called <u>proper</u>.

Using (2.1) one can construct iterative processes by setting

(2.3) $$Mx_{k+1} = Qx_k + b.$$

Since M is singular in general, one must first define what is to be understood as x_{k+1} in (2.3). Second, the obtaining x_{k+1} from (2.3) should be in a sense "cheap".

An explicit form of x_{k+1} can be expressed by using the <u>iteration operator</u> T, i.e. (2.3) is equivalent with

(2.4) $$x_{k+1} = Tx_k + M^g b_k ,$$

where M^g is a generalized inverse of M.

3. Generalized inverse operators

To resolve (2.3) at each iteration step one has to introduce appropriate tools. As most relevant we show two kinds of generalized inverses: the <u>Moore-Penrose pseudoinverse</u> and <u>Drazin generalized inverse</u>.

Let A $B(Y_1, Y_2)$ and let us consider system of relations for determing Z $B(Y_2, Y_1)$

(3.1) $\quad$ (i) AZA, (ii) ZAZ = Z,

$\quad$ (iii) (AZ) = AZ, (iv) (ZA) = ZA.

It can shown that there is at most one Z $B(Y_2, Y_1)$ fulfilling (3.1).

Let

$$A^* A = \sum_{j=1}^{s} \lambda_j P_j$$

be the spectral representation of A A with projections P_j, $j = 0, 1, \ldots$ for which

(3.2) $$P_j P_k = P_k P_j = \delta_{jk} P_j, \quad P_j^2 = P_j = P_j^*, \sum_{j=1}^{s}{}' P_j = I - P_o,$$

where

$$A^* A P_j = \lambda_j P_j, \quad \lambda_j > 0, \quad j = 1, \ldots, s$$

and $A^* A P_o = 0$, i.e. $0 = \lambda_o$ is in the spectrum $\sigma(A^* A)$ only if $P_o \neq 0$.

A unique solution to (3.1) is called <u>Moore-Penrose pseudoinverse</u> and is denoted by A^+.

It is easy to see that

(3.3) $$A^+ = \sum_{j=1}^{s} {}_j^{-1} P_j A \quad .$$

Let $Y_1 = Y_2 = Y$ and let $A \in B(Y)$ be such that 0 is a pole of the resolvent operator $(\lambda I - A)^{-1}$ of order $q \geq 0$. Let us consider relations

(3.4) (i) $AZ = ZA$, (ii) $ZAZ = Z$, (iii) $A^{q+1} Z = A^q$.

Again, there is at most one $Z \in B(Y)$ satisfying (3.4). A unique solution to (3.4) is called the <u>Drazin generalized inverse</u> and is denoted by A^D.

It can be shown that $A^D = f(A)$, where $f(z) = \frac{1}{z}$ for z with $|z| > a$ and $f(z) = 0$ for z with $|z| \leq a$, where a is such that $0 = \lambda \in \sigma(A)$ implies that $|\lambda| > a$.

A particular case of A^D if $q \leq 1$ is the <u>group inverse</u> denoted by $A^{\#}$, i.e.

(3.5) $AA^{\#} = A^{\#}A, \quad A^{\#}AA^{\#} = A^{\#}, \quad AA^{\#}A = A.$

<u>Remark</u>. It is easy to see that if $Y_1 = Y_2 = Y$ and $A \in B(Y)$ is an operator for which 0 is not in the spectrum $\sigma(A)$, then $A^{-1} = A^{+} = A^{D} = A^{\#}$.

4. <u>Convergence</u>

We derive a convergence result under some rather abstract hypotheses and then will show how this result applies in some particular situations.

<u>Convergence hypotheses</u>. It is assumed that a subproper splitting of A and a sort of generalized inverse M^g have been chosen such that the following assumptions 1°-4° are fulfilled:

1° $T = M^gQ = P + S$,

where $P^2 = P$, $PS = SP = 0$ and the spectral radius of S $r(S) = \sup \{ |\lambda| : \lambda \in \sigma(S) \} < 1$, $\sigma(T)$ being the spectrum of T. We call such T <u>convergent operator</u>.

<u>Remark</u>. In the literature one may meet a concept <u>s-convergent</u> (semiconvergent) for the above concept and convergent for the <u>zero-convergent</u> operator, i.e. $T^k \rightarrow 0$. In general, for a convergent T in 1° the limit $\lim T^k = P \neq 0$.

2° $AP = 0$.

3° Y_2 is a direct sum (not necessarily orthogonal) of the subspaces $R(A)$ and $\ker(AM^g)$, i.e. $Y_2 = R(A) \oplus \ker(AM^g)$.

4° For $b = b_1 + b_2$ in the sense of 3° it holds that

$M^g b_1 = (I - P)b_1$ and $M^g b_2 = PM^g b$.

Proposition 4.1 Let $A = M-Q$ be a subproper splitting of A. Then under the hypotheses $1^o - 4^o$ the sequence $\{y_k\}$, where

$$(4.1) \qquad y_{k+1} = x_{k+1} - kPM^g b,$$

is convergent and its limit $\bar{x} = \lim_{k\to\infty} y_k$ is a solution to the system $Ax = b_1$.

Remark. Each vector $x \in Y$, for which $Ax = b_1$ is called generalized solution to the system $Ax = b$.

Proposition 4.2 Let $A = M-Q$ be a subproper splitting and let either (a) $M^g = M^+$ or (b) $M^g = M^{\#}$. If $T = M^g Q$ is convergent then the hypotheses $2^o - 4^o$ hold true.

As corollary we obtain

Theorem 4.1 If $A = M-Q$ is a subproper splitting and $T = M^g Q$ with either $M^g = M^+$ or $M^g = M^{\#}$, then the process (2.3), i.e., the sequence (4.1) is convergent if and only if T is convergent: $\bar{x} = \lim_{k\to\infty} y_k$. The speed of convergence is given by the estimate

$$\|\bar{x} - y_k\|_1 = c\,[r(S)]^k,$$

where c is a constant independent of k.

Remark. As a rule, the kernel of A is unknown and so are the operator P and the vector $b_2 = PM^g b$. However, we have that

$$x_{k+1} - x_k = (S^{k+1} - S^k)\,x_o + \sum_{p=0}^{k} (S^{p+1} - S^p)M^g b + PM^g b$$

and thus, $PM^g b = \lim_{k\to\infty}(x_{k+1} - x_k) = \lim_{k\to\infty}(y_{k+1} - y_k)$.

This fact has a quite important impact on the practical computations; for large values of k the difference $x_{k+1}-x_k$ is a good approximation to $c_2 = PM^g b = M^g b_2$.

It should also be noted that $PM^g b = 0$ if the system $Ax = b$ is consistent, i.e. if $b = b_1 \in R(A)$.

Proposition 4.3 Let $A = M-Q$ be a proper splitting where iteration operator $T = M^g b$ is convergent. The T is also zero-convergent.

Theorem 4.2 Let $A = M-Q$ be a proper splitting and let $T = M^g b$ be convergent, where either $M^g = M^+$ or $M^g = M^{\#}$. Then the iteration process (2.3) is convergent and $\bar{x}=\lim_k x_k$ is a generalized solution to (1.1).

5. Partial ordering and constructing of subproper splittings

Let Y_1 contain a set K such that

(i) $K + K \subset K$, (ii) $aK \subset K$ for $a \in R^1$, $a \geq 0$, (iii) $K \cap (-K) = \{0\}$, (iv) $\bar{K} = K$, where $\bar{K}$ denotes the norm closure of K, (v) $Y = K-K$, (vi) there is a $d > 0$ such that $\|x + y\| \geq d \operatorname{Max}(\|x\|, \|y\|)$ for $x, y \in K$.

A set K fulfilling (i)-(iii) is called cone. A cone K is closed, generating and normal if (iv), (v) and (vi) holds respectively (see Krein-Rutman, 1948).

An operator T B(Y) is called K-positive, if $TK \subset K$. A K-positive operator T is called K-irreducible, if for every pair $x \in K$, $x' \in K'$, there is an index $p = p(x, x')$ such that $(T^p x, x') > 0$.

Remark. Note that this definition is equivalent with the classical Frobenius definition (see Varga 1962 and Marek-Žitný 1984).

A partial order can be introduced in Y by setting $x \leq y$ (or $y \geq x$) $\Longleftrightarrow$ $(y-x)$ K. Similarly for T, S $\in$ B(Y), $S \leq T$ (or $T \geq S$) $\Longleftrightarrow$ $(T-S)\,K \subset K$.

An operator $A \in B(Y)$ is called <u>M(K)-operator,</u> if there is a number $a \in R^1$ and an operator $B \in B(Y)$ such that $BK \subset K$ and $a \geq r(B)$, where $\geq r(B)$ is the spectral radius of B. An M(K)-operator is said to have "property c", if $A = aI-B$ is convergent.

If K is a closed generating and normal cone in Y, then the set $K' = \{ x' \in Y : (x, x') \geq 0$ for all $x \in K \}$ forms also a closed generating and normal cone in Y - the so called <u>dual cone</u>.

Let Int K denote the norm-topological interior of K. We see that

$$\text{Int } K = K^d = \{ x \in K : (x, x') > 0 \text{ for all } x' \in K', x' \neq 0 \}.$$

The next results concern the convergence of iteration operators corresponding to subproper splittings. Let M^g be either M^+ or $M^{\#}$.

<u>Theorem 5.1</u> Let $A = M-Q$ be a subproper splitting. Let $M^gATK \subset K$ and $TK \subset K$, where $T = M^gQ$ and K is a closed generating normal cone. Then T is convergent.

<u>Theorem 5.2</u> Let $A \in B(Y)$ be symmetric ($A^* = A$) and let $(Ax, x) = w\,(Mx, x)$ for all $x \in Y$, $0 \leq w < 2$ and $A = M-Q$. If $T = M^gQ = 0$ in the sense of some closed generating normal cone $K \subset Y$ and if there is a vector $u \in K$ such that $M^gMu \in K^d$ and also $M^gAu \in K$, then T is convergent.

<u>Theorem 5.3</u> Let $A = M-Q$ be such that $TK \subset K$ with some closed generating normal cone $K \subset Y$, where $T = M^gQ$. Let A be positively semidefinite and M be positively definite. Then $r(T) \leq 1$.

Let $K_j \subset Y_j$, $j = 1,2$ be closed generating normal cones in Y_j respectively. A splitting $A = M-Q$ is called <u>weakly regular</u>, if $M^gK_2 \subset K_1$ and $QK_1 \subset K_2$.

If $Y_1 = Y_2 = Y$, $K_1 = K_2$ and $M^{-1}K \subset K$, $QK \subset K$, then the splitting $A = M-Q$ is called <u>regular</u>.

<u>Theorem 5.4</u> Let A be an $M(K)$-operator such that $Ae \in K$ for some $e \in K^d$ and let $A = M-Q$ be a weakly regular subproper splitting such that $TK \subset K$, $T = M^gQ$. Then $r(T) \leq 1$ and if $\lambda = 1 \in \sigma(T)$, then 1 is a simple pole of $(\lambda I - T)^{-1}$.

<u>Remark</u>. As shows the example of the Jacobi operator corresponding to the discrete Neumann problem the operator T in Theorems 5.3 and 5.4 in general is not convergent.

On the other hand, an application of Theorem 5.3 to the Gauss-Seidel iteration operator and the SOR-operator (w) $0 < w \leq 1$, for the same discrete Neumann problem yields convergence of these operators.

Let us assume that $A \in B(Y)$ is an $M(K)$-operator, where $K = R^n_+$. Let

$$P_n = \{ (j,k) : 1 \leq j, k \leq n, j \neq k \} .$$

A rather effective method of solving systems $Ax = b$ with particular A's is based on the following result.

<u>Proposition 5.1</u> Let $A = bI-B$ be an $M(R)$-matrix with K-irreducible B. Then, there exist for every $P \subset P_n$ a lower

triangular matrix $L = (l_{jk})$ with $l_{jj}=1$ and an upper triangular matrix $U = (u_{jk})$ and a matrix $Q = (q_{jk})$ such that the splitting $A = LU-R$ is weakly regular and

$$l_{jk} = 0 = u_{jk} \text{ for } (j,k) \in P \text{ and } q_{jk} = 0 \text{ for } (j,k) \notin P.$$

The factors L and U are unique and if $Q \neq 0$ then $LU - Q = A$ is a regular splitting.

If $A^* = A$ then there exist for every $P \subset P_n$ a lower triangular matrix $L = (l_{jk})$ and a matrix $Q = (q_{jk})$ such that the splitting $A = LL^* - R$ is weakly regular (if $Q \neq 0$ regular) and $l_{jk} = 0$ for $(j,k) \in P$.

<u>Theorem 5.5</u> If A, L, U, Q are defined as in Proposition 5.1, where $A = LU-Q$ is a regular splitting, then the process

$$(5.1) \qquad LUx_{k+1} = Qx_k + b, \; k = 0,1,\dots.$$

is convergent in the sense of Theorem 4.1 as soon as $\lambda \in \sigma(T)$, $|\lambda| = 1$, implies that $\lambda = 1$; here $T = (LU)^{-1}Q$.

<u>Remark</u>. If $A^* = A$ and $T = (LL^*)^{-1}Q$ fulfils the hypotheses of Theorem 5.5 then it is advantageous to solve (5.1) by a conjugate gradient method. This combination produces a very effective method the properties of which are similar to those described for the case $\det A \neq 0$ in (Meijerink, Van der Vorst 1977).

6. <u>Semiiterative methods</u>

Since the structure of the spectrum of a convergent operator is rather special we would like to broaden the applicability of iterative methods. A reasonable way is to consider <u>semiiterative methods</u> (see Eiermann, Niethammer, Varga 1985).

Let

$$p_k(z) = \sum_{t=0}^{k} p_{kt} z, \; p_{kt} \in C$$

be a polynomial such that

(6.1) $$p_{kk} \neq 0$$

and

(6.2) $$p_k(1) = 1, \; k = 0, 1, \ldots$$

Define a new iteration scheme by setting

(6.3) $$y_k = p_{k0} x_0 + \sum_{t=1}^{k} p_{kt} (x_t - (t-1)PM^g b),$$

where x_t is defined by (2.4).

Theorem 6.1 Let $A = M-Q$ be a subproper splitting such that $T = M^g Q = P+R$, $P^2 = P$, $PR = RP = 0$, $1 \notin \sigma(R)$. Let $\{p_k\}$, $k = 0, 1, \ldots$ be a sequence of polynomials such that (6.1) and (6.2) hold and that the sequence $\{p_k(R)\}$ is zero-convergent. Then the iteration process (6.3) is convergent and its limit $\bar{x} = \lim y_k$ is a generalized solution to $Ax = b$.

In particular we derive the following result as a combination of Theorems 5.5 and 6.1.

Theorem 6.2 Let $a \in (0, 1)$ and $p(z) = az + 1-a$. Let $A = M-Q$ be a weakly regular subproper splitting such that $T = M^g Q$ is K-irreducible, then the iterative process (6.3) with $p_k(z) = [p(z)]^k$ is convergent.

Remark. Theorem 6.2 applies for A being an M(K)-operator by combining a regular splitting LU-Q as described in Theorem 5.5.

We observe that the results of this section apply to convergent iterative processes of type (2.3). In that case we may try to find polynomials p_k in order to accelerate the rate of convergence of the semiiterations. However, a more important fact is that using even some simple polynomials like in Theorem 6.2 we are able to broaden an applicability of iterative methods essentially by choosing the polynomials p_k appropriately. Some methods of constructing suitable sequences p_k are presented in (Eiermann, Niethammer, Varga 1985) for the case of regular systems. As shown above the same methods apply to singular systems as well.

References

1. Ben Israel A., Greville T.N.E. Generalized Matrix Inverses. Theory and Applications. J.Wiley Publ., New York 1973.
2. Berman A., Neumann M. Proper splittings and rectangular matrices. SIAM J.Appl.Math. 31 (1976), 307-312.
3. Berman A., Plemmons R.J. Cones and iterative methods for least squares solutions. SIAM J.Numer Anal. 11(1977), 145-154.
4. Eiermann M., Niethammer W., Varga R.S. A study of semiiterative methods for nonsymmetric systems of linear equations. To be published in Numer.Math. 1985.
5. Krein M.G., Rutman M.A. Linear operators leaving invariant a cone in a Banach space. Usp.mat.nauk III (1948),

N 1, 3-95 (Russian). English translation in Translations of the Amer.Math.Soc. 26 (1950), 128 pp.

6. Marchouk G.I., Kuznetsov Yu.A. Iterative Methods and Quadratic Functionals. Nauka, Siberian branch, Novosibirsk (1972) (Russian).

7. Marek I. Iterative methods of solving linear systems with a rectangular matrix. Report 8(32), 1981, Mathematisch Instituut Katholieke Universiteit, Nijmegen, The Netherlands.

8. Marek I., Žitný K. An equivalence of K irreducibility concepts. Comment.Math.Univ.Carol. 25 (1984), 61-72.

9. Meijerink J.A., van der Vorst H.A. An iterative solution method for linear systems of which the coefficient matrix is a symmetrix M-matrix. Math.Comp. 31 (1977), 148-162.

10. Nashed Z. Generalized inverses, normal solvability and iteration for singular operator equations. In Nonlinear Functional Analysis and Applications. Academic Press, New York 1971, 311-359.

11. Neumann M. Some applications of partial orderings to iterative methods for rectangular linear systems. Linear Algebra Appl. 19 (1978), 95-116.

12. Neumann M. Subproper splittings of rectangular matrices. Linear Algebra Appl. 14 (1976), 41-51.

13. Neumann M., Plemmons R.J. Convergent nonnegative matrices and iterative methods for consistent linear systems. Numer.Math. 31 (1978), 265-279.

14. Neumann M., Plemmons R.J. Generalized inverse positivity and splittings of M-matrices. Linear Algebra Appl. 23 (1979), 21-35.

15. Ortega J., Rheinboldt W. Monotone iterations for non-linear equations with application to Gauss-Seidel methods. SIAM J.Numer.Anal. 4 (1967), 171-190.

16. Plemmons R.J. M-matrices leading to semiconvergent splitting. Linear Algebra Appl. 15 (1976), 243-252.

17. Plemmons R.J. Regular splittings and the discrete Neumann problem. Numer.Math. 25 (1976), 153-161.

18. Varga R.S. Matrix Iterative Analysis. Prentice Hall, Englewood, New Jersey 1962.

Numerical Approximation of Partial Differential Equations
E.L. Ortiz (Editor)

ON DIFFERENT NUMERICAL METHODS TO SOLVE SINGULAR BOUNDARY PROBLEMS.

Francisco Michavila
Departamento de Cálculo Numérico e Informática
E.T.S. de Ingenieros de Minas
Universidad Politécnica de Madrid

Abstract:

Two model problems are introduced, one bidimensional with mixed boundary conditions and other tridimensional linear elasticity case with crack, in order to analyze the types of possible singularities in boundary problems in general, the alterations due to these in error bounds, and the numerical methods for their solution. The singular isoparametric degenerate elements and those of the quadratic Akin type will be extensively shown thereafter, as well as the triangular ones of six nodes, and the quadrilateral ones of eight nodes, together with their corresponding shape functions, modified, radial behaviour, compatibility, etc. Comparative numerical results are obtained in different cases.

1 PRELIMINARIES

Let us consider both of the following elliptical problems:

Model problem P1: Be that $\Omega \subset \mathbb{R}^2$(Figure 1), and the boundary problem:

$$-\Delta u = f \text{ in } \Omega$$

$$u = g_1 \text{ in } \partial\Omega_1$$

$$\frac{\partial u}{\partial n} = g_2 \text{ in } \partial\Omega_2$$

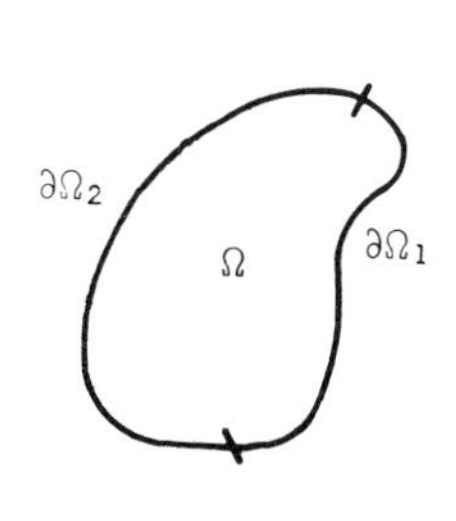

Figure 1

Its weak formulation is:

"Find $u \in H^1(\Omega)$, $u=g_1$ in $\partial\Omega_1$:

$$\int_\Omega \nabla u \nabla v \, dx\, dy = \int_\Omega fv \, dx\, dy + \int_{\partial\Omega_2} g_2 \, v \, ds \qquad (1.1)$$

$$\forall \; v \in H^1(\Omega),\; v=0 \text{ in } \partial\Omega_1\text{"}$$

Model problem P2: Be that $\Omega \subset \mathbb{R}^3$ (Figure 2) and let us consider the problem of elasticity:

$$[D]^t \{\sigma\} + \{f\} = \{0\} \text{ in } \Omega$$

$$[N]\{\sigma\} = \{\hat{\sigma}\} \text{ in } \partial\Omega$$

whose weak formulation is:

"Find $\{u\}\in\hat{H}^1(\Omega)$:

$$\int_\Omega ([D]\{v\})^t [E][D]\{u\}\, dxdydz = \int_\Omega \{v\}^t\{f\}\, dxdydz + \int_{\partial\Omega}\{v\}^t\{\hat{\sigma}\}ds \quad (1.2)$$

$$\forall \{v\}\in\hat{H}^1(\Omega)"$$

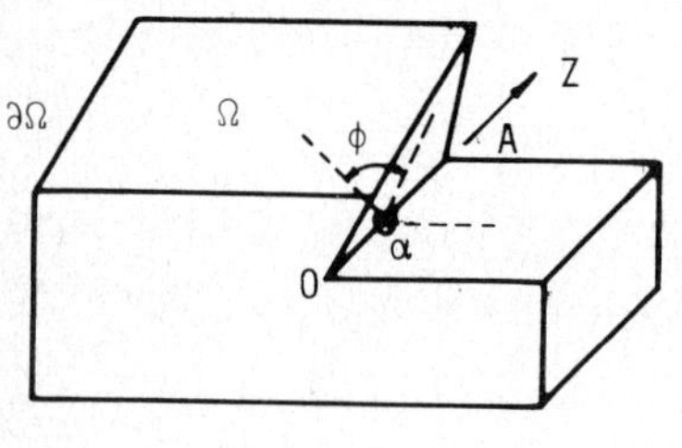

Figure 2

where $\hat{H}^1(\Omega) = \{\{v\} \mid v_x, v_y, v_z \in H^1(\Omega)\}$ and where the expressions of the constitutive law have been employed for an -- elastic, linear homogeneous and isotropic material, and the relation strain - displacement:

$$\{\sigma\} = [E]\{\varepsilon\}$$

$$\{\varepsilon\} = [D]\{u\}$$

Being Ω_h the geometrical appoximation - of Ω by an inscribed polygonal, and being Ω_e each one of the finite elements of the mesh constituting a Ω_h partition, if Ω_e is triangular -- with 6 nodes (Figure 3), the local approximation is made through a polynomial of this way:

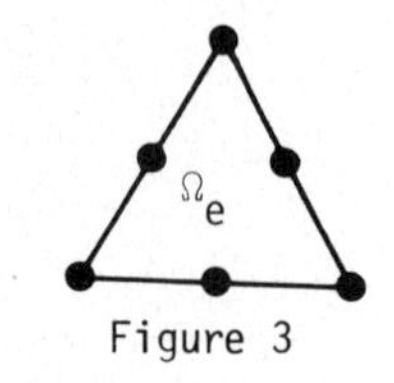

Figure 3

$$u_h|_{\Omega_e} = a_0+a_1x+a_2y+a_3x^2+a_4y^2+a_5xy \quad (1.3)$$

which is a complete one of order 2. If Ω_e is quadrilateral with 8 nodes (Figure 4), the local approximation - is:

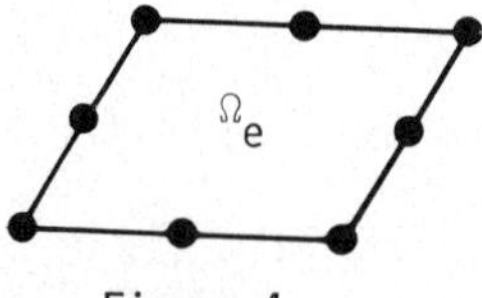

Figure 4

$$u_h|_{\Omega_e} = a_0+a_1x+a_2y+a_3x^2+a_4y^2+a_5xy + a_6x^2y+a_7xy^2 \quad (1.4)$$

whose order of completitude is also 2. For the discretization of P1, -- should it be that:

$$H^h = \{v_h\in H^1(\Omega_h),\ v_h|_{\Omega_e} \text{ is of the form (1.3) or (1.4)}\}$$

while the approximation to the problem will be:

"Find $u_h \in H^h$, $u_h = g_1$ in $\partial\Omega_{1h}$:

$$\int_{\Omega_h} \nabla u_h \nabla v_h \, dxdy = \int_{\Omega_h} f \, v_h \, dxdy + \int_{\partial\Omega_{2h}} g_2 \, v_h \, ds$$

$$\forall \, v_h \in H^h, \; v_h = 0 \text{ in } \partial\Omega_{1h}"$$

Being $\partial\Omega_{1h}$ and $\partial\Omega_{2h}$ the parts of $\partial\Omega_h$ respectively approaching $\partial\Omega_1$ and $\partial\Omega_2$. As the first member of (1-1) is a continous bilinear form and H^1-elliptic, acoording to CIARLET-RAVIART [1], we get to this:

$$||u-u_h||_1 \leqslant \frac{c}{\alpha} \, h^{\mu} \, ||u||_r \tag{1.5}$$

$$\mu = \min(2, r-1)$$

If $r \geqslant 3$:

$$||u-u_h||_1 \leqslant K.||u||_r.h^2$$

By which it is defined the convergence order for the standard approxima-- tion by finite elements of P1.

In order to discretize P2, we shall use a mesh of parallelepiped finite elements with 20 nodes (Figure 5), whose form functions are:

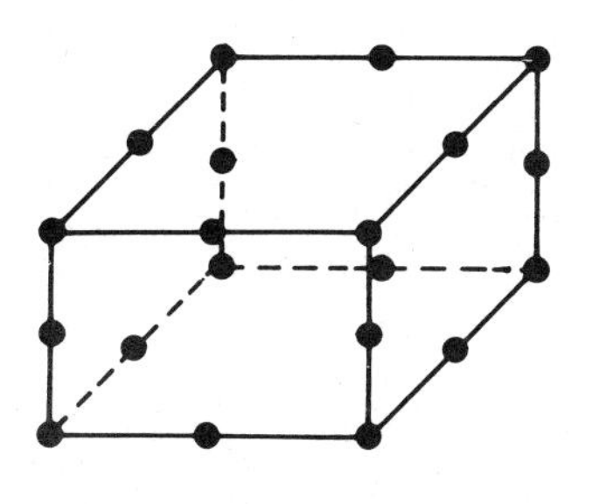

Figure 5

$$\hat{\psi}_i(\xi,\eta,\zeta) = \frac{1}{8}(1+\xi\xi_i)(1+\eta\eta_i)(1+\zeta\zeta_i) \,.$$
$$.\,(\xi\xi_i+\eta\eta_i+\zeta\zeta_i-2)\xi_i^2\,\eta_i^2\,\zeta_i^2 \;+$$
$$+\frac{1}{4}(1-\xi^2)(1+\eta\eta_i)(1+\zeta\zeta_i)(1-\xi_i^2) \;+$$
$$+\frac{1}{4}(1-\eta^2)(1+\xi\xi_i)(1+\zeta\zeta_i)(1-\eta_i^2) \;+$$
$$+\frac{1}{4}(1-\zeta^2)(1+\xi\xi_i)(1+\eta\eta_i)(1-\zeta_i^2)$$

$$i=1,2,\ldots,20$$

As (1.2) has a first member which is an $\hat{H}^1_-$ elliptic form, being - continous and linear, by (1.5) and if $r \geq 3$:

$$||u_x - u_{xh}||_1 \leq C.h^2$$

and the same thing goes for the u_y, u_z components.

2. SINGULARITIES

The singularities in boundary problems may be due to:

a) Irregular form of the boundary $\partial\Omega$.

b) Discontinuities of the boundary conditions.

c) Other different causes.

The singular behaviour of a boundary problem's solution is referred to the fact that either this or its partial derivatives may take infinitely big values in points of $\bar{\Omega}$. Such problems appear in crack mechanics, in heat transmission, or, in general, in a boundary problem presenting irregularities in its data.

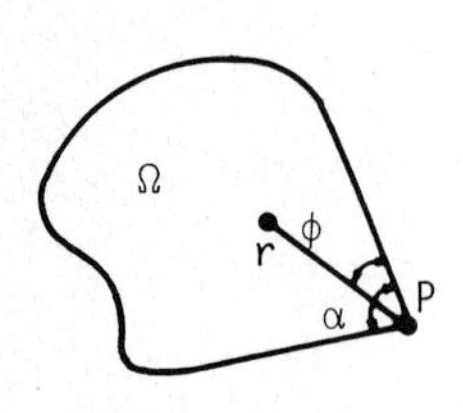

Figure 6

If $\partial\Omega$ shows an angular P point (Fig.6) with α inner angle, ODEN [2] establish that its solution is this way:

$$u(r,\theta) = \sum_{n=1}^{\infty} A_n . r^{n\pi/\alpha} \psi_n(\theta) + \sum_{n=1}^{\infty} \sum_{m=1}^{\infty} f_{mn} \chi_{mn}(r) \psi_n(\theta) \quad (2.1)$$

in P is neighbourhood.

The dominating singular term is $r^{\pi/\alpha} . \sin \frac{\pi\theta}{\alpha}$. In MICHAVILA and -- GAVETE [3] , the crack problem is studied in detail, for which $\alpha \simeq 2\pi$.

The singularity of P1, corresponding to discontinuity in boundary conditions, has been established by EMERY [4] and BIRKHOFF [5], its solution being:

$$u(r,\theta) = \tilde{u}\,(r,\theta) + \sum_{j=1}^{\infty} k_j . r^{j/2}\,(\sin j\,\frac{\theta}{2} + \frac{\sin j\,\frac{\pi}{2}}{\cos j\,\frac{\pi}{2}}\,\cos j\,\frac{\theta}{2})$$

where $\tilde{u}(r,\theta)$, is the solution's regular part, whose main singular term -- (2) is that of the $Kr^{1/2}$ form.

In the edge's neighbourhood OA of Figure 2, the solution of P2 -- behaves in a singular way, being the main term of the form $r^{\pi/\alpha} . \sin \frac{\pi\phi}{\alpha} Z$. In STEPHAN and WHITEMAN [6] , the problem of crack in 3D is analyzed in detail.

In P1, if $f \in C^{\infty}(\Omega)$, when Ω is of polygonal boundary (Figure 7) -- non convex calling:

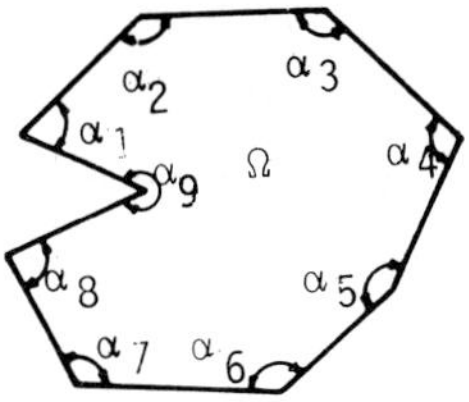

$$\beta_j = \frac{\pi}{\alpha_j}$$

$$\beta_M = \operatorname*{Sup}_{j} \beta_j \qquad (2.2)$$

Figure 7

one gets that:

$$u \in H^{1+\beta_M-\varepsilon}(\Omega), \qquad \varepsilon > 0$$

In particular, in rectangular domains without or with a crack -- (Figures 8 and 9) for the quadratic elements of Figures 3 and 4, the following boundary marks of error are obtained in $H^1(\Omega)$, respectively:

$\alpha_M = \frac{\pi}{2}$

Figure 8

$$||u-u_h||_{1,\Omega} \leqslant C\, h^{2-\varepsilon} ||u||_{3-\varepsilon,\,\Omega}$$

and

$$||u-u_h||_{1,\Omega} \leqslant C\, h^{1/2-\varepsilon} ||u||_{3/2-\varepsilon,\Omega}$$

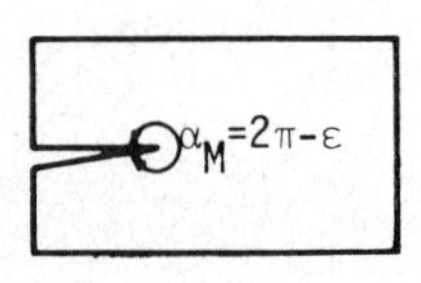

Figure 9

Also, for the seeked boundary mark of -- error, a different technique must be - - used. It consists of a local estimate in the neighbourhood of the singular points, in such a way that in the Figure 7, some zones should be established $\{\Omega_j\}_{j=1}^{9}$ (Figure 10), where the norm of the error is given by an expre- - ssion, different form the rest of domain $\tilde{\Omega} = \Omega \setminus (\bigcup_{j=1}^{9} \Omega_j)$. So,

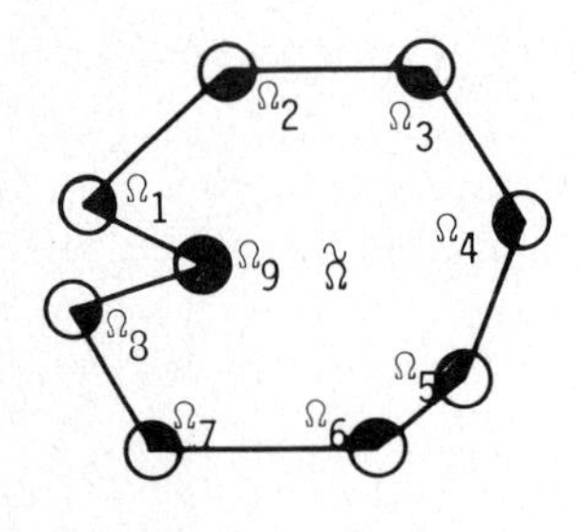

Figure 10

$$||u-u_h||_{1,\Omega_j} \leqslant C.h^{\min(2,\beta_j)-\varepsilon}.||u||_{\beta_j+1-\varepsilon,\Omega_j} \tag{2.3}$$

$$||u-u_h||_{1,\tilde{\Omega}} \leqslant C.h^2\, ||u||_{3,\tilde{\Omega}}$$

Another different technique proposed by SCHARTZ and WAHLBIN [7] allows to establish L^∞ boundary marks. Be that $\hat{\Omega}$ and Ω_0 so that:

$$\Omega_0 \subset\subset \hat{\Omega} \subset\subset \Omega$$

then, for quadratic elements, one gets:

$$||u-u_h||_{\infty,\Omega_0} \leqslant C.|||u-u_h|||_{-s,\hat{\Omega}}$$

where $|||.|||_{-s,\Omega}$ is the norm in $[H_0^s(\Omega)]'$, beings an arbitrary positive integer. The term $|||u-u_h|||_{-s,\hat{\Omega}}$ shows the influence of the irregularities local error, out of $\hat{\Omega}$.

If $s>\beta_M-1$:

$$|||u-u_h|||_{-s,\hat{\Omega}} \leqslant C(u)\ h^{2\beta M-\varepsilon} \tag{2.5}$$

Consequently, $||u-u_h||_{\infty,\Omega_0}$ is of the order of $h^{2\beta M-\varepsilon}$ if $1/2\leqslant \beta_M\leqslant 1$ that is to say, if Ω has a crack, or is L-formed.

For the crack and quadratic finite elements case, in other words, with complete polynomials of 2 grade for each element's approximation, following the results obtained by Schartz and Wahlbin, one gets:

$$\beta_M = \frac{1}{2} + \varepsilon'$$

$$||u-u_h||_{\infty,\Omega_0} \leqslant C\ |||u-u_h|||_{-s,\hat{\Omega}}$$

$$|||u-u_h|||_{-s,\hat{\Omega}} \leqslant C(u).h^{\min(4,1)-\varepsilon}$$

Therefore:

$$||u-u_h||_{\infty,\Omega_0} \leqslant C(u).h^{1-\varepsilon} \tag{2.6}$$

Analogically, in the case of a L-form, one would obtain:

$$||u-u_h||_{\infty,\Omega_0} \leqslant C(u).h^{4/3-\varepsilon}$$

3. NUMERICAL METHODS EMPLOYED IN THE TREATMENT OF SINGULARITIES.

The various kinds of approximation techniques are collected in an exhaustive manner by WHITEMAN and AKIN [8] , and, particularly, the use of degenerate isoparametric elements has been introduced by BARSOUM[9], as -- well as its extension to tridimensional problems whose singularity consists of the $K r^{-1/2}$ for to be systematized by HUSSAIN, COFFIN and ZALESKI [10].

We can group those numerical methods in the following manner:

a) Methods employing a local refining of mesh.
b) Methods consisting on an increase of test functions space (through the incorporation of singular functions to the base of approximating subs- pace).
c) Methods based upon the use of singular elements which, on their turn, can be classified like this:
   c-1) Use of degenerate isoparametric elements obtained through node -- displacement.
   c-2) Modification of base functions which are thus transformed into ra tionale polynomials.

The methods for mesh local refining consist of the use of sma- - ller diameter elements in singular points neighbourhood. This technique - does not require of any special modification in standard programas of fi- nite elements, but it is very costly. The number of used refining levels is automatically controlled by predetermined laws. The determination of elements size in the nearness of singular points, and the error local es- timates are established in SCHARTZ WAHLBIN [11].

In those methods based upon an increase of test functions space, the additional base functions are defined in a singular way, near the studied point and the globaly hold upon Ω themselves the continuity sought - after . If the solution $u(r,\theta)$ is written in the following way:

$$u(r,\theta) = \sum_{j=1}^{N} A_j \, \rho_j(r,\theta) + w(r,\theta) \qquad (3.1)$$

where $w \in H^2(\Omega)$ and the first term represent the solution's singular part, $\{A_j\}_{j=1}^{N}$ are constant, obtaining (Figure 11) that:

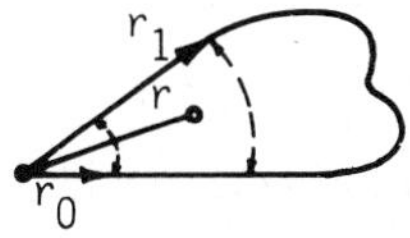

Figure 11

$$\rho_j(r,\theta) = \begin{cases} r^{\mu_j} \sin \mu_j\theta & 0 \leqslant r \leqslant r_0 \\ \gamma_j(r) \sin \mu_j\theta & r_0 \leqslant r \leqslant r_1 \\ 0 & r_1 \leqslant r \end{cases} \qquad (3.2)$$

whenever ρ_j is known, by a finite elements standard method, we obtain w_h, and one gets that:

$$u_h = \sum_{j=1}^{N} A_j^h \, \rho_j + w_h$$

in $H^h \subset H_0^1(\Omega)$ by means of appoximation by k grade polynomials upon each element of mesh. The increased space base is:

$$\{\rho_1, \rho_2, \ldots, \rho_N, \phi_1, \phi_2, \ldots, \phi_M\}$$

while the boundary mark for error is given by:

$$||u-u_h||_{1,\Omega} \leqslant C.h^k ||w||_k$$

This method is seldom used due to the fact that singular functions are not usually known in an explicit manner for elliptic problems - in general, and, besides, when they are known, it presents the additional inconveniance that the band estructure of the stiffness matrix gets destroyed.

4. SINGULAR ELEMENTS.

The degenerate isoparametrical elements, which were introduced by BARSOUM [9] and HENSHELL and SHAW [12] can be formed in a standard programme of finite elements, using isoparametric elements, through an adequate - localization of nodes. Various results and applications of these elements can be found in MICHAVILA, GAVETE and CONDE [13], CONDE [14] and MICHAVILA and GAVETE [15] . Said degenerate isoparametric elements have a quadrangular or a triangular shape. In MICHAVILA [16] , the conclusion is reached, for one side (1.2) of the generic element Ω_e of the mesh so that the point 1(x=0) is singular, that:

$$u_h^e(x,y)\Big|_{1-2} = -\frac{1}{2}\left(2\sqrt{\frac{x}{\ell}} - 1\right)\left(2-2\sqrt{\frac{x}{\ell}}\right)u_1^e +$$

$$+ \frac{1}{2}\left(-1+2\sqrt{\frac{x}{\ell}}\right) 2\sqrt{\frac{x}{\ell}}\, u_2^e + 4\left(\sqrt{\frac{x}{\ell}} - \frac{x}{\ell}\right) u_5^e$$

whose derivative presents a behaviour $O(r^{-1/2})$ in the neighbourhood of -- the singular point.

The degenerate isoparametric triangular element of six nodes is obtained directly out of the quadrilateral of eight nodes degenerate isoparametric collapsing one side.

In tridimensional problems, from an parallelepiped element of 20 nodes, a degenerate can be obtained by collapsing one of its sides. In -- this case, the jacobian of the standard element $\hat{\Omega}(\xi,\eta,\zeta)$ transformation, into one of the mesh $\Omega e(x,y,z)$ with nodes displaced to a 1/4 and a collapsed side is:

$$J = \begin{vmatrix} \ell/4\ (\xi+1)\ \{(1+\zeta)\cos\alpha+(1-\zeta)\} & 0 & \frac{\ell}{8}(\xi+1)^2(\cos\alpha-1) \\ 0 & \ell/2 & 0 \\ \ell/4\ (\xi+1)\ \text{sen}\ \alpha(1+\zeta) & 0 & \frac{\ell}{8}(\xi+1)^2\ \text{sen}\ \alpha \end{vmatrix} =$$

$$= \frac{\ell^3}{32}\ \text{sen}\ \alpha(\xi+1)^2$$

where α is the angle formed by both sides coinciding over the side of singular points.

A method's variant consists of transition elements use between - the singular element and the normal elements. In said transition elements, the nodes placed on the middle points of the sides are displaced to intermediate positions ranged between 1/2 and 1/4(normal position and "degenerate" position). This technique was introduced by LYNN and INGRAFFEA [17] , and several results about the approximation improvement due to said ele- - ments are presented in MICHAVILA, GAVETE and VICENTE [18] , MICHAVILA and GAVETE [15] , and MICHAVILA and GAVETE [19] .

Another method for singular elements generation, introduced by - AKIN [20] for two dimensions consists of shape functions modification without altering the nodes placement. With these singular elements, it is possible to approximate singularities of from $O(r^{-\lambda})$, $0<\lambda<1$(differently - - from degenerate isoparametric elements, where it is only possible to analyze the case $\lambda= 1/2$).

For bidimensional problems, shape functions for AKIN type elements get generated in the following fashion:

The singularity $r^{-\lambda}$ in a $\hat{\Omega}$ node (let us suppose $\hat{1}$ node) is obtained replacing standard form functions $\{\hat{\psi}_j\}_{j=1}^{N}$ by some other new ones $\{\tilde{\psi}_j\}_{j=1}^{N}$ generated form those through the following method. Be that:

$$\begin{aligned} W(\xi,\eta) &= 1-\hat{\psi}_1(\xi,\eta) \\ R(\xi,\eta) &= (W(\xi,\eta))^{\lambda} \end{aligned} \tag{4.1}$$

where (ξ,η) are the coordinates in $\hat{\Omega}$. The new shape functions are given by:

$$\begin{aligned} \tilde{\psi}_1(\xi,\eta) &= 1 - \frac{W(\xi,\eta)}{R(\xi,\eta)} \\ \tilde{\psi}_j(\xi,\eta) &= \frac{\hat{\psi}_j(\xi,\eta)}{R(\xi,\eta)} \qquad j = 2,3,....,N \end{aligned} \tag{4.2}$$

that is, rationale functions. With them, the approximation to the solution in the singular element is:

$$\hat{u}_h(\xi,\eta) = \sum_{j=1}^{N} u_j^e \, \tilde{\psi}_j(\xi,\eta)$$

The $\{\tilde{\psi}_j\}_{j=1}^{N}$ has the following properties:

a) $\sum_{j=1}^{N} \tilde{\psi}_j(\xi,\eta) = 1$

b) $\{\tilde{\psi}_j\}_{j=1}^{N}$ are compatibles.

In other words, the singular elements of AKIN type are conformed with the normal elements connected with them.

Both elements of this type, which are exhaustively studied in -- this work are, in conformity with P1 presentation, the six nodes triangular and the eight nodes quadrilateral:

a) Akin type six nodes triangular element.

The singularity is found in 1^e node (Figure 12). The shape functions corresponding to the standard element are:

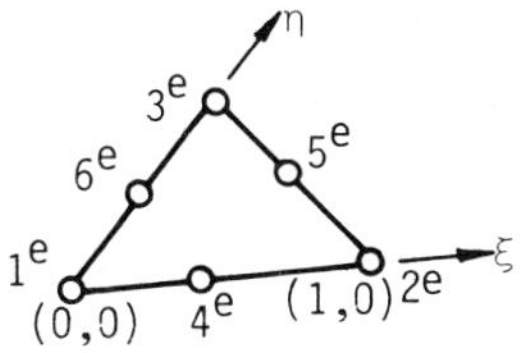

$$\hat{\psi}_1 = (1-\xi-\eta)(1-2\xi-2\eta)$$

$$\hat{\psi}_2 = \xi(1-2\xi)$$

$$\hat{\psi}_3 = \eta(1-2\eta)$$

$$\hat{\psi}_4 = 4\xi(1-\xi-\eta)$$

$$\hat{\psi}_5 = 4\xi\eta$$

$$\hat{\psi}_6 = 4\eta\,(1-\xi-\eta)$$

Figure 12

They should be:

$$W(\xi,\eta) = \xi(2\xi + 4\eta - 3) + \eta(2\eta-3)$$

$$R(\xi,\eta) = \left[\xi(2\xi + 4\eta - 3) + \eta(2\eta-3)\right]^{\lambda}$$

The modified shape functions corresponding the singular element are:

$$\tilde{\psi}_1 = 1- \left[\xi(2\xi+4\eta-3) + \eta(2\eta-3)\right]^{1-\lambda}$$

$$\tilde{\psi}_2 = \xi(1-2\xi)\ \left[\xi(2\xi+4\eta-3)+\eta\ (2\eta-3)\right]^{-\lambda}$$

$$\tilde{\psi}_3 = \eta(1-2\eta)\ \left[\xi(2\xi+4\eta-3) + \eta(2\eta-3)\right]^{-\lambda}$$

$$\tilde{\psi}_4 = 4\xi\ (1-\xi-\eta)\left[\xi(2\xi+4\eta-3) +\eta\ (2\eta-3)\right]^{\lambda}$$

$$\tilde{\psi}_5 = 4\xi\eta\ \left[\xi(2\xi+4\eta-3)+ \eta(2\eta-3)\right]^{-\lambda}$$

$$\tilde{\psi}_6 = 4\eta\ (1-\xi-\eta)\ \left[\xi(2\xi+4\eta-3)+ \eta(2\eta-3)\right]^{-\lambda}$$

The expression of the approximation in the singular element in the case $\lambda=1/2$ is:

$$u_h = u_1^e \left[1-\sqrt{\xi(2\xi+4\eta-3) + \eta(2\eta-3)}\right] + \left[\xi(2\xi+4\eta-3)+\eta(2\eta-3)\right]^{-1/2} \cdot$$

$$\cdot \left[u_2^e\, \xi(1-2\xi) + u_3^e\, \eta(1-2\eta) + u_4^e\, 4\xi(1-\xi-\eta) + u_5^e\, 4\xi\eta + u_6^e\, 4\eta(1-\xi-\eta)\right]$$

Its radial behaviour for $\lambda = 1/2$ is (with $\xi = ar$, $\eta = br$ where r is the distance to the singularity) the following:

$$u_h = u_1^e \{1-\left[(2a^2+4ab+2b^2)r^2+(-3a-3b)r\right]^{1/2}\} + \left[(2a^2+4ab+2b^2)r^2 +\right.$$

$$\left. + (-3a-3b)r\right]^{-1/2} \cdot \left[u_2^e(ar-2a^2r^2) + u_3^e\,(br-2b^2r^2) +\right.$$

$$\left. + u_4^e(4ar - 4a^2r^2 - 4abr^2) + u_5^e\, 4\,abr^2 + u_6^e(4\,br - 4abr^2-4b^2r^2)\right] =$$

$$= u_1^e \left[1-r^{1/2}.(c_1+c_2\,r)^{1/2}\right] + \frac{r^{1/2}}{(c_1+c_2\,r)^{1/2}}\,(c_3+c_4\,r) =$$

$$= u_1^e + k_1\,r^{1/2} + k_2\,r^{3/2} + k_3\,r^{5/2} + \ldots..$$

which fits the successive potencies of the solution singular part.

The solution behaviour in every one of the sides of the singular element is as follows:

In 1^e-2^e ($\eta = 0$):

$$u_h\Big|_{\eta=0} = u_1^e\,\{1-[\xi(2\xi-3)]^{1-\lambda}\} + u_2^e\,\frac{\xi(1-2\xi)}{[\xi(2\xi-3)]^{\lambda}} + u_4^e\,\frac{4\,\xi(1-\xi)}{[\xi(2\xi-3)]^{\lambda}}$$

In 1^e-3^e ($\xi = 0$):

$$u_h\Big|_{\xi=0} = u_1^e\,\{1-[\eta(2\eta-3)]^{1-\lambda}\} + u_3^e\,\frac{\eta(1-2\eta)}{[\eta(2\eta-3)]^{\lambda}} + u_6^e\,\frac{4\,\eta(1-\eta)}{[\eta(2\eta-3)]^{\lambda}}$$

In 2^e-3^e $(\xi+\eta=1)$:

$$u_h\big|_{\xi+\eta=1} = u_2^e\ \xi(1-2\xi)+u_3^e(1-\xi)(2\xi-1)+u_5^e\ 4\xi\ (1-\xi)$$

Consequently, its behaviour on this side is quadratic and the element is -- compatible with others of same number of nodes, but of the standard type in the mesh.

b) Akin type eight nodes quadrilateral element.

The singularity is found in 1^e node (Figure 13).Shape functions corresponding the standard element are:

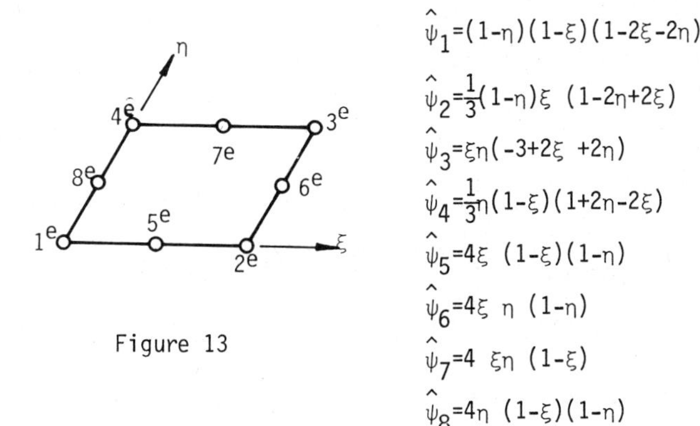

Figure 13

$$\hat{\psi}_1=(1-\eta)(1-\xi)(1-2\xi-2\eta)$$

$$\hat{\psi}_2=\frac{1}{3}(1-\eta)\xi\ (1-2\eta+2\xi)$$

$$\hat{\psi}_3=\xi\eta(-3+2\xi\ +2\eta)$$

$$\hat{\psi}_4=\frac{1}{3}\eta(1-\xi)(1+2\eta-2\xi)$$

$$\hat{\psi}_5=4\xi\ (1-\xi)(1-\eta)$$

$$\hat{\psi}_6=4\xi\ \eta\ (1-\eta)$$

$$\hat{\psi}_7=4\ \xi\eta\ (1-\xi)$$

$$\hat{\psi}_8=4\eta\ (1-\xi)(1-\eta)$$

Let us form the functions:

$$W(\xi,\eta)=3(\xi+\eta)+5\ \xi\eta+\ 2\ (\xi^2+\eta^2)-2\ \xi\eta(\xi+\eta)$$

$$R(\xi,\eta)=[3(\xi+\eta)+5\xi\eta\ +\ 2\ (\xi^2+\eta^2)-2\ \xi\eta(\xi+\eta)]^\lambda$$

shape functions modified for the singular element are:

$$\tilde{\psi}_1 = 1 - [3(\xi+\eta) + 5\xi\eta + 2(\xi^2+\eta^2) - 2\xi\eta(\xi+\eta)]^{1-\lambda}$$

$$\tilde{\psi}_2 = \frac{1}{3}\xi(1-\eta)(1+2\xi-2\eta).R(\xi,\eta)^{-1}$$

$$\tilde{\psi}_3 = \xi\eta\,(-3+2\xi+2\eta).R(\xi,\eta)^{-1}$$

$$\tilde{\psi}_4 = \frac{1}{3}\eta(1-\xi)(1+2\eta-2\xi).R(\xi,\eta)^{-1}$$

$$\tilde{\psi}_5 = 4\,\xi(1-\xi)(1-\eta).R(\xi,\eta)^{-1}$$

$$\tilde{\psi}_6 = 4\xi\eta(1-\eta).R(\xi,\eta)^{-1}$$

$$\tilde{\psi}_7 = 4\xi\eta(1-\xi).R(\xi,\eta)^{-1}$$

$$\tilde{\psi}_8 = 4\eta\,(1-\xi)(1-\eta).R(\xi,\eta)^{-1}$$

The expression of approximation in the singular element of case $\lambda = 1/2$ is:

$$u_h = [u_1^e\,\{1 - [3(\xi+\eta)+5\,\xi\eta + 2(\xi^2+\eta^2) - 2\,\xi\eta(\xi+\eta)]^{1/2}\} +$$

$$+ [3(\xi+\eta)+5\,\xi\eta+2\,(\xi^2+\eta^2)-2\,\xi\eta(\xi+\eta)]^{-1/2}\,.\,[u_2^e.\frac{\xi}{3}(1-\eta)(1+2\xi-2\eta) +$$

$$+ u_3^e\,\xi\eta(-3+2\xi+2\eta)+u_4^e\,\frac{\eta}{3}\,(1-\xi)(1+2\eta-2\xi)+u_5^e\,4\,\xi(1-\xi)(1-\eta) +$$

$$+ u_6^e\,4\,\xi\eta(1-\eta)+u_7^e\,4\,\xi\eta(1-\xi) + u_8^e\,4\eta(1-\xi)(1-\eta)]$$

Its radial behaviour for $\lambda = 1/2$ (with $\xi = ar$, $\eta = br$) is:

$$u_h = u_1^e\,\{1-r^{1/2}.\,[3(a+b)+5abr+2(a^2r+b^2r)-2abr(ar+br)]^{1/2}\}+$$

$$+ r^{1/2}\,.[3(a+b)+5abr+2(a^2r+b^2r)-2abr(ar+br)]^{-1/2}\,.[u_2^e\,\frac{a}{3}\,(1-br)\,.$$

$$(1+2ar-2br)+u_3^e\,abr\,(-3+2ar+2br)+u_4^e\,\frac{b}{3}\,(1-ar)(1+2b-2ar) +$$

$$+ u_5^e\,4a\,(1-ar)(1-br)+u_6^e\,4abr\,(1-br) + u_7^e\,4\,abr\,(1-ar) +$$

$$+ u_8^e\,4b\,(1-ar)(1-br)] = u_1^e + K_1\,r^{1/2} + K_2\,r^{3/2} + K_3\,r^{5/2}+\ldots$$

what indicates that, when r tends to zero, the singular solution approxima tion is correct.

The solution behaviour on singular element sides is:

In 1^e-2^e (η= 0):

$$u_h\Big|_{\eta=0} = u_1^e \left[1-(3\xi+2\xi^2)^{1-\lambda}\right] +$$

$$+ \left[3\,\xi+ 2\xi^2\right]^{-\lambda} \left[u_2^e \frac{\xi}{3} \,.\, (1+2\xi)+u_5^e \;\; 4\xi\;(1-\xi)\right]$$

In 1^e-4^e(ξ= 0):

$$u_h\Big|_{\xi=0} = u_1^e \left[1-(3\eta+2\eta^2)^{1-\lambda}\right]+ \left[3\eta+2\eta^2\right]^{\lambda}.\left[u_4^e \frac{\eta}{3}\,(1+2\eta)+u_8^e\; 4\eta\;(1-\eta)\right]$$

On sides 2^e-3^e and 3^e-4^e the behaviour is quadratic, and consequently, the element is compatible with a mesh of standard elements of same form and sa me number of nodes.

5. NUMERICAL RESULTS

As an application of singular elements, let us solve an ellipti-- cal, order 2 problem with mixed boundary conditions introducing not adequa tely treatable singularities through a finite element method H^1- conformed classical.

Let us consider the rectangular domain Ω= $\{(x,y)\,|\,0\leqslant x\leqslant 1,\ 0\leqslant y\leqslant 0{,}5\}$ (Figure 14) and let us define the problem on Ω :

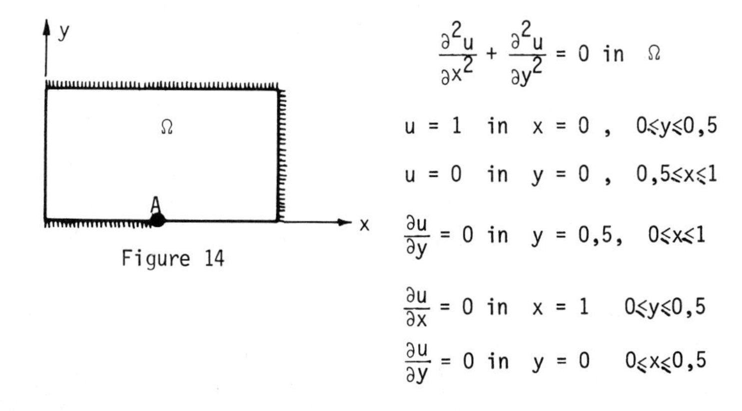

Figure 14

$$\frac{\partial^2 u}{\partial x^2} + \frac{\partial^2 u}{\partial y^2} = 0 \text{ in } \Omega$$

$u = 1$ in $x = 0$, $0\leqslant y\leqslant 0{,}5$

$u = 0$ in $y = 0$, $0{,}5\leqslant x\leqslant 1$

$\frac{\partial u}{\partial y} = 0$ in $y = 0{,}5$, $0\leqslant x\leqslant 1$

$\frac{\partial u}{\partial x} = 0$ in $x = 1$ $0\leqslant y\leqslant 0{,}5$

$\frac{\partial u}{\partial y} = 0$ in $y = 0$ $0\leqslant x\leqslant 0{,}5$

By that has been above said, there is a singularity of the type $r^{-1/2}$ in A, as $u=k\, r^{1/2}$ being K a constant, and by deriving, we shall obtain:

$$\frac{du}{dr} = \frac{K}{2}\,\frac{1}{r^{1/2}}$$

This problem has been solved by using a model (Fig. 15) with degenerate isoparametric elements, and quadratic Akin type elements. The near by zone to the singular point has been (Figures 16 and 17) to study the influence of a transition ele-- ment.

In order to establish a comparaison allowing us determine the model effect in the singularity's approximation the problem - has been also solved by using C° quadrilateral elements with 4 nodes, progress - - ively increasing mesh of elements taking as lengths of sides of squares 1/4, 1/8, 1/12, 1/16 and 1/20. On figure 18 the obtained results can be observed.

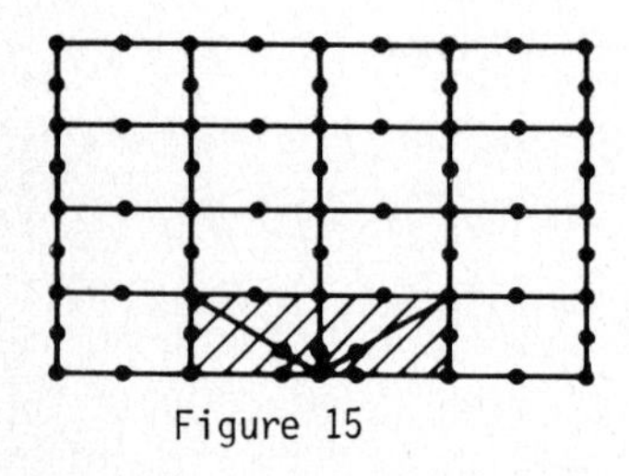
Figure 15

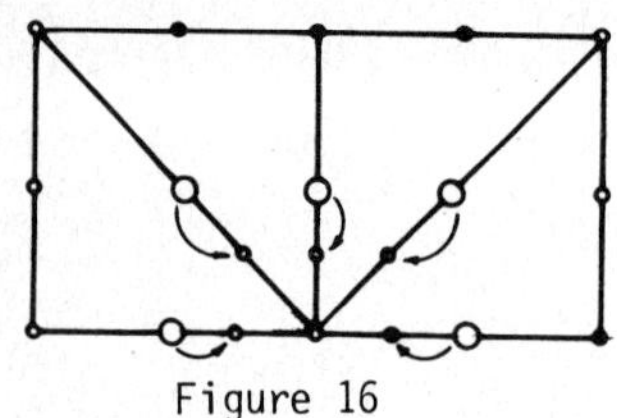
Figure 16

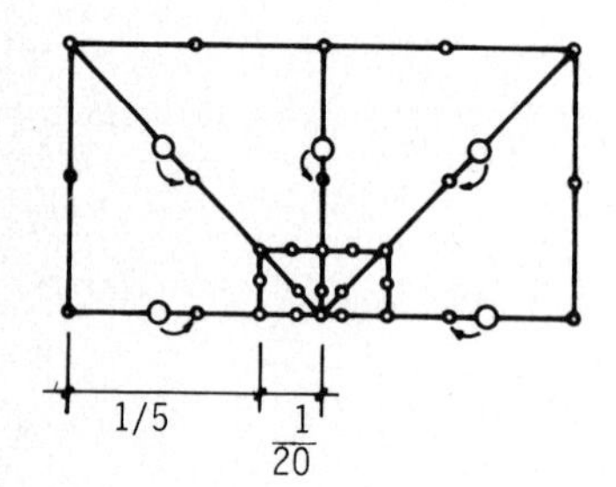

Figure 17

Has been represented $\left(\frac{\partial u}{\partial y}\right)_{y=0}$ for $0<x<0,5$ by using central finite differences to approximate the derivative. One can see how the approxima-- tion improves by refining the model and how, in its turn, improves still - more by using the quadratic degenerate isoparametric element.

On figures 19 and 20 the $\left(\frac{\partial u}{\partial y}\right)_{x=0,5}$ and $\left(\frac{\partial u}{\partial y}\right)_{y=0,125}$ have been res-- pectively represented for the two cases of 6 node triangular Akin type elements, and 8 node quadrilateral elements and the triangular degenerate isoparametric element.

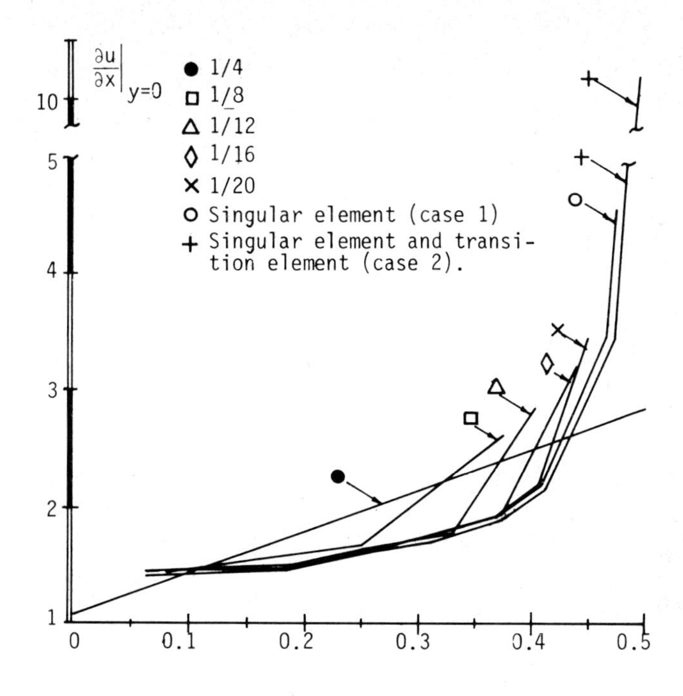

Figure 18

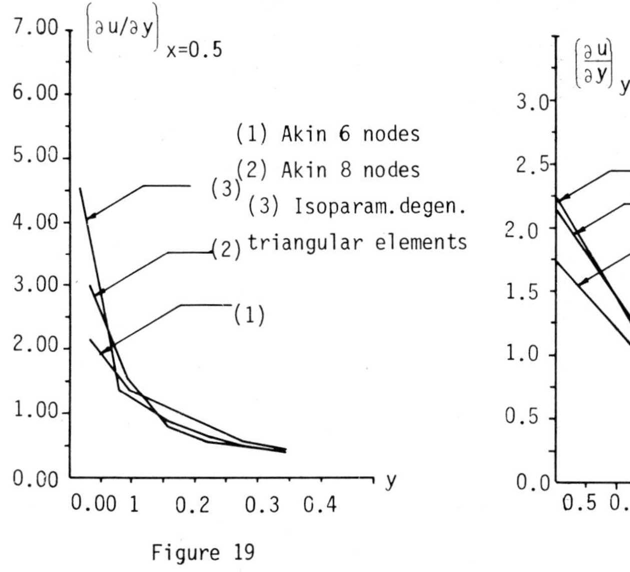

Figure 19

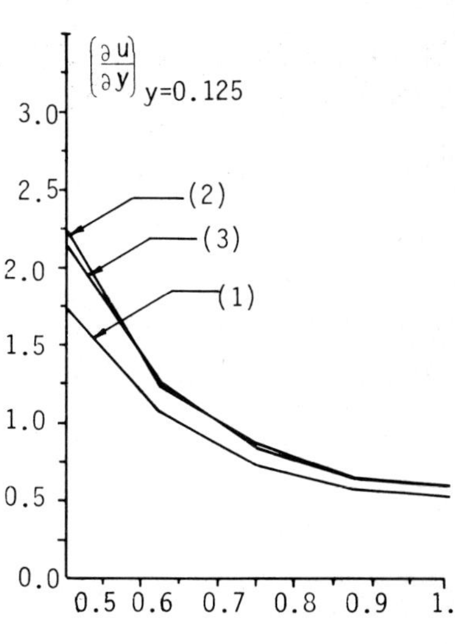

Figure 20

The 3-Dimension crack models make a very urgent problem nowadays, in mechanical engineering, considering that pipe and pressure vessel cracks can be detected, beforte the said pressure broken. Width and length intervene very directly in this models, and even the crack form is important to it.

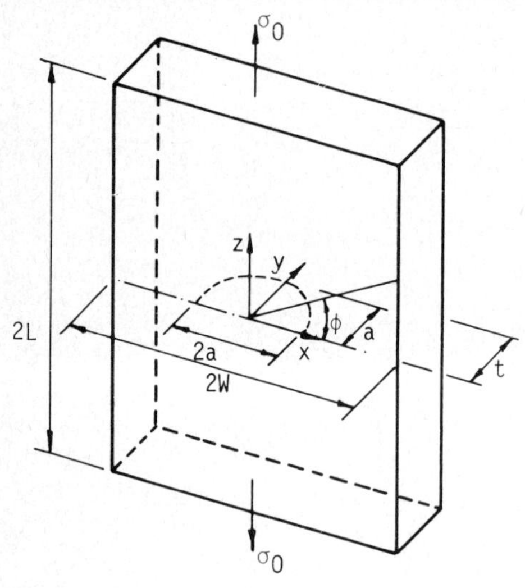

Figure 21

Figure 21, shows , subject to a uniform stress field and having a circular center crack Data are a/t = 0,4, L/t=2.5 and a/w = 0.2.

Figure 22 model has been car-- ried out.

Degenerate singular elements use in 3 dimensions when model contains crack curved edges,it requires certain additional conditions due to suplementary conditions appearance in - the singular element jacobian estimate. -- That implies, on Figure 22 model, to modelize crack with straight elements, which - provokes to eliminate the advantage implied in the use of isoparametric elements, considering that the discretization error increases on boundary, as it can't curved the sides. The obtained results are shown on - Figure 23.

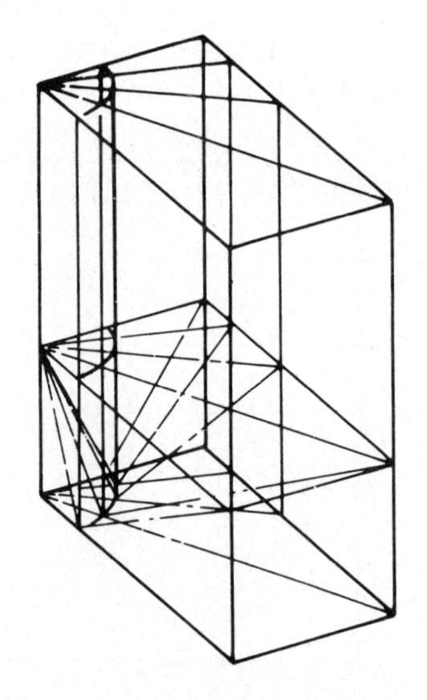

Figure 22

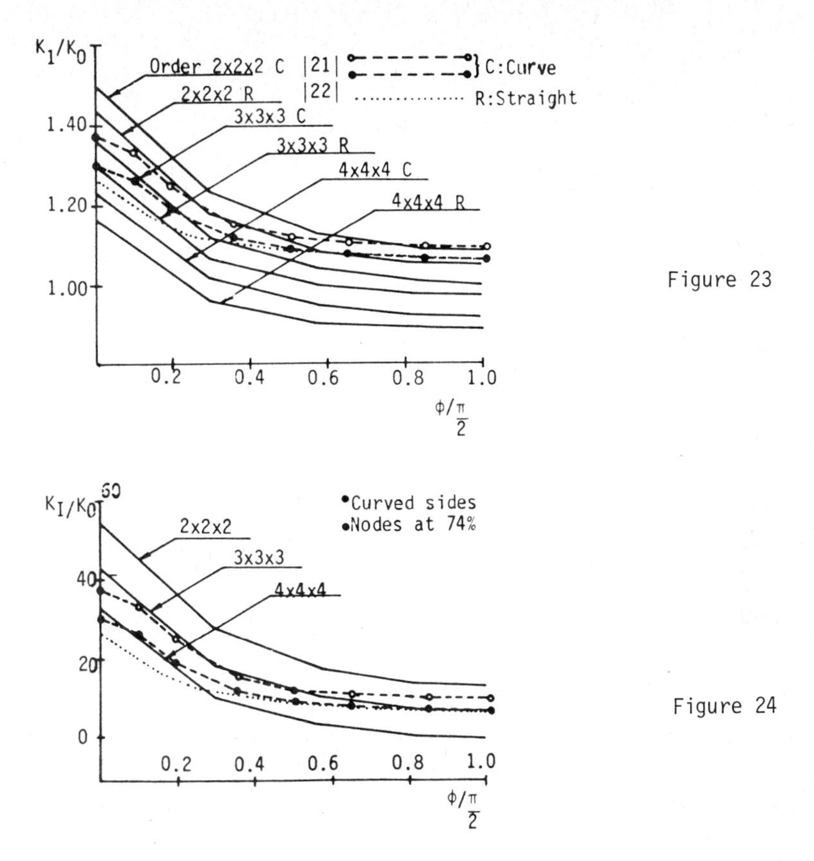

Figure 23

Figure 24

The performed errors can reach up to 10% and even lower, which represents an acceptable error level, if we consider that reference models of MIYAZAKI, -- WATANABE and YAGAWA [21] and YAGAWA, ICHIMIYA and ANDO [22] have more elements. In general, these errors depend on the use of straight crack edges or - - curb ones, on the numerical integration order, on the diameter and on the elements form of model.

In some cases, a negative pivot is obtained during the -- factorization which is carried out in the equations system solution, this occurs in some curb elements in which the jacobian may become negative.

A simple and effective numerical method, according to PEANO and PASINI [23] that avoids numerical difficulties, is locating the nodes of degenerate singular elements, very near by but not coinciding with the -- side's fourth part. This has been used by moving nodes 1% on Figure 24.

BIBLIOGRAPHY

[1] CIARLET, RAVIART:"General and Hermite Interpolation in $\mathbb{R}^n$ with applications to finite elements". Arch. Rat. Mech. Anal. Vol. 46, 177-199 (1972).

[2] ODEN: "Finite elements, vol II" Ed. Prentice Hall (1983).

[3] MICHAVILA, GAVETE: "On the use of quadratic transition element proceedings of Thrid Int. Conf. Num. Meth. Fract. Mech. (1984).

[4] EMERY: "The use of singularity programming in Finite-Difference and Finite-Element Computations of temperature". Trans. ASME, J. of Heat-transfer, 344-351 (1973).

[5] BIRKHOFF: "Angular singularities of Elliptic Problems".Jour. of Approx. Theory 6, 215-230 (1972).

[6] STEPHAN, WHITEMAN: "Singularities of the Laplacian at corners and edges of three-dimensional domains and their treatment with finite elements methods". Tech. Rep. Bicom 81/1. Ins. Comp. Math. Brunel (1981).

[7] SCHARTZ, WAHLBIN: "Maximum norm estimates in the finite element method on plane polygonal domains". Part I, Math. Comp 32, 73-109 (1978).

[8] WHITEMAN, AKIN: "Finite elements singularities and fracture" MAFELAP 1978, 35-54, Academic Press (1979).

[9] BARSOUM: "On the use of isoparametric finite elements in linear fracture mechanics". Int. J. of Num. Meth. in Eng. 10, 25-37 (1976).

[10] HUSSAIN, COFFIN, ZALESKI: "Three dimensional singular element" Comp. and Struct. Vol. 13, 595-599 (1981).

[11] SCHARTZ, WAHLBIN: "Maximum Norm Estimates in the finite element method on Plane Polygonal Domains. Part 2, Refinements". Math. Comp. 465-492 (1979).

[12] HENSHELL, SHAW: "Crack tip finite elements are unnecesary" Int. J. Num. Meth. Eng 12, 1031-1036 (1978).

[13] MICHAVILA, GAVETE, CONDE: "Una aplicación de elementos finitos isoparamétricos degenerados en la modelización de ciertas singularidades que aparecen en problemas de grietas". Anal. Ing. Mec. 1, 104--113 (1982).

[14] CONDE: "Mecánica de fractura de materiales metálicos" Unv. Ver. Gijón, (1984).

[15] MICHAVILA, GAVETE: "Aproximación de un problema elíptico singular - mediante elementos finitos isoparamétricos degenerados". Rev. Real Acad. Ciencias (to appear).

[16] MICHAVILA: "Técnicas de elementos finitos para la resolución de problemas elípticos bidimensionales con singularidades en el contorno" Conf. Cong. Iberoam. Met. Num. y Comp. (to appear).

[17] LYNN, INGRAFFEA: "Transition elements to be used with quarter-point crack-tip elements". Int. J. Num. Meth. Eng. 12, 1031-1036 (1978).

[18] MICHAVILA, GAVETE, VICENTE: "Elementos finitos singulares y de transición, empleando la aproximación cuadrática en el estudio de fractu_ ras en 2 dimensiones" Anal. Ing. Mec. 2, 118-127 (1982).

[19] MICHAVILA, GAVETE: "Sobre el tratamiento numérico de singularidades tridimensionales" Rev. Real Acad. Ciencias (to appear).

[20] AKIN: "The generation of elements with singularities" Int. J. Num. Meth. Vol. 10, 1249-1259 (1976).

[21] MIYAZAKI, WATANABE, YAGAWA: "Calculation of stress intensification factors of surface cracks in complex structures: applications of -- efficient computer program EPAS-J1". Num. Eng. and Design. Vol 68, 71-85 (1981).

[22] YAGAWA, ICHIMIYA, ANDO: "Two and three-dimensional analysis of stress intensity factors based on discretization error in finite elements" Num. Meth. in Fracture Mech. 149-167. Pineridge Press (1978).

[23] PEANO, PASINI: "A warning against misuse of quarter point elements" Int. J. Num. Meth. Eng. 314-319 (1982).

Numerical Approximation of Partial Differential Equations
E.L. Ortiz (Editor)

SOME NUMERICAL TECHNIQUES FOR THE SOLUTION OF PROBLEMS RELATED TO SEMICONDUCTOR DEVICES

John J. H. Miller
Numerical Analysis Group, Trinity College
Dublin, Ireland

In this paper we derive the basic equations governing the behaviour of semiconductor devices. We then examine several reformulations of the equations in terms of different dependent variables. The necessity of generating, automatically, local mesh refinements is then discussed, and the advantages of a Delaunay triangulation are described. In the final section a method is given for discretising the equations in a systematic manner, which will produce the standard generalisation of the Scharfetter-Gummel discretisation to two dimensions currently in use. The paper concludes with a brief annotated bibliography.

1. Introduction

The electrical and thermal behaviour of semiconductor devices under various operating conditions are of great importance to electrical engineers, because such devices are the basic building blocks in the construction of integrated circuits. Since it is generally not feasible to produce prototypes of these devices, it is necessary to seek either computational or analytic models. As we shall see below, the nonlinear nature of the governing equations and the complexity of the device geometries means that analytic solutions can be obtained only under very restrictive assumptions about the operating conditions and the types of device. For this reason considerable efforts have been made during the last 20 years to develop computer codes to model devices computationally. The principal uses of such codes are the interactive computer aided design of new devices and the derivation of accurate device parameters for use in circuit modelling codes.

2. Derivation of the Basic Equations

The governing equations for the electromagnetic and thermal field in a semiconductor device are obtained from Maxwell's equations, the heat flow equation and the constitutive relations. The former are

$$\dot{\underline{D}} - \nabla \times \underline{H} = -\underline{J}$$

$$\dot{\underline{B}} + \nabla \times \underline{E} = 0$$

$$\nabla \cdot \underline{D} = \rho$$

$$\nabla \cdot \underline{B} = 0$$

where the dot denotes partial differentiation with respect to time, $\underline{B}$ and $\underline{H}$ are respectively the magnetic induction and field, $\underline{D}$ and $\underline{E}$ are respectively the electric displacement and field, $\underline{J}$ is the conduction current density and ρ is the charge density. The electric displacement and field are linked by the relations

$$\underline{D} = \varepsilon \underline{E}$$

where ε is the permittivity. This is in general a tensor of rank two, but semiconductor materials in normal use are isotropic and homogeneous. Thus, in each separate region of a device ε is a scalar constant.

From the above equations we see that $\underline{B}$ is divergence free, which means that we can introduce a vector potential $\underline{A}$ satisfying

$$\underline{B} = \nabla \times \underline{A} \quad , \quad \nabla \cdot \underline{A} = 0$$

Plugging this expression for $\underline{B}$ into the second Maxwell equation we obtain

$$\nabla \times (\dot{\underline{A}} + \underline{E}) = 0$$

which shows that $\dot{\underline{A}} + \underline{E}$ is rotation free. It follows that there is a scalar electrostatic potential ψ satisfying

$$\dot{\underline{A}} + \underline{E} = -\nabla \psi$$

We can now express $\underline{D}$ in terms of these potentials as

$$\underline{D} = \varepsilon \underline{E} = -(\dot{\underline{A}} + \nabla \psi)$$

Plugging this expression into the third Maxwell equation we obtain

$$\nabla \cdot (\varepsilon \dot{\underline{A}} + \varepsilon \nabla \psi) = -\rho$$

But ε is homgeneous so

$$\nabla \cdot (\varepsilon \dot{\underline{A}}) = \varepsilon \nabla \cdot \dot{\underline{A}} = \varepsilon \frac{\partial}{\partial t}(\nabla \cdot \underline{A}) = 0$$

since $\underline{A}$ is divergence free. From this we conclude that in each region of a device (defined by the requirement that ε is homogeneous there)

$$\nabla \cdot (\varepsilon \nabla \psi) = -\rho$$

This is Poisson's equation for the electrostatic potential. We now take the divergence of both sides of the first Maxwell equation and use the third equation to eliminate $\underline{D}$, thus obtaining the current continuity equation

$$\nabla \cdot \underline{J} = -\dot{\rho}$$

The total charge ρ in a semiconductor device can be expressed by

$$\rho = q(p - n + N)$$

where q denotes the elementary charge, p is the concentration of holes, n is the concentration of electrons and N is the net concentration of impurities from the doping. Plugging this expression for ρ into Poisson's equation gives the first of our basic equations

$$\nabla \cdot (\varepsilon \nabla \psi) = q(n - p - N) \tag{2.1}$$

The total conduction current density $\underline{J}$ in a semiconductor device can be decomposed into the sum of the carrier current densities $\underline{J}_p$ and $\underline{J}_n$ due to the conduction band holes and electrons respectively, so

$$\underline{J} = \underline{J}_p + \underline{J}_n$$

Assuming that N does not change with time we have

$$\dot{\rho} = q(\dot{p} - \dot{n})$$

Using these expressions to eliminate $\underline{J}$ and $\dot{\rho}$ from the current continuity equation we obtain

$$\dot{p} + (1/q)\nabla \cdot \underline{J}_p = \dot{n} - (1/q)\nabla \cdot \underline{J}_n$$

This suggests introducing the new quantity R equal to the common value on each side of this relation and called the net recombination rate. Thus

$$\dot{n} - (1/q)\nabla \cdot \underline{J}_n = -R \tag{2.2}$$

$$\dot{p} + (1/q)\nabla \cdot \underline{J}_p = -R \tag{2.3}$$

To obtain expressions for the carrier current densities $\underline{J}_n$ and $\underline{J}_p$ we recall that, from irreversible thermodynamics, these are proportional to the gradients of the corresponding electrochemical potentials or quasi-Fermi potentials ϕ_n and ϕ_p respectively. This means that

$$\underline{J}_n = -q\mu_n n\nabla\phi_n \quad , \quad \underline{J}_p = -q\mu_p p\nabla\phi_p$$

where μ_n and μ_p are the mobilities of the appropriate carriers.

To relate these quasi-Fermi potentials to the carrier concentrations and the electrostatic potential we use the classical Boltzmann approximation to Fermi-Dirac statistics and we obtain the equations

$$n = n_i e^{\alpha(\psi-\phi_n)} \quad , \quad p = n_i e^{\alpha(\phi_p-\psi)}$$

where $\alpha = q/(kT)$, k is the Boltzmann constant, T is the absolute temperature and n_i is the intrinsic concentration.

Using these expressions to eliminate the quasi-Fermi potentials, we obtain the following expressions for the carrier currents

$$\underline{J}_n = -q\mu_n(n\nabla\psi - (1/\alpha)\nabla n) \tag{2.4}$$

$$\underline{J}_p = -q\mu_p(p\nabla\psi + (1/\alpha)\nabla p) \tag{2.5}$$

Finally, the heat flow equation is

$$\rho c\dot{T} - \nabla \cdot (\kappa\nabla T) = H \tag{2.6}$$

where ρ, c and κ denote respectively the mass density, the specific heat density and the thermal conductivity of the semiconductor material and H is the locally generated heat.

The set of equations (2.1) – (2.6) form the basis for the computational modelling of semiconductor devices. It is normal to eliminate the carrier current densities from the other equations using (2.4) and (2.5). This gives the following system of four scalar equations for the four quantities ψ, n, p and T

$$\nabla \cdot (\varepsilon \nabla \psi) = q(n - p - N) \tag{2.7}$$

$$\dot{n} + \nabla \cdot (\mu_n (n \nabla \psi - (1/\alpha) \nabla n)) = -R \tag{2.8}$$

$$\dot{p} - \nabla \cdot (\mu_p (p \nabla \psi + (1/\alpha) \nabla p)) = -R \tag{2.9}$$

$$\rho c \dot{T} - \nabla \cdot (\kappa \nabla T) = H \tag{2.10}$$

We note that ε, μ_n, μ_p, ρ, c and κ are always positive and so the first of these is an elliptic equation while the remainder are parabolic equations, provided of course that R and H do not contain partial derivitives of the dependent variables with respect to time or higher than first order with respect to space.

In what follows we ignore the equation of heat flow, so we consider only the system of equations (2.7) – (2.9) and take T as a constant.

3. Reformulation of the Equations

In the system (2.7) – (2.9) the dependent variables are ψ, n and p. If we now change to the dependent variables ψ, ϕ_n and ϕ_p, where ϕ_n and ϕ_p are the quasi-Fermi potentials, the system becomes

$$\nabla \cdot (\varepsilon \nabla \psi) = q(n_i e^{\alpha(\psi - \phi_n)} - n_i e^{\alpha(\phi_p - \psi)} - N) \tag{3.1}$$

$$\alpha n_i e^{\alpha(\psi - \phi_n)} (\dot{\phi}_n - \dot{\psi}) - \nabla \cdot (\mu_n n_i e^{\alpha(\psi - \phi_n)} \nabla \phi_n) = R \tag{3.2}$$

$$\alpha n_i e^{\alpha(\phi_p - \psi)} (\dot{\phi}_p - \dot{\psi}) - \nabla \cdot (\mu_p n_i e^{\alpha(\phi_p - \psi)} \nabla \phi_p) = -R \tag{3.3}$$

These variables have been used successfully in numerical computations. One of their advantages is that they are all of the same order of magnitude. Another is that we are guaranteed positive carrier concentrations, even in the presence of numerical errors, because of the relations

$$n = n_i e^{\alpha(\psi - \phi_n)} \quad , \quad p = n_i e^{\alpha(\phi_p - \psi)}$$

A disadvantage is the exponential nonlinearity in all three equations (3.1) – (3.3).

Another commonly used set of dependent variables is ψ, η and ρ where η and ρ are the variables

$$\eta = e^{-\alpha\phi_n} \quad , \quad \rho = e^{\alpha\phi_p}$$

In terms of these variables the system of equations becomes

$$\nabla \cdot (\varepsilon \nabla \psi) = q(n_i e^{\alpha\psi}\eta - n_i e^{-\alpha\psi}\rho - N) \tag{3.4}$$

$$n_i e^{\alpha\psi}(\dot{\eta} + \alpha\dot{\psi}\eta) - \nabla \cdot ((1/\alpha) n_i \mu_n e^{\alpha\psi} \nabla \eta) = -R \tag{3.5}$$

$$n_i e^{-\alpha\psi}(\dot{\rho} - \alpha\dot{\psi}\rho) - \nabla \cdot ((1/\alpha) n_i \mu_p e^{-\alpha\psi} \nabla \rho) = -R \tag{3.6}$$

and the relations between these variables and the carrier concentrations are

$$\eta = \frac{n}{n_i} e^{-\alpha\psi} \quad , \quad \rho = \frac{p}{n_i} e^{\alpha\psi}$$

At room temperature the approximate value of α is $\alpha \approx 40$. This means that, for the electrostatic potential ψ varying from $+10$ to -10, both η and ρ are multiplied by factors ranging between e^{400} and e^{-400}. The main disadvantage therefore of these variables is the problem of overflow and underflow in their floating point representation on most computers, except for problems where the electrostatic potentials are very small in magnitude. The main advantage of these variables is that in the stationary case ($\dot{\psi} = \dot{\eta} = \dot{\rho} = 0$) the equations reduce to

$$\nabla \cdot (\varepsilon \nabla \psi) = q(n_i e^{\alpha\psi}\eta - n_i e^{-\alpha\psi}\rho - N) \tag{3.7}$$

$$\nabla \cdot ((1/\alpha) n_i \mu_n e^{\alpha\psi} \nabla \eta) = R \tag{3.8}$$

$$\nabla \cdot ((1/\alpha) n_i \mu_p e^{-\alpha\psi} \nabla \rho) = R \tag{3.9}$$

which are formally self adjoint differential equations. If these are then discretised in an appropriate way the numerical approximation to their solution may be obtained by solving symmetric systems of linear equations, rather than the unsymmetric systems that must be solved for the other choices of the variables.

Another choice of variables arises if we adopt the stream function approach. This appears to be satisfactory in two dimensions but may be unwieldly in three dimensions. Details of this may be found in the monograph of Mock(1983)

We remark now that if restrict ourselves to the stationary case and we use the variables ψ, η and ρ, each of the equations (3.7) – (3.9) is in the conservation form

$$\nabla \cdot (c \nabla u) = f$$

where u is ψ, η and ρ ; c is ε, $(1/\alpha)n_i\mu_n e^{\alpha\psi}$ and $(1/\alpha)n_i\mu_p e^{-\alpha\psi}$; f is $q(n_i e^{\alpha\psi}\eta - n_i e^{-\alpha\psi}\rho - N)$, R and R for (3.7), (3.8) and (3.9) respectively.

4. Meshes for Finite Difference and Finite Element Methods

In a typical semiconductor device there are regions where the dependent variables change extremely rapidly, and other regions where they vary only gradually. For this reason it is essential to construct meshes which are fine in some regions and coarse in others. It should be noted that the option of constructing a sufficiently fine uniform mesh is not available, because this would lead to an impossibly large number of mesh points, especially for two and three dimensional problems. The conclusion therefore is that the spatial mesh must be non-uniform, and this is true for both finite difference and finite element methods.

The next important consideration about the required non-uniform mesh is that normally we do not know in advance where the different regions are located. Some information will be available from the physical insight of the device designer, but the full picture will be available only after a satisfactory numerical approximation has been computed. This apparent paradox is resolved by allowing the mesh to change as the solution process develops. This means that we must construct an adaptive mesh, and that this must be done in an automatic way, since a manual technique would be impossibly slow.

In order to achieve the required automatic mesh refinement a computable criterion must be specified, which can be used to determine whether or not additional mesh points should be added to the current mesh. Such criteria usually start from some variant of the principle that the local truncation error of the discretisation should be of the same order of magnitude at all points of the mesh. Some simplification of this requirement is then made in order to obtain an economical criterion that can be used in practice.

When a local refinement is required in a given mesh, it must be done in such a way that appropriate changes are induced in the discretisation of the problem. It is well known that this is not an easy problem for finite difference methods on a non-uniform mesh, and it may not be completely natural even for finite element discretisations involving rectangles. In both of these cases an appropriately refined mesh is constructed, which often makes use of some triangular elements. However, one case in which local refinement can be achieved in a completely natural way is with a triangular mesh. This is because an additional mesh point can be placed in the interior of any triangle and an acceptable new mesh is obtained by simply joining this point to the vertices of the triangle in which it lies. The original triangle is replaced therefore by three new subtriangles, and a local refinement of the mesh is achieved.

A particularly elegant triangulation of a region is the Delaunay triangulation. Given N points this triangulation maximises the minimum angle in the triangulation. Any triangulation having this property is called a Delaunay triangulation. To investigate the conditions under which such a triangulation is unique we consider the set of all pairs of triangles having a common edge. If the four distinct vertices belonging to any pair of triangles in this set lie on a circle, then the Delaunay triangulation is not unique, otherwise it is unique. The four vertices of the pair of triangles form a quadrilateral having the common edge of the triangles as one of the diagonals. We can form a different pair of triangles from this quadrilateral simply by annihilating the given diagonal and replacing it with the other diagonal. The new triangulation obtained is still a Delaunay triangulation, but this would not be the case if the four vertices had not lain on a circle. We conclude that the Delaunay triangulation is unique apart from the possibility of switching diagonals as explained above.

Given a set of N points $\{x_i\}_1^N$ in the plane we can characterise their Delaunay triangulation as follows. For each point x_i we introduce its Dirichlet polygon D_i, defined to be the set of all points in the plane that are closer to x_i than to any other point x_j, $j \neq i$. The set of all such polygons is called the Dirichlet tessellation of the plane. The dual of the Dirichlet tessellation is constructed in the following way. Any two points x_i and x_j are joined by a straight line segment if and only if their Dirichlet polygons have an edge in common. The result is a triangulation whose edges are perpendicular to the edges of the Dirichlet polygons. This triangulation is the Delaunay triangulation of the N points. Note that we can recover the Dirichlet polygons simply by constructing the perpendicular bisectors of the edges of the triangles. Because we started from a Delaunay triangulation there are no problems with self intersecting polygons as is the case with arbitrary triangulations.

5. Discretisation Techniques

We recall from §3 that when we are using the dependent variables ψ, η and ρ in the stationary case all three equations in the system may be written in the conservative form

$$\nabla \cdot (c\nabla u) = f$$

We now multiply both sides of this equation by a smooth test function ψ, on Ω, which vanishes on the boundary $\partial\Omega$ of Ω. Letting S and T denote respectively the solution and test space we have the following weak form of the problem

Find $u \in S$ such that

$$\int_\Omega \nabla \cdot (c\nabla u)\psi = \int_\Omega f\psi \quad \text{for all } \psi \in T$$

Integrating by parts and using the fact that ψ vanishes on $\partial\Omega$ we get

$$-\int_\Omega c\nabla u \cdot \nabla\psi = \int_\Omega f\psi$$

To discretise this we choose N nodes $\{x_i\}_1^N$ in Ω and on its boundary $\partial\Omega$. We join the nodes on $\partial\Omega$ by straight line segments and we approximate Ω by Ω_h, the polygonal domain determined by these line segments. We assume that the remaining nodes are all in Ω_h and we let $\{D_i\}_1^N$ be the Dirichlet polygonals of the nodes. Using these N nodes we now construct finite dimensional solution and test spaces S_h and T_h of dimension N corresponding to the Delaunay triangulation of Ω_h. We take S_h to be a space of elements corresponding to this triangulation, while T_h is some other space with a basis $\{\psi_i\}_1^N$ say. The discretised problem may then be written in the form

Find $u^h \in S_h$ such that

$$-\int_{\Omega_h} c\nabla u^h \cdot \nabla\psi_i = \int_{\Omega_h} f\psi_i \quad \text{for } i = 1, \ldots, N$$

Various choices of T_h are possible. One possibility is the following using an idea of P. Mole. We take T_h to be the space of piecewise constants obtained from the basis functions $\{\psi_i\}_1^N$, where $\psi_i(x) = 1$ if $x \in D_i$ and $\psi_i(x) = 0$ otherwise. Then $\nabla\psi_i(x) = 0$ unless $x \in \partial D_i$ and the equations become

$$-\int_{R_i} c\nabla u^h \cdot \nabla\psi_i = \int_{D_i} f$$

where R_i denotes a small domain containing ∂D_i. Integrating by parts we get

$$-\int_{\partial R_i} c\frac{\partial u^h}{\partial \nu}\psi_i + \int_{R_i} \nabla \cdot (c\nabla u^h)\psi_i = \int_{D_i} f$$

where $\frac{\partial}{\partial \nu}$ denotes differentiation along the outward normal to ∂R_i. Noting that $\partial R_i = \partial R_i^+ \cup \partial R_i^-$, where ∂R_i^+ is outside D_i and ∂R_i^- is inside D_i, we can write the above equation as

$$-\int_{\partial R_i^-} c\frac{\partial u^h}{\partial \nu} + \int_{R_i} \nabla \cdot (c\nabla u^h)\psi_i = \int_{D_i} f$$

Noting that $\nabla \cdot (c\nabla u^h)$ and ψ_i are bounded in R_i, we see that the second integral vanishes when we let area $R_i \to 0$. We have in the limit

$$-\int_{\partial R_i^-} c\frac{\partial u^h}{\partial \nu} = \int_{D_i} f$$

or

$$\int_{\partial D_i} c\frac{\partial u^h}{\partial \nu} = \int_{D_i} f$$

where for consistency $\frac{\partial}{\partial \nu}$ now denotes differentiation along the outward normal to ∂D_i. This choice of T_h has given us an integral form of the equation, which generalises to Delaunay triangulations the well known box integration method for rectangular meshes.

We now consider the approximation of the integrals. We let e denote a typical edge of the polygon ∂D_i and we introduce the following notation :

x_e the neighbouring node to x_i on the line perpendicular to e

ℓ_e the length of e

u_i the value of u at x_i

u_e the value of u at x_e

A_i the area of D_i

We rewrite the integral equation as

$$\sum_{e\in\partial D_i} \int_e c\frac{\partial u^h}{\partial \nu} = \int_{D_i} f$$

Then we lump the values of the first integrand at the point $(x_i + x_e)/2$ and the values of the second at the node x_i. This leads to the approximate equation

$$\sum_{e\in\partial D_i} \ell_e\left(c\frac{\partial u^h}{\partial \nu}\right)\left(\frac{x_i + x_e}{2}\right) = A_i f(x_i)$$

We now need to express the value of $\frac{\partial u^h}{\partial \nu}$ at the point $(x_i + x_e)/2$, by its values at the nodes x_i and x_e. From physical reasoning we expect $c\frac{\partial u^h}{\partial \nu}$ to be more regular than $\frac{\partial u^h}{\partial \nu}$. The simplest assumption we can make is that on the line segment joining x_i and x_e

$$c\frac{\partial u^h}{\partial \nu} = J$$

where J is some constant. Dividing by c and integrating from x_i to x_e we obtain

$$u_e - u_i = J \int_{x_i}^{x_e} \frac{1}{c}$$

and so

$$J = \frac{u_e - u_i}{\int_{x_i}^{x_e} \frac{1}{c}}$$

We conclude that, with the above assumption,

$$(c\frac{\partial u^h}{\partial \nu})(\frac{x_i + x_e}{2}) = \frac{u_e - u_i}{\int_{x_i}^{x_e} \frac{1}{c}}$$

Plugging this into the approximate equation we obtain the difference equation

$$\sum_{e \in \partial D_i} \frac{\ell_e}{\int_{x_i}^{x_e} \frac{1}{c}} (u_e - u_i) = A_i f(x_i)$$

This is a generalisation to conservation laws in two dimensions and Delaunay triangulations of the discretisation of conservation laws in one dimension which is examined in a paper of Miller. Using the methods described there the standard extension of the Scharfetter-Gummel method to two dimensions may be obtained from the above, and new discretisations may also be derived. Details of these will be published later.

Bibliography

Monographs :

Mamoru Kurata, "Numerical Analysis for Semiconductor Devices" Lexington Books, Massachusetts (1982)

M. S. Mock, "Analysis of Mathematical Models of Semiconductor Devices" Boole Press, Dublin (1983)

Siegfried Selberherr, "Analysis and Simulation of Semiconductor Devices" Springer-Verlag, Vienna (1984)

The first and third of these books are written by electrical engineers and they provide excellent introductions to the subject. The second is written by a mathematican and it contains practically all of the mathematical analysis in the subject to date.

Journals :

Solid-State Electronics – An International Journal, Pergamon Press, Oxford

Emphasizes the association of theory with practice in the borderland between solid-state physics and circuit engineering

IEEE Transactions on Electron Devices, IEEE, New York

Publishes original contributions to the theory, design and performance of active electron and ion devices, solid-state devices, integrated electronic devices, and energy sources.

COMPEL – The International Journal for Computation and Mathematics in Electrical and Electronic Engineering, Boole Press, Dublin

A journal devoted exclusively to numerical and analytical methods in all areas of electrical and electronic engineering.

Proceedings of Conferences :

"Numerical Analysis of Semiconductor Devices – Proceedings of the NASECODE I Conference", eds. B. T. Browne and J. J. H. Miller, Boole Press, Dublin (1979)

"Numerical Analysis of Semiconductor Devices and Integrated Circuits – Proceedings of the NASECODE II Conference", eds. B. T. Browne and J. J. H. Miller, Boole Press, Dublin (1981)

"NASECODE III – Proceedings of the Third International Conference on the Numerical Analysis of Semiconductor Devices and Integrated Circuits", ed. J. J. .H. Miller, Boole Press, Dublin (1983)

"Joint Special Issue on Numerical Simulation of VLSI Devices", eds. W. Fichtner and D. J. Rose, Transactions on Electron Devices, ED-30, 9, (1983) also SIAM Journal of Scientific and Statistical Computing , 4, 3 , (1983)

"Simulation of Semiconductor Devices and Processes – Proceedings of an International Conference" eds. K. Board and D. R. J. Owen, Pineridge Press, Swansea (1984)

"NASECODE IV – Proceedings of the Fourth International Conference on the Numerical Analysis of Semiconductor Devices and Integrated Circuits", ed. J. J. H. Miller, Boole Press, Dublin (1985)

Lecture Notes :

"An Introduction to the Numerical Analysis of Semiconductor Devices and Integrated Circuits – Lecture Notes of a Short Course", ed. J. J. H. Miller, Boole Press, Dublin (1981)

"Finite Element Programming with Special Emphasis on Semiconductor Device and Process Modelling – Lecture Notes of a Short Course", ed. J. J. H. Miller, Boole Press, Dublin (1983)

"New Problems and New Solutions for Device and Process Modelling – Lecture Notes of a Short Course", ed. J. J. H. Miller, Boole Press, Dublin (1985)

Selected Papers :

W. van Roosbroeck, Bell System Technical Journal, 29, 560 – 607, (1950)

The first statement of the basic semiconductor equations in essentially the form used nowadays.

H. K. Gummel, "A Self-Consistent Iterative Scheme for One-Dimensional Steady State Transistor Calculations", IEEE Trans. Electron Devices, ED-11, 455 – 465, (1964)

The classic paper on the numerical solution of the basic equations.

D. L. Scharfetter and H. K. Gummel, "Large-Signal Analysis of a Silicon Read Diode Oscillator" IEEE Trans. Electron Devices, ED-16, 64 – 77, (1969)

The original description of the most successful discretisation known to date.

W. L. Engl, H. K. Dirks and B. Meinerzhagen, "Device Modelling", Procs. IEEE, 71, 1, 10 – 33, (1983)

An excellent modern review article.

B. J. McCartin, "Discretization of the Semiconductor Device Equations" in New Problems and New Solutions for Device and Process Modelling, Boole Press, Dublin, (1985)

An excellent tutorial on Delaunay triangulation and related matters.

J. J. H. Miller, "On the Inclusion of the Recombination Term in Discretisations of the Semiconductor Device Equations", Mathematics and Computers in Simulation (to appear)

Numerical Approximation of Partial Differential Equations
E.L. Ortiz (Editor)

RECENT PROGRESS IN THE NUMERICAL TREATMENT OF SINGULAR PROBLEMS FOR PARTIAL DIFFERENTIAL EQUATIONS WITH TECHNIQUES BASED ON THE TAU METHOD

Eduardo L. Ortiz
Imperial College
London SW7 2BZ - England

A B S T R A C T

In this paper we report on recent research work on the use of new formulations of the Tau Method for the numerical solution of singular problems defined by partial differential equations. The problems considered here are relevant to the field of computational fracture mechanics and have attracted considerable attention in the recent literature. Results relate to Laplace´s and the biharmonic equations.

Comparisons of results obtained by using the Tau Method and other well established methods, such as specificly designed techniques based on the Finite Element Method, the Boundary Integral Method and Collocation, are reported in this paper. These comparisons show that the Tau Method is a versatile technique capable of giving results of a very high accuracy near the singular point with relatively low degrees of approximation.

Three formulations of the Tau method are used in this paper for the numerical treatment of singular problems. The first of them, the Tau-Lines approach is a hybrid technique. The second is based on a multidimensional formulation of the Tau Method in terms of Tau elements. In the third one the basis of representation currently used in the Tau Method is augmented with nonpolynomial elements to take into account the singular behaviour of the solution.

1.- Introduction

In recent years the classical Tau Method, a high accuracy approximation technique proposed by Lanczos in 1938 (see [1]) and largely ignored for a long period, has been given different new formulations. They made possible its application to the numerical solution of complex problems for ordinary and systems of ordinary differential equations and related eigenvalue problems in the presence of stiffness and singular behaviour.

Recent numerical and theoretical work has shown that the Tau Method is an optimal technique in the sense that for a given degree n, an approximation constructed by using the Tau Method and a best uniform approximation by algebraic polynomials have an error of approximation of the same order (see Ortiz [2], Freilich and Ortiz [3] and Onumanyi and Ortiz [4].)

In this paper we discuss several formulations of the Tau Method for the numerical treatment of partial differential equations. The first of them is based on the reduction of the given problem to one for a system of ordinary differential equations. It takes advantage of the efficiency of Tau Method as a boundary value problem solver. Two other formulations are briefly sketched in this paper; with them it is possible to treat partial differential equations preserving unmodified the given differential operator; therefore, taking full advantage of the Tau Method approach. Theoretical results (see Namasivayam and Ortiz [5]) show that for some classes of partial differential equations the optimality results indicated before are still valid (see Ortiz and Samara [6] for numerical examples).

Results obtained with the Tau Method are compared with those reported by other authors and obtained by using several of today standard techniques: Finite Elements, the Boundary Integral Method and Collocation. In most examples the techniques used for comparison were specificly designed or adapted to the problems considered. Numerical results reported in this paper show that the Tau Method can achieve comparable accuracy with relatively low degrees of approximation and with remarkable simplicity in its implementation.

The research reported here has been carried out and discussed at the Numerical Approximation Seminar, Imperial College, London, mainly by Drs. J.H. Freilich, E.M. El Misiery, K.M. Liu, S. Namasivayam, P. Onumanyi, A. Pham-Ngoc-Dinh, K.-S. Pun, H. Samara and by the author of this paper.

2.- Discrete-analytic techniques

The Method of Lines is a technique for the reduction of the numerical solution of a given partial differential equation to that of a partial differential equation with a smaller number of independent variables. This is achieved by the discretization of one (or more) of the variables and the replacement of the partial derivatives taken with respect to such variable by a finite difference estimate. Such process obviously introduces a restriction of the original domain to the union of a number of subdomains defined on varieties of a lower dimension. The number of subdomains is related to the number of points in the discretization scheme.

In the case of a partial differential equation involving two variables, discretization of one of them reduces the original domain, defined in R^2, to a series of one dimensional lines. On each of such lines a parametric function of the non-discretized variable is defined. Such reduction of the given domain into a collection of lines gives the method its name. However, it does not do full justice to it since the method can be applied in far more general situations.

Since the parametric functions defined on different lines are interrelated by the discretization formulae, the differential equations satisfied on each of such lines form a system of coupled ordinary differential equations, which is a model of the original one.

The Method of Lines is a very attractive approach and has lead to particularly interesting applications. If the resulting system of ordinary differential equations can be integrated analytically, it leads to a semi-analytic technique. This is, however, rarely the case in concrete applications.

Therefore, the numerical approximation of such system of ordinary differential equations must in general be considered. However, further use of discretization destroyes the semi-analytic character of the method, which is one of the attractive features of the Method of Lines and differentiates it from ordinary Finite Difference techniques based on discretization of all variables.

3.- A hybrid approach: the Tau-Lines Method

The optimality results mentioned before suggest that the Tau Method, an accurate boundary value solver, could be used in association with the Method of Lines. Its use preserves, without further alteration, the differential operator defined by the non-discretized variables. That is, it preserves the semi-analytical character of the model. Furthermore, the Tau approximations are continuously differentiable functions on each of the lines of the model. Such approach is known as the Tau-Lines Method.

The Tau-Lines Method can be developped by using either the recursive formulation of the Tau Method (see Ortiz [7]) or the operational approach (see Ortiz and Samara [8]-[9]). The first possibility has been discussed in a series of recent papers by El Misiery and Ortiz (see [10]-[11]). These authors applied it to the numerical treatment of singular partial differential equations related to crack problems, in the field of fracture mechanics. They have considered problems governed by either Laplace´s or the biharmonic equation. The second approach, based on the operational formulation of the Tau Method, has been followed by Liu and Ortiz who have used it extensively in connection with eigenvalue problems for partial differential equations (see [12]). This last approach, which is well documented in [12], will not be discussed here.

4.- The recursive approach

We shall illustrate briefly the key steps involved in the recursive formulation of the Tau-Lines Method by considering a problem defined by Laplace's equation in the square R_k:

$$R_k := \{ (x,y) \subset \mathbf{R}^2 : -k \leq x,y \leq k\}, \text{ with } k := 1.$$

Let us construct first a semi-analytic model cf Laplace's equation. Let

$$\nabla^2 u(x,y) = 0, \ (x,y) \subset R_1 \subset \mathbf{R}^2,$$

be given in R_1 together with the boundary conditions

$$u(-1,y) = g_1(y);\ u(1,y) = g_2(y);\ u(x,-1) = g_3(x);\ u(x,1) = g_4(x). \tag{1}$$

which we shall assume to be polynomials or polynomial approximations of degree $\leq \nu$ of given functions.

Let $P_y := \{y_j\}$, $j = 0(1)m$, be a partition of the interval $-1 \leq y_j \leq 1$ and let $L = \{L_j\}$, $j = 0(1)m$, be a system of lines which, for simplicity, we shall assume to be parallel to the direction of the x-axis and at a constant distance $h = 2/m$ from each other. On each of the m+1 lines L_j we define a function $U_j(x)$ for $j = 0(1)m$, such that $U_j(x) := u(x,jh-1)$. If we introduce a discretization scheme S for the second derivative of u with respect of y, in terms of the parametric functions $U_j(x)$, then equation (1) can be modelled by a system of coupled ordinary differential equations. In particular, if we take

$$U_0(x) := g_3(x);\ U_m(x) := g_4(x),$$

and replace the second order derivative u_{yy} by, say, the $O(h^2)$-accurate discretization estimate $[U_{j+1}(x) - 2U_j(x) + U_{j-1}(x)]/h^2$, we obtain the system of ordinary differential equations defined by:

$$U''_j(x) - (2/h^2)\, U_j(x) = -\,[U_{j+1}(x) + U_{j-1}(x)]/h^2,\ -1 \leq x \leq 1,\ j = 1(1)m-1,$$

with the boundary conditions

$$U_j(-1) = g_1(jh-1), \text{ and } U_j(1) = g_2(jh-1),\ j = 1(1)m-1. \tag{2}$$

The vector function $\underline{U}(x) := (U_0(x), U_1(x), \ldots, U_m(x))^T$ is an approximation of $u(x,y)$ relative to the chosen system of lines L and the discretization scheme S.

Our next step is the construction of a Tau Method approximation of $\underline{U}(x)$, more precisely of its m-1 unknown components: $\underline{V}(x) := \{U_1(x), \ldots, U_{m-1}(x)\}^T$. Let us write (2) in terms of $\underline{V}$ in a more concise vectorial form:

$$D\,\underline{V}(x) = \underline{F}(x), \text{ for } -1 \le x \le +1,$$

with the boundary conditions

$$\underline{V}(-1) = \underline{A} \text{ and } \underline{V}(1) = \underline{B}, \tag{3}$$

where D stand for the (m-1) x (m-1) matrix differential operator

$$\begin{vmatrix} d & 1 & & & \\ 1 & d & 1 & & \\ & 1 & d & 1 & \\ & & \cdots & \cdots & \\ & & & 1 & d \end{vmatrix};$$

$\underline{F}(x) := (-g_3(x), 0, \ldots, 0, -g_4(x))^T$,

$d := h^2[d^2/dx^2] - 2$,

and $\underline{A}$, $\underline{B}$ are constant vectors with m-1 components which contain the values of $g_k(jh-1)$, $j = 1(1)m-1$, for $k = 1,2$ respectively.

With the system of ordinary differential equations (3) we shall associate the <u>Tau Problem</u>

$$D\,\underline{V}_n(x) = \underline{F}(x) + \underline{H}_n(x), \text{ for } -1 \le x \le +1,$$

with the boundary conditions

$$\underline{V}_n(-1) = \underline{A} \text{ and } \underline{V}_n(1) = \underline{B}, \tag{4}$$

where $\underline{H}_n(x)$ is a polynomial vector, the m-1 components of which are polynomials of degree not greater that n. $\underline{H}_n$ is chosen for the <u>exact</u> solution $\underline{V}_n(x)$ of (4) to be a polynomial vector with m-1 components; the degree of the components of $\underline{V}_n(x)$ is trivially related to n.

We remark that the <u>Tau approximation</u> $\underline{V}_n(x)$ is defined by the same matrix differential operator of (2)-(3) and that it satisfies the same boundary conditions as $\underline{V}(x)$. The choice of the perturbation term $\underline{H}_n(x)$ depends on the norm in which we wish $\underline{V}_n(x)$ to approximate $\underline{U}(x)$. For details on this point see Namasivayam and Ortiz [13].

5.- Vector canonical polynomials

Let us assume that a sequence of strings of polynomial vectors ($Q^{(1)}{}_k(x)$, ..., $\mathbf{Q}^{(m-1)}{}_k(x)$) can be associated uniquely with an (m-1) x (m-1) matrix linear differential operator **D** and that they are such that

$$\mathbf{D}\,Q^{(1)}{}_k(x) = (x^k, 0, 0, \ldots, 0)^T;$$

....

$$\mathbf{D}\,Q^{(m-1)}{}_k(x) = (0, 0, \ldots, 0, x^k)^T;$$

these polynomials are called the sequence of canonical polynomial vectors associated with the matrix differential operator **D.** For a more precise definition of such sequence, taking into account possible residuals, see Ortiz [7].

Let us write for the components $\underline{H}_{nj}(x)$ of $\underline{H}_n(x)$, j = 1(1)m-1:

$$\underline{H}_{nj}(x) := \Big(\sum_{k=1}^{k=t} \tau^n{}_{kj} \sum_{i=0}^{n+1-k} c^{n+1-k}{}_i x^i \Big)^T,$$

where $\tau^n{}_{kj}$ are t free constant vectors which will be used to satisfy the supplementary conditions of our problem. In our case t = 2, since only two boundary conditions must be satisfied. Let us write for the components $\underline{F}_j(x)$ of the right hand side $\underline{F}$:

$$\underline{F}_j(x) := \Big(\sum_{i=0}^{i=v} f^{(j)}{}_i\, x^i \Big)^T, \; j = 1(1)m-1,$$

where we choose $n \geq v$.

Therefore, if the sequence of canonical polynomial vectors is available, the exact solution of the Tau Problem is given by:

$$\underline{H}_{nj}(x) := \left(\sum_{k=1}^{k=t} \tau^{n}_{kj} \sum_{i=0}^{n+1-k} c^{n+1-k}_{i} Q^{(j)}_{i}(x)\right)^{T} +$$

$$+ \left(\sum_{i=0}^{i=v} f^{(j)}_{i} Q^{(j)}_{i}(x)\right)^{T}, \text{ for } j = 1(1)m-1,$$

where the two free constant vectors τ^{n}_{kj} are fixed by using the two sets of boundary conditions for the vector $\underline{V}(x)$.

For the construction of the recursive relation between canonical polynomial vectors associated with the matrix operator D we proceed as in the one dimensional case (see Ortiz [7]): we begin by testing D on a simple monomial vectors (the so-called generating polynomials) and from these expressions deduce the recursive relation.

Let us sketch such procedure for equation (4): applying **D** to the monomial vector $((m-1), (m-2), \ldots, 2, 1)^T x^n$ we find the polynomial vector:

$$\begin{vmatrix} -m\,x^{n} + [(m-1)h^{2}n(n-1)]x^{n-2} \\ [(m-2)h^{2}n(n-1)]x^{n-2} \\ [(m-3)h^{2}n(n-1)]x^{n-2} \\ \cdots\cdots\cdots\cdots \\ \cdots\cdots\cdots\cdots \\ [\ (1)h^{2}n(n-1)]x^{n-2} \end{vmatrix}$$

Therefore, to isolate the vector $(x^n, 0, 0, \ldots, 0)^T$ in the last expression, which would give $\mathbf{Q}^{(1)}_{n}(x)$, we only have to introduce further terms in the testing monomial vector to cancel the terms of degree x^{n-2}. This is done by using canonical polynomial vectors of order n-2. A recursive expression then follows for $Q^{(1)}_{n}(x)$:

$$Q^{(1)}_{n}(x) := [\underline{p}_{m-1}\, x^{n} + Z \sum_{i=1}^{m-1} (m-i) Q^{(i)}_{n-2}(x)]/(-\,m)$$

where $\underline{p}_k$ stands for the vector $(k, k-1, \ldots, 2, 1)^T$ and $Z := h^2 n(n-1)$. An entirely similar argument gives $\mathbf{Q}^{(j)}_{n}(x)$, for $j = 2(1)m-1$. After the order of approximation n_1 is chosen, the recursive relation is run for $n = 0(1)\, n_1$.

Once the system of lines L and the discretization scheme S have been chosen, the recursive expressions for the canonical polynomial vectors are also fixed, independently of the particular choice of the boundary conditions in problems (1) or (4). We then have a kind of approximate recursive formula to solve a class of similar problems for partial differential equations with the Tau-Lines approach. If a more accurate calculation, with $n = n_2 > n_1$ is required, to construct the new Tau approximation it is only necessary to add the canonical polynomial vectors corresponding to $n = (n_1+1)(1)n_2$ to the sequence already constructed by recurrence and to update the value of the Tau-parameters.

6.- Laplace´s singular problems

Let us consider the problem

$$\nabla^2 u(x,y) = 0, \; (x,y) \subset R_k \subset \mathbf{R}^2, \tag{5}$$

with $k := 7$, where the square R_k has a slit along the segment $y = 0$, $-7 \leq x \leq 0$. The boundary conditions are:

$u(x,y) := 0$ on $x = -7$, $0 \leq y \leq 7$;

$u(x,y) := u_0$ on $x = -7$, $-7 \leq y \leq 0$;

$u_n(x,y) = 0$ on $y = \pm 7$, $-7 \leq x \leq 7$; $x = 7$, $-7 \leq y \leq 7$ and along the crack line; u_n stands for the normal derivative.

By symmetry considerations the problem can be solved in the rectangle:

$R^+_7 := \{ (x,y) \subset \mathbf{R}^2 : 0 \leq |x|, y \leq 7\}$,

with the additional boundary condition:

$u(x,y) := u_0/2$ on $y = 0$, $0 \leq x \leq 7$

El Misiery and Ortiz report in [11] numerical results obtained with Tau approximations of degrees $n = 7$ and $n = 9$ and by using 47 and 71 internal lines respectively, for $u_0 := 1\,000$. These Tau-Lines estimates are particularly interesting near the crack tip, where they are compared with results obtained by using the Boundary Integral Method with 96 boundary elements and either ignoring the crack tip singularity or with singularity treatment and then giving only approximate answers for displacements and stresses (see Symm [14], Papamichael and Symm [15] and Xanthis et al. [16]). The Tau Method results do not include singularity treatment, to make the comparison more clear.

Nevertheless, they are considerably more accurate than those obtained with a basic boundary element method and of a comparable accuracy with those obtained with the Boundary Integral Method when the presence of the singularity is taken into account. Details of these comparisons and other examples are given in El Misiery and Ortiz [11].

7.- Biharmonic regular and singular problems

Let us consider the biharmonic problem defined by:

$$\nabla^4 u(x,y) = f(x,y),\ (x,y) \subset R_k \subset R^2,$$

with the boundary conditions $u(x,y) = u_n(x,y) = 0$ on the boundary of R_k, where u_n stands for the normal derivative of u and $k = 1$. The first problem tested is a regular one which has the exact solution $u(x,y) = (1 - x^2)^2(1 - y^2)^2$.

This choice was made in [10] to compare Tau Method results with those reported by Amara and Destuynder [17] by using an interesting Finite Element Method devised to deal specificly with the biharmonic equation. The objective in [17] was to produce a Finite Element Method formulation leading to a small matrix, even at the price of a difficult assemble; numerical results are reported for the function and first and second partial derivatives. The number of triangular elements used is 72 and the corresponding matrix has order 81.

The number of lines required by a Tau-Lines approximation to reach an error not greater than the one reported for the Finite Element Method of [17] is 5, leading to a matrix of order equal to only 20.

In [11] El Misiery and Ortiz compared the Tau-Lines technique with two other well tested techniques pioneered by J. Whiteman and collaborators in connection with the singular biharmonic problem defined by:

$$\nabla^4 u(x,y) = 0,$$

in the rectangle $R_{kh} \subset R^2$:

$R_{kh} := \{(x,y) \in R^2:\ -0.4 \leq x \leq 0.4;\ -0.7 \leq y \leq 0.7\}$, which has a slit on the segment $y = 0,\ -0.4 \leq x \leq 0$.

The symmetric boundary conditions are:

$u(x,y) = u_n(x,y) := 0$ on $y = 0,\ -0.4 \leq x \leq 0$ and $x = -0.4,\ 0 \leq y \leq 0.7$;

$u(x,y) := 5\,000\ (x^2 + 0.8\,x + 0.16)$ and $u_n(x,y) := 0$ on $y = 0.7$ and $-0.4 \leq x \leq 0.4$;

$u(x,y) := 1\,600$ and $u_n(x,y) := 8\,000$ on $x = 0.4,\ 0 \leq y \leq 0.7$.

(6)

The Tau Method approximation is compared with two techniques: (a) Finite Elements with conformal mesh refinement (see Schiff, Fishelov and Whiteman [18]) and (b) Collocation of N terms of the known series expansion of the solution (given in polar coordinates centered at the crack tip) at M points, where M is much larger than N. A Linear Programming technique is used to solve the overdetermined collocation problem (see Mangasarian [19]). Whiteman [20] used this technique for problem (6).

The accuracy of results reported for the Finite Element Method with conformal mesh refinement (7 to 14 elements) and Collocation with Linear Programming (N = 23, M = 47) in [18] and [20] is matched by using the Tau-Lines Method _without singularity treatment_ with 10 to 15 internal lines.

8.- _A fully analytic approach: the multidimensional Tau Method_

The canonical polynomials´ approach discussed in the previous sections can be used for partial differential equations. A trivial modification is necessary: linear partial differential operators **D** with polynomial coefficients must now be tested on $x^i y^j$. From these generating polynomials , a recursive expression for the canonical polynomials

$$Q_{ij}(x,y): \mathbf{D}\, Q_{ij}(x,y) = x^i y^j$$

associated with **D** is deduced. Exact polynomial solutions, not necessarily satisfying the given boundary conditions, must be taken into account, as in the one dimensional case. In the general case residuals, in the sense of Ortiz [7], must be introduced.

We shall discuss here a direct multidimensional formulation of the Tau Method which is based on the operational approach of Ortiz and Samara [8]-[9].

These authors remarked in [6] that the effect of multiplication by powers of the variables and of partial differentiation on the coefficients of a multivariate polynomial can be reduced to elementary operations involving matrices in which only one line, parallel to the main diagonal, is different from zero.

In the two dimensional case this observation can be formulated in the form:

$$x^i y^j \,[\partial^{r+s} a(x,y)/\partial x^r \partial y^s] = \underline{x}^T (\eta^r \mu^i)^T A (\eta^s \mu^j)\, \underline{y},$$

where

$a(x,y) := \underline{x}^T A\, \underline{y}$, and A is the coefficient matrix of the polynomial a(x,y) and $\underline{t}^T := (1, t, t^2, \ldots)$. The matrices η and μ are:

$$\eta := \begin{vmatrix} 0 & & & & \\ 1 & 0 & & & \\ & 2 & 0 & & \\ & & & \cdots & \\ & & & \cdots & \end{vmatrix}, \qquad \mu := \begin{vmatrix} 0 & 1 & & & \\ & 0 & 1 & & \\ & & 0 & 1 & \\ & & & \cdots & \\ & & & & \cdots \end{vmatrix}$$

This result can be used to express the effect of any linear differential operator **D** with polynomial coefficients on the coefficients of a bivariate polynomial:

$$u_{nm}(x,y) := \underline{x}^T U_{nm} \underline{y}.$$

In the case of Laplace´s equation

$$\nabla^2 u(x,y) = F(x,y), \ (x,y) \subset R_1 \subset \mathbf{R}^2,$$

with the boundary conditions

$$u(-1,y) = g_1(y);\ u(1,y) = g_2(y);\ u(x,-1) = g_3(x);\ u(x,1) = g_4(x).$$

it follows that the coefficients of a polynomial approximation of degree n in x and m in y satisfy the following linear system of matrix equations:

$$\underline{x}^T [(\eta^2)^T A + A \ \eta^2] \underline{y} = \underline{x}^T F \underline{y},$$

where $\underline{x}^T F \underline{y}$ is the polynomial representation of F(x,y). The boundary conditions can be expressed by the equations:

$$\underline{x}^T|_{x=-1} U_{nm} \underline{y} = g_1 \underline{y};$$
$$\underline{x}^T|_{x=1} U_{nm} \underline{y} = g_2 \underline{y};$$
$$\underline{x}^T U_{nm} \underline{y}|_{y=-1} = g_3 \underline{x};$$
$$\underline{x}^T U_{nm} \underline{y}|_{y=1} = g_4 \underline{x};$$

(7)

These equations can be collected in the general form bX + Xc = d, where X is a matrix which contains the matrix U_{nm} of the unknown coefficients of $u_{nm}(x,y)$. Ortiz and Samara have shown in [6] that, except for a similarity transformation, the algebraic problem defined by (7) is equivalent to a bivariate Tau Method approximation of u(x,y). Such similarity transformation depends on the choice of perturbation term, in most cases Chebyshev or Legendre polynomials. The associated linear algebraic problem is solved with standard software.

9.- Laplace´s singular problems

Let us return to the singular problem defined by Laplace´s equation (5).

Ortiz and Pun [21] have used different approaches based on the multidimensional formulation of the Tau Method to find approximate solutions to this problem.

We shall take advantage of the natural symmetry of this problem to consider it on R^+_7, the subdomain of R_7 which is situated over the line $y = 0$.

Since the boundary conditions given on the segments $I_1 := \{ y = 0 \text{ and } -7 \leq x < 0\}$ and $I_2 := \{ y = 0 \text{ and } 0 < x \leq 7\}$ have different forms, we shall segment R^+_7 into two subdomains (or Tau elements) symmetrical about the line $x = 0$ and impose a C^1 continuity condition along that line.

Ortiz and Pun have compared results obtained without singularity treatment with results reported by Whiteman and Papamichael in [22] and obtained by using an accurate Conformal Transformation method. Their results for a Tau approximation of degree $n = m = 5$ agree to within 0.5 o/o outside the crack line, where differences reach up to 8.2 o/o by the crack tip. If the degree of the Tau approximation is increased to $n = m = 9$ the error by the crack tip is reduced to about a half of the previous value.

A further symmetric segmentation of each subdomain, by $x = \pm 3.5$, gives errors of 1.0 and 4.8 o/o respectively for $n = m = 6$, if compared with the values given in [22]. If n is increased to 9, while m remains equal to 6, the erors are reduced to 0.6 and 3.4 o/o respectively.

These results show that with small values of the degrees n and m an accurate solution can be constructed with the Tau Method away from the crack tip, but not in the proximity of it. In all cases the accuracy increases with n,m and with segmentation of the domain. Such increase in accuracy is slow near the crack tip.

10.- Singularity treatment in the context of the Tau Method

The situation changes very dramatically if singularity treatment is introduced in the Tau Method. Let us consider the singular series expansion of the exact solution, expressed in polar coordinates (r,θ) centered at the crack tip (see Kondrat´ev [23]):

$$u(r,\theta) := u_0/2 + \sum_{i=0}^{\infty} a_i \, g_i(r,\theta), \tag{8}$$

where

$$g_i(r,\theta) := r^{(4i+1)/2} \cos \{[(4i+1)/2]\, \theta\}.$$

The numerical estimation of the parameters a_i of expansion (8) is of a considerable interest in engineering applications. The first of them, a_0, is trivially related to the <u>stress intensity factor</u>.

Ortiz and Pun introduced singularity treatment in the context of the Tau Method in [21] by using a representation of the approximate solution which includes nonpolynomial terms. These terms are taken from the first few terms of (8).

Symm reported in [14] a detailed calculation of approximate values of u(x,y) obtained by using the Boundary Integral Method at two sets of points. The first of them, P_1, is defined by the points $x_i := -6(2)6$, $y_j := 0(2)6$. The second one, P_2, is a refinement of the previous one near the crack tip, defined by $x_i := -1(0.25)1$, $y_j := 0(0.25)1$.

In [21] Ortiz and Pun compare Tau Method results obtained with singularity treatment with Symm´s estimates on P_1 and P_2. The first two terms of (8) are retained for the treatment of the singularity in both methods. For a Tau approximation of degree n = m = 7, and using a Legendre perturbation term, a maximum difference of 0.05 is reported. It reduces to about a half if n = m = 9 is taken for the Tau Method approximation.

Estimates of the singular parameters a_i in expansion (8) have also been compared. For a_0 and a_1 the Tau Method, with n = m = 9, gives 151.61 and 4.73, while Symm reported 151.63 and 4.72 for n = m = 13. More detailed calculations, taking the first five terms of (8) and n = m = 7, gives 151.622 and 4.719 respectively.

By using different values of n,m up to 15; Chebyshev or Legendre perturbation terms and the first five coefficients of the singular expansion (8), Ortiz and Pun have reported in [21] the following Tau Method estimates:

$$a_0 = 151.625;$$
$$a_1 = 4.733;$$
$$a_2 = 0.133;$$
$$a_3 = -0.00888;$$
$$a_4 = 0.000282,$$

which probably are the most accurate estimates available in the literature on this problem. In [16] Xanthis et al. have given estimates which agree in absolute value with the latter up to 0.005; 0.0033; 0.00005; 0.00001 and 0.000062 respectively.

11.-Final remarks

The approach just sketched for the Tau Method has been applied by Ortiz and Pun to L-shaped domains [24]. Comparisons were made there with the accurate results reported by Whiteman and Papamichael [22] by using the Conformal Transformation method and by Symm [14] by using the Boundary Integral Method in the case of two L-shaped Laplace´s problems of different proportions. The results of these authors were reproduced with a low order Tau approximation (n = m = 6). Onumanyi and Ortiz have recently developed, in [25], a different approach, based on the differentiation of canonical polynomial with respect to the index, which they have successfully applied to the treatment of singular problems for ordinary differential equations.

Ortiz and Pun have also used their techniques in the numerical treatment of nonlinear partial differential equations (see [26]-[27]).

R e f e r e n c e s

1.- C. Lanczos, Trigonometric interpolation of empirical and analytical functions, J. Math. Phys, **17**, pp.123-199 (1938).

2.- E.L. Ortiz, Canonical polynomials in Lanczos´ Tau Method, in STUDIES IN NUMERICAL ANALYSIS, B.K.P. Scaife, Ed., Academic Press, New York, pp. 73-93 (1979).

3.- J.H. Freilich and E.L. Ortiz, Numerical solution of systems of differential equations with the Tau Method: an error analysis, Maths. Comp. **39**, pp. 189-203, 1984.

4.- P. Onumanyi and E.L. Ortiz, Numerical solution of stiff and singularly perturbed boundary value problems with a segmented-adaptive formulation of the Tau Method, Math. Comput., **43**, pp. 189-203, 1984.

5.- S. Namasivayam and E.L. Ortiz, Best approximation and the numerical solution of partial differential equations, Portugaliae Mathematica, **40**, pp. 97-119 (1985).

6.- E.L. Ortiz and H. Samara, Numerical solution of partial differential equations with variable coefficients with an operational approach to the Tau Method, Comp. and Maths. with Appli., **31**, pp. 95-103, 1984.

7.- E.L. Ortiz, The Tau Method, SIAM J. Numer. Analysis, **6**, pp. 480-491, 1969.

8.- E.L. Ortiz and H. Samara, An operational approach to the Tau method for the numerical solution of nonlinear differential equations, Computing, **27**, pp. 15-25 (1981).

9.- E.L. Ortiz and H. Samara, Numerical solution of differential eigenvalue problems with an operational approach to the Tau method, Computing, **31**, pp. 95-103 (1983).

10.- A.E.M. El Misiery and E.L. Ortiz, Numerical solution of regular and singular biharmonic problems with the Tau-Lines Method, Comm. in Appl. Numer. Methods, **1**, pp. 281-285, 1985.

11.- A.E.M. El Misiery and E.L. Ortiz, Tau-Lines: a new hybrid approach to the numerical treatment of crack problems based on the Tau Method, Comp. Meth. in Appl. Mech. and Engng., in press, 1986.

12.- K.M. Liu and E.L. Ortiz, Numerical solution of eigenvalue problems for partial differential equations with the Tau-Lines method, Computers and Maths. with Appli. (in press) (1985).

13.- S. Namasivayam and E.L. Ortiz, Dependence of the local truncation error on the choice of perturbation term in the step by step Tau Method for systems of differential equations, NAS Res. Rep. Imperial College, London (1984).

14.- G.T. Symm, Treatment of singularities in the numerical solution of Laplace´s equation by an integral equation method, Res. Rep. NAC 31.01.1973, Nat. Phys. Lab. (U.K.), 1973.

15.- N. Papamichael and G.T. Symm, Numerical techniques for two-dimensional Laplacian problems, Comput. Meths. Appl. Mech. Engrg. **6**, pp. 175-194, 1975.

16.- L. Xanthis, M. Bernal and C. Atkinson, The treatment of singularities in the calculation of stress intensity factors using the boundary integral method, Comput. Meths. Appl. Mech. Engrg. **26**, pp. 285-304, 1981.

17.- M. Amara and P. Destuynder, A numerical method for the biharmonic problem, Int. J. Numer. Methods Eng, **17**, pp. 1515-1523, 1981.

18.- B. Schiff, D. Fishelov and J.R. Whiteman, Determination of a stress intensity factor using local mesh refinement, in THE MATHEMATICS OF FINITE ELEMENTS AND APPLICATIONS III, J.R.Whiteman, Ed., Academic Press, London (1979).

19.- O.L. Mangasarian, Numerical solution of the first biharmonic problem by linear programming, Internat. J. Engng. Sci., **1**, pp.231-240 (1963).

20.- J.R. Whiteman, Numerical treatment of a problem from linear fracture mechanics, in NUMERICAL METHODS IN FRACTURE MECHANICS, D.R.Owens and A.R. Luxmore, Eds., University of Wales, Swansea, pp. 128-138 (1978).

21.- E.L. Ortiz and K.-S.Pun, Numerical solution of Laplace´s equation with a bidimensional formulation of the Tau Method, Invited Lectures, Proc. Second Int. Conference on Computational Mathematics, Boole Press, Dublin, in press, 1985.

22.- J.R. Whiteman and N. Papamichael, Numerical solution of two dimensional harmonic boundary value problems containing singularities by Conformal Transformation Methods, Res. Rep. No. TR/2, Mathematics Department, Brunel University, 1971.

23.- V.A. Kondrat´ev, Boundary problems for elliptic equations in domains with conical or angular points, Trans. Moscow Math. Soc., pp. 227-313, 1967.

24.- E.L. Ortiz and K.S. Pun, Numerical solution of Laplace´s equation in L-shaped domains with the Tau Method, Research Report NAS 08-85, Imperial College, London, 1985.

25.- P. Onumanyi and E.L. Ortiz, A method of finite series approximation of Bessel´s differential equation, Abacus, **17**, pp. 74-90, 1986.

26.- E.L. Ortiz and K.-S.Pun, Numerical solution of nonlinear partial differential equations with the Tau method, J. Comp. and Appl. Math., 12 & 13, pp. 511-516 (1985).

27.- E.L. Ortiz and K.-S.Pun, Numerical solution of Burgers´ nonlinear partial differential equation with a multi-dimensional formulation of the Tau method, Computers and Maths. with Appli., (in press) (1986).

Numerical Approximation of Partial Differential Equations
E.L. Ortiz (Editor)

PRESENT STATE AND NEW TRENDS IN PARALLEL COMPUTATION

Rafael Portaencasa and Carlos Vega
Polytechnic University of Madrid
Madrid, Spain

A B S T R A C T

This is a sketch of the origins and development of parallel computation from the point of view of computer architecture, software and applications, including those related to numerical methods. It is used as a basis for the evaluation of a policy for the development of parallel computation in Spain.

The recent evolution of computing machines

The history of parallel computation overlaps with that of computers. The first generation of computers, in the decade of the 50's, was based on the use of electronic tubes; their components gave response times far higher than the ones achieved at present. The time required for a signal to pass from the entrance of a logical gate to the entrance of the following one was approximatively 1 μ s. Such time was reduced to 0.3 μ s by 1960, in second generation computers such as the IBM 7079, through the use of Germanium transistors.

The silicon integrated circuits of 1965, with very few gates per chip, registered times of about 10 ns. By 1975 such times were reduced to under 1 ns. The MOS alternative allowed for these times to be reduced again, from 5 to 19 times. The microprocessors of the present decade made it possible to reduce such times even further.

If we consider the period between 1950 and 1975, we remark that the basic speed of components, measured as the inverse of the times considered before, was multiplied by a factor of 10^3, while computer efficiency, measured as the inverse of the time required for a multiplication, increased by a factor of 10^5. This additional increment in speed has been possible because of novelties introduced in systems architecture. In particular, through the introduction of parallelism.

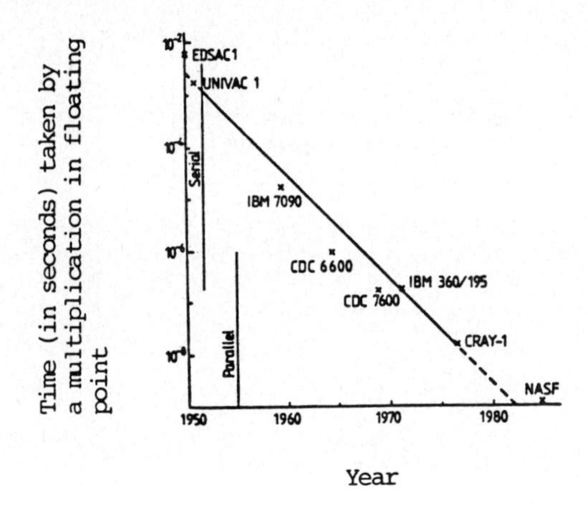

The impact of these technological developments on the design of computers, particularly on those devoted to high speed scientific computing, has been discussed in detail since 1982 at several international scientific meetings and by many different authors,in particular, by D. Parkinson (see References).

The comparisons made before take no account of the superposition of different arithmetical and logical operations. Alternatively, we can compare the computer clock period, regarded as a measure of the speed of technology, with the number of basic arithmetic operations which can be performed in one second. That is taken as a measure of the efficiency of a computer.

In 1951 the execution time required to evaluate a series of n terms with 2n arithmetical operations was 18n s for a clock with a period of 2 && s; that is, with an average speed of 100 arithmetical operations per second. This figure should be compared today with an efficiency of 130 millions of arithmetical operations per second for matrix multiplications in computers such as CRAY-1, where the clock period is 12.5 ns. Thus, the increment in efficiency over three decades is about 10^6, while the technological factor accounts for only 160. Therefore, it could be said that the introduction of vector and array processors in scientific computation has opened a new era.

These advances required the deveolpment of new languages and algorithms, which were necessary to take advantage of progress made in computer architecture. The use of parallel computers has given further impulse to the analysis of vector and asynchronic processes.

We shall consider now those architectures which have been designed to take advantage of parallelism in large-scale scientific computing and of some new prototype machines which were specificly designed for the analysis of parallel process.

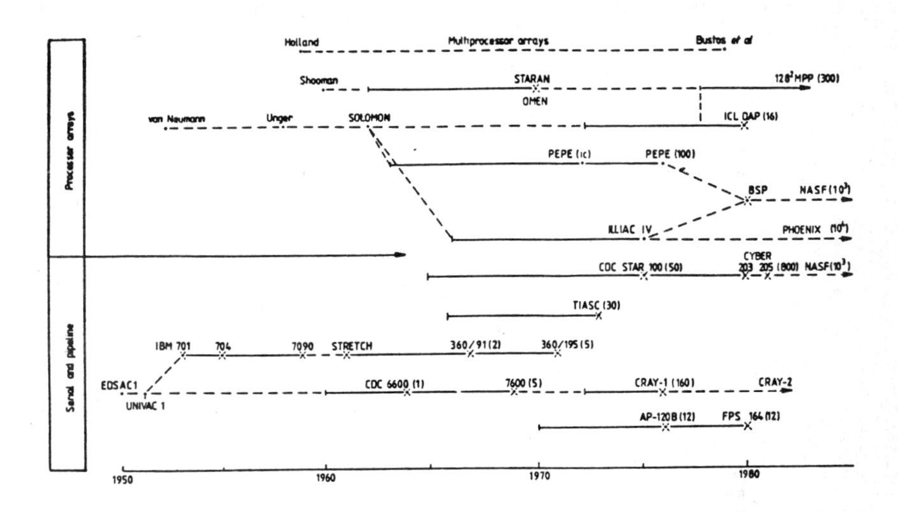

Parallel architectures

A typical classification of parallel architectures is the following:

a) Pipelining: It allows for an increment of the efficiency of a control or arithmetic unit by lines assembly.

b) Functionalism: Several independent units help increasing the efficiency of different functions, such as the logical or arithmetical (addition or multiplication) one, allowing them to operate simultaneously on different sets of data.

c) Array systems: A set of identical processing elements operate under a unified control improving the efficiency of a given operation which is simultaneously performed on different data stored locally.

d) Multiprocessing: Uses several processors, each one with its own instructions and usually interconnected through a common storage.

These elements can be combined in many different ways.

Until 1980 designs which made use of multiprocessing were mainly used for connecting independent computers with the purpose of increasing the efficiency of a given installation.

History shows that parallelism has been used to increase the efficiency of computers from the very early designs and at different levels. We can classify such levels of parallelism in the following way:

1.- Parallelism at job level:

(i) between jobs

(ii) between phases of a job.

2.- Parallelism at programming level:

(i) between parts of a program

(ii) between loops of a DO.

3.- Parallelism at instruction level:

(i) between different phases of the execution of an instruction.

4.- Arithmetical parallelism and at bit level:

(i) between the elements of a vector operation

(ii) inside logical-arithmetical circuits.

Basically, each job can be regarded as made up of different sequential phases. Each of them requires a different system´s program and different resources from the system. Because of that, the principal objective of a computer installation is to maximize the speed at which jobs can be processed and to try to use parallelism at any possible level.

Parallelism and hardware

Parallel computers can also be classified, according hardware considerations, as SIMD and MIMD. The first group, Single Instruction Multiple Data, can be vector or array processors. The latter differ in the way in which data is communicated within the system. Vector computing is guided by pipeline and synchronic multiprocessing. In a matrix or array computer, processors receive information from their own memories or from those of close neighbours.

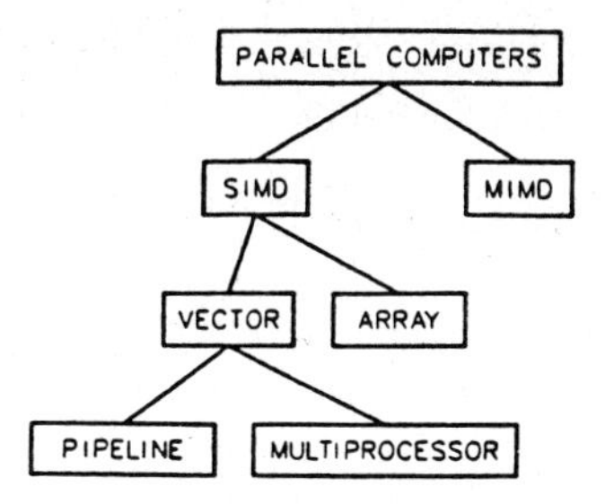

Computers of MIMD Multiple Instruction Multiple Data type do not give rise to further classification; their hardware characteristics were described by Jordan in 1979.

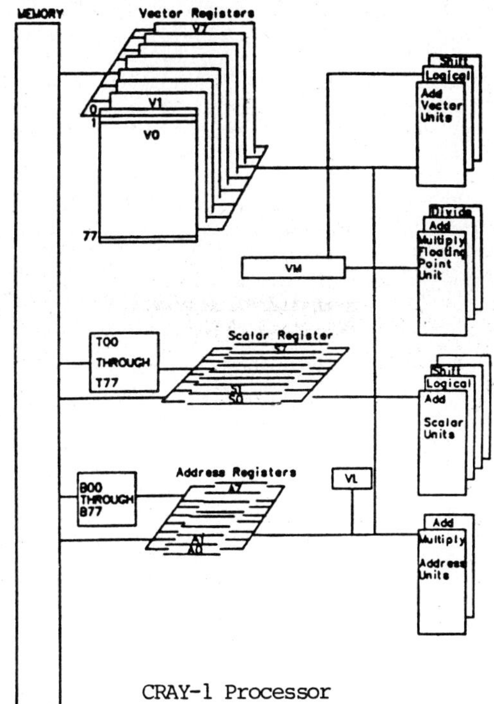

CRAY-1 Processor

Typical SIMD architectures for parallel processors are used in the following systems:

Vector pipeline: Cyber 203; and 205; CRAY-1;

Vector multiprocessor: Bouroughs BSPYPEPE;

Matrix: Bouroughs ILLIAC IV and ICL DAP;

while MIMD processor units are present in:

Carnegie Mellon´s CM and University of Texas´ TRAC.

Independently of the new problems parallelism presents to architecture design, the development of languages and algorithms for an optimal use of computers in a parallel environment have to be considered. This is because parallel architectures do not execute computer instructions in the same way as conventional machines. Conversely, to achieve high performances, it is necessary to adapt the computer programs to computers´ architecture.

The efficiency of a computer program is directly related to the numerical procedure that is being use, that is, to the algorithm; to the tools used by the programmer to implement it in the computer and to the compiler, which codifies it. I we define the efficiency of a computer program as inversely proportional to the CPU time required to perform it, our objective will be to minimize such time. In computers of the first generation the objective was to make a minimum use of the memory; algorithms design was dominated by this condition, A typical example is Gill´s algorithm, developed in 1951, for the computer implementation of Runge-Kutta´s method for the numerical solution of ordinary differential equations. In the case of parallel computers the minimization of the execution time is not necessarily equivalent to the use of a minimum number of arithmetical operations.

In the case of a parallel architecture, the parallelism of an algorithm is defined as the number of independent arithmetical operations which can be performed concurrently or simultaneously. In a pipelined computer data for the operations is defined by vectors and the operations could be interpreted as vector instructions; parallelism is the same irrespectively of the size of the vector. In an array processor data for each operation is to be found in each processing element of a matrix. Therefore, as a result of the interpretation of an instruction from the master control unit, operations on all elements are performed simultaneously. Parallelism is measured by the number of elements which operate along a given path. It can remain constant through different stages of a given process, as in the case of matrix multiplication, or change from one stage to the next.

These ideas, which we have sketched only superficially, have made possible important advances in different research areas in the present decade. The large number of international meetings, new books and journals and courses to update personnel, related to parallel computing, are a testimony of their impact.

New numerical algorithms are being developed to try to solve more efficiently complex problems which required a considerable amount of effort with the approaches of the past. At the same time parallelism is affecting new areas cf research which could not be treated before. There is also a considerable effort devoted to the development of new projects related to the design of parallel computers and processors.

The MIMD prototype

An example of such work is the large capacity parallel processesor prototype which is being developed at the Watson Research Center, in New York. It will make possible large scale analysis of both hardware and software. An interesting feature of this project is the fact that the prototype is a research tool for parallel processes rather than a design fcr an specific computer.

This prototype is built as a MIMD machine in which shared stores of up to 512 conventional 32 bits processors can be supported.

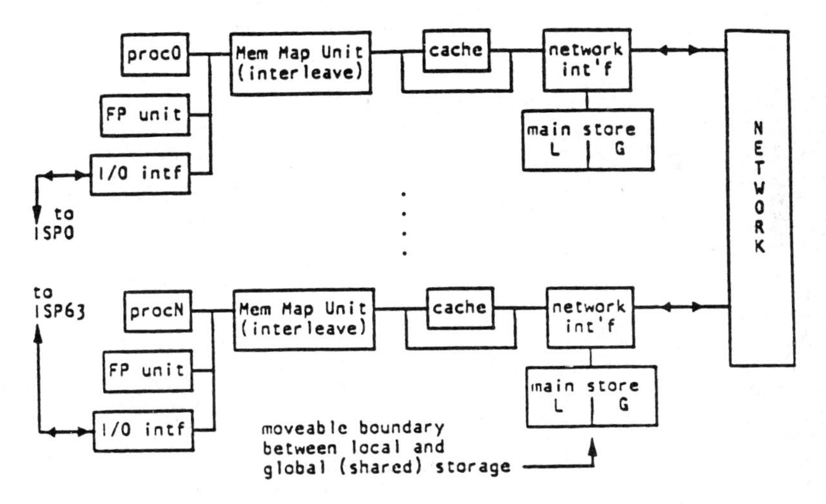

Architecture of the prototype

Processor memory elements (PME) and their connections are shown in the figure above; the memory organization is schematically displayed in the figure belcw.

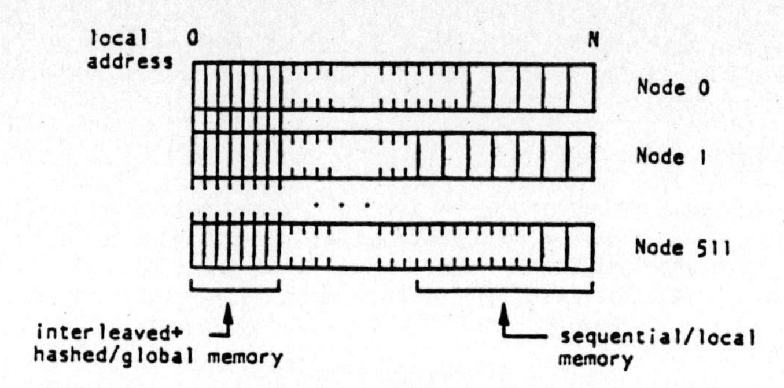

Memory organization

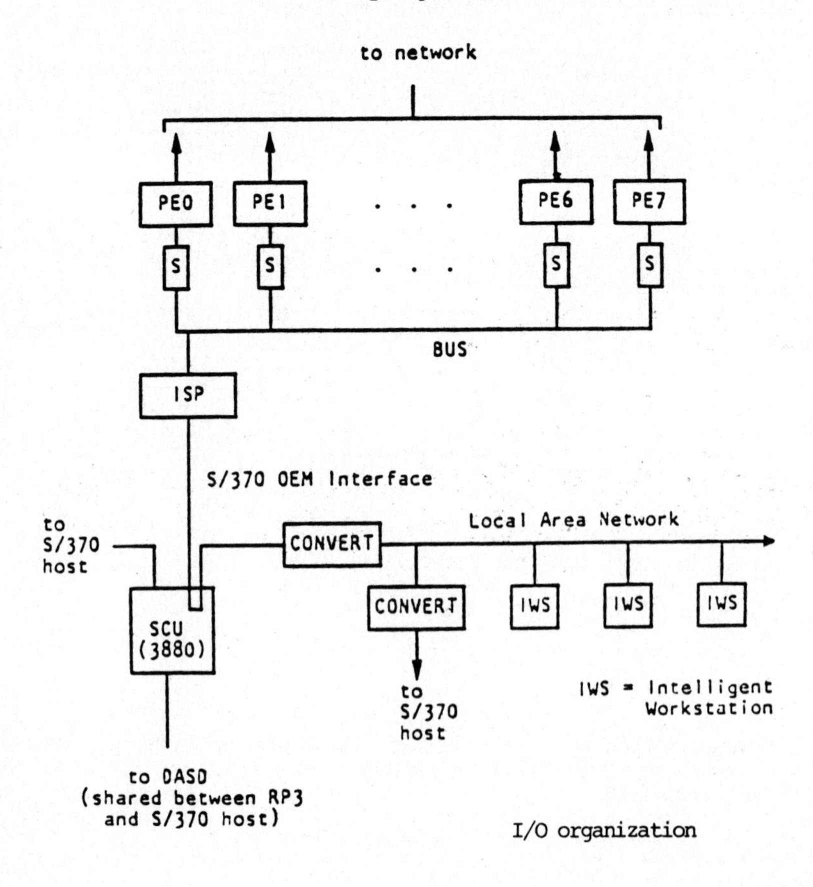

I/O organization

Software is not initially designed to make effective operational procedures for a parallel system but, rather, to enable an experienced programmer to use it to develop specific applications. It provides the basic infrastructure for gaining experience on the use of programming methods in highly efficient parallel systems. It will use an operating system based on a modified version of UNIX which will be able to support internally highly efficient parallel operations, and with the necessary additions to deal with a shared memory. Languages available for it will be language C, Fortran and Pascal with token modifications. It is hoped that on this basis more radical concepts and approaches will evolve.

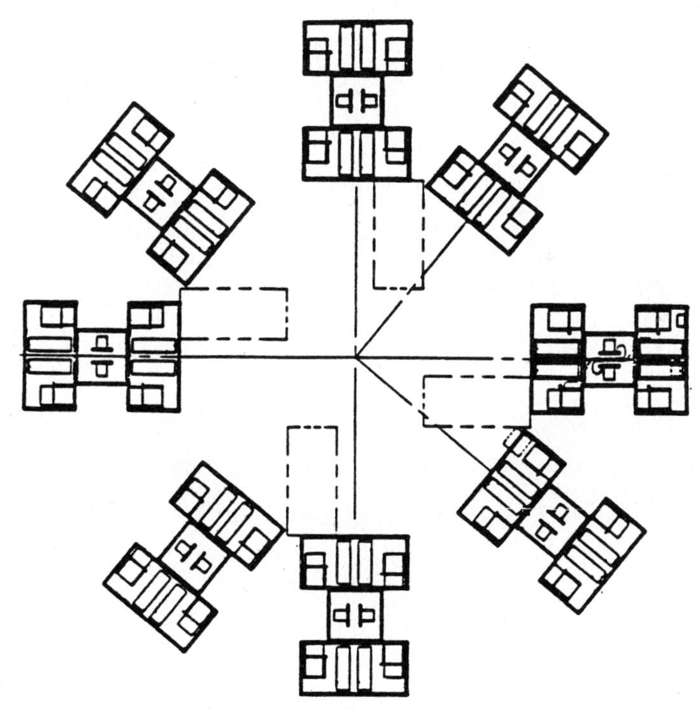

Memory coupling in the prototype

One of the advantages offered by this prototype is that it will make possible to gain considerable efficiency with a certain degree of flexibility. It is hoped that it will promote the use of advanced parallel systems in the solution of concrete problems. Because of this last objective, a high degree of efficiency was required from its design, otherwise the testing of parallel techniques would have no meaning. This may account for a choice of 512 processors.

Parallel computing in Spain

We would like to end this paper with some remarks on the situation of parallel computing in Spain. Needless to say that the group of scientists and research departments with parallel computing requirements is far from being large in Spain today. There are small collective groups working on parallel design techniques; a more numerous group of research workers would require them for the development of their present computing requirements.

However, the necessity to be aware of and prepared to take advantage of new developments in this area is clearly felt. With this end in mind, the Spanish Ministry of Education and Science has prepared a project, called Projecto CDIN, the main purpose of which is to find ways of solving the requirements of Spanish scientists which cannot be solved today by using large conventional computers or the small vector installations they posses. This project aims at discussing the needs of those involved in the solution of real life large computer problem with parallel computing in view. In this respect we find some affinity with the prototype approach discussed earlier, which, at a later stage, could be implemented with the necessary accessories for effective use in one of our main research centers, preferably in the university sector.

REFERENCES

Hockney and Jesshope, Parallel Computers, Adam Hilger Ltd., 1981.

Rodrigue, Parallel Computations, Academic Press, 1982.

Parallel Computing 83, North Holland, 1984.

Numerical Approximation of Partial Differential Equations
E.L. Ortiz (Editor)

FINITE ELEMENT METHODS FOR TREATING PROBLEMS INVOLVING SINGULARITIES, WITH APPLICATIONS TO LINEAR ELASTIC FRACTURE.

J. R. Whiteman

Brunel University, Uxbridge, England.

1. INTRODUCTION

The need for effective finite element methods for treating singularities in elliptic boundary value problems is well recognised. As a result mathematicians and engineers have developed special finite element techniques which produce accurate approximations to the solutions of problems involving singularities and which can exhibit good asymptotic convergence properties. A major driving force behind this work has been the need for such techniques in fracture mechanics.

The work of mathematicians has been mainly in the context of model two dimensional Poisson problems. For such problems many theoretical results are known concerning the regularity of solutions of weak forms in Sobolev space settings. As a result significant finite element error estimates for this limited class of problems have been produced. Mainly on account of the lack of comparable theoretical results, much less progress has been made with three-dimensional problems of this type. A similar situation exists for the case of linear elastic fracture.

In this paper we discuss in detail the finite element treatment of singularities for a model two-dimensional Poisson problem, giving estimates for various norms of the error. Some finite element techniques for singularities are then described. In problems with singularities the approximation of secondary (derived) quantities by retrieval from approximations to the solutions (primary quantities) is of great importance. Path integral retrieval methods are discussed. The state-of-the-art for treating problems with singularities in three-dimensions is then presented and contrasted with that for two dimensions. Finally the finite element treatment of problems of linear elastic fracture is discussed, with reference back to the Sections on Poisson problems.

2. POISSON PROBLEMS INVOLVING SINGULARITIES

2.1 Two Dimensional Poisson Problems. Let $\Omega \subset \mathbb{R}^2$ be a simply connected polygonal domain with boundary $\partial\Omega$. We consider first the much studied model problem in which the scalar function $u(\underline{x})$ satisfies

$$\begin{aligned} -\Delta[u(\underline{x})] &= f(\underline{x}), \quad & \underline{x} \in \Omega, \\ u(\underline{x}) &= 0 \quad , & \underline{x} \in \partial\Omega, \end{aligned} \tag{2.1}$$

where $f \in L_2(\Omega)$. A weak form of (2.1) is defined in the usual Sobolev space $\overset{\circ}{H}{}^1(\Omega)$, and for this $u \in \overset{\circ}{H}{}^1(\Omega)$ satisfies

$$a(u,v) = F(v), \ \forall \ v \in \overset{\circ}{H}{}^1(\Omega), \tag{2.2}$$

where

$$a(u,v) \equiv \int_\Omega \nabla u \ \nabla v \ d\underline{x}, \qquad u, \ v \in \overset{\circ}{H}{}^1(\Omega), \tag{2.3}$$

and

$$F(v) \equiv \int_\Omega f \ v \ d\underline{x} \ , \quad v \in \overset{\circ}{H}{}^1(\Omega). \tag{2.4}$$

Problem (2.1) is treated by considering the weak form (2.2), where the bilinear form has the important properties that it is continuous, symmetric and elliptic on $\overset{\circ}{H}{}^1(\Omega)$, see Ciarlet [1].

For the finite element solution of (2.1) the region Ω is partitioned quasi uniformly into triangular elements Ω^e in the usual manner and the Galerkin method is applied to (2.2). Conforming trial and test functions are employed and the solution $u \in \overset{\circ}{H}{}^1(\Omega)$ is approximated by $u_h \in S^h$, where $S^h \subset \overset{\circ}{H}{}^1(\Omega)$ is a finite dimensional space of piecewise polynomial functions of degree p, $(p \geqq 1)$, and u_h satisfies

$$a(u_h, v_h) = F(v_h) \quad \forall \ v_h \in S^h. \tag{2.5}$$

The well known best approximation property of the Galerkin solution gives the inequality

$$\| u - u_h \|_{1,\Omega} \leqq \| u - w_h \|_{1,\Omega} \quad \forall \ w_h \in S^h, \tag{2.6}$$

where $\| v \|_{1,\Omega}$ is the energy norm $\| \nabla v \|_{L_2(\Omega)}$. Since (2.6) holds for all $w_h \in S^h$, we may take the interpolant $\tilde{u}_h \in S^h$ to u for w_h in (2.6) and, using approximation theory, it follows that

$$\| u - u_h \|_{1,\Omega} \leqq C \ h^\mu \ |u|_{k+1,\Omega} \ , \tag{2.7}$$

where $\mu = \min(p,k)$, whilst C is a constant. Throughout the paper all constants in the estimates are denoted by C .

The actual value of μ is thus dependent both on the choice of p and on the regularity of the solution u of (2.2). Under the condition that $f \in L_2(\Omega)$ the regularity of u is determined by the shape of $\partial\Omega$. If Ω is a convex polygon, then $u \in \mathring{H}^1(\Omega) \cap H^2(\Omega)$, so that $k = 1$ in (2.7) and

$$\| u - u_h \|_{1,\Omega} \leq C\, h\, |u|_{2,\Omega} \tag{2.8}$$

In this case, see Schatz [2],

$$\| u - u_h \|_{L_2(\Omega)} \leq C\, h^2\, |u|_{2,\Omega}\ , \tag{2.9}$$

so that there is an $O(h)$ convergence gain through changing from the 1-norm to the L_2-norm. The above two estimates are *optimal* in that they are the best that can be obtained by approximating from S^h a function with the regularity of u . It has also been shown, see Nitsche [3] and Ciarlet [1], that for this case the L_∞-norm of the error has $O(h^2)$ convergence.

As has been stated in Section 1, problems with boundaries having re-entrant corners, and thus containing boundary singularities, are of main interest here. We thus consider again problem (2.1), but now in the situation where Ω is a non-convex polygonal domain with interior angles α_j, $1 \leq j \leq M$, where

$$0 < \alpha_1 \leq \alpha_2 \leq \ldots \leq \pi < \alpha_m \leq \ldots \leq \alpha_M \leq 2\pi .$$

In this case the solution u of (2.2) is such that $u \in \mathring{H}^1(\Omega) - H^2(\Omega)$, and it has been shown by Grisvard [4] that over Ω u can be written as

$$u = \sum_{j=m}^{M} a_j\, \chi_j(r_j)\, u_j(r_j,\theta_j) + w, \tag{2.10}$$

where (r_j,θ_j) are local polar coordinates centred on the jth corner of $\partial\Omega$, the χ_j are smooth cut-off functions for the corners, $w \in H^2(\Omega)$ and

$$u_j(r_j,\theta_j) = r_j^{\pi/\alpha_j} \sin \frac{\pi\,\theta_j}{\alpha_j} .$$

The regularity of u is clearly determined by the term in the summation in (2.10) associated with the Mth corner. In fact $u \in H^{1+\pi/\alpha_M-\varepsilon}(\Omega)$ for every $\varepsilon > 0$, see also Schatz and Wahlbin [5].

Since $\alpha_M > \pi$ and $u \in H^{1+\pi/\alpha_M-\varepsilon}(\Omega)$, it follows from (2.7) that

$$\| u - u_h \|_{1,\Omega} \leq C\, h^{(\pi/\alpha_M-\varepsilon)}\, |u|_{1+\pi/\alpha_M-\varepsilon} . \tag{2.11}$$

and, see Schatz [2], that

$$\| u - u_h \|_{L_2(\Omega)} \leq C\, h^{2(\pi/\alpha_M-\varepsilon)}\, |u|_{1+\pi/\alpha_M-\varepsilon} . \tag{2.12}$$

Whereas the convergence gain in the changing from the 1-norm to the L_2-norm is $O(h)$ for the case where Ω is a convex polygon, (2.8),(2.9), the gain is less for the re-entrant case.

Estimates of the type (2.11) and (2.12), being global, reflect the worst behaviour of the solution over Ω. The situation may not be so bad locally, in particular away from the corners where from (2.10) $u \in H^2$. Thus we now consider L_∞-estimates. Suppose that at the jth vertex z_j of $\partial\Omega$ the intersection of Ω with a disc centred at z_j and containing no other corner is Ω_j and that $\Omega_0 \equiv \Omega \setminus \left(\bigcup_{j=1}^{M} \Omega_j\right)$. It has been shown by Schatz and Wahlbin [5] that

$$\| u - u_h \|_{L_\infty(\Omega_0)} \leq C\, h^{\min(p+1,\, 2\pi/\alpha_M) - \varepsilon} \tag{2.13}$$

$$\| u - u_h \|_{L_\infty(\Omega_M)} \leq C\, h^{\pi/\alpha_M - \varepsilon} \tag{2.14}$$

Similar estimates were discussed by Oden and O'Leary [6]. It should be emphasised that all the above estimates are based on quasi-uniform meshes and piecewise p^{th} order polynomials.

Specific examples of estimates (2.11) - (2.14) are those where the region Ω contains a slit, $\alpha_M = 2\pi$, for which

$$u \in H^{3/2-\varepsilon}(\Omega),\quad \| u - u_h \|_{1,\Omega} = O(h^{\frac{1}{2}-\varepsilon}),\quad \| u - u_h \|_{L_2(\Omega)} = O(h^{1-\varepsilon})$$

$$\| u - u_h \|_{L_\infty(\Omega_0)} = O(h^{1-\varepsilon}),\qquad \| u - u_h \|_{L_\infty(\Omega_M)} = O(h^{\frac{1}{2}-\varepsilon})$$

and where the region is L-shaped, $\alpha_M = 3\pi/2$, for which

$$u \in H^{5/3-\varepsilon}(\Omega),\quad \| u - u_h \|_{1,\Omega} = O(h^{2/3-\varepsilon}),\quad \| u - u_h \|_{L_2(\Omega)} = O(h^{4/3-\varepsilon})$$

$$\| u - u_h \|_{L_\infty(\Omega_0)} = O(h^{4/3-\varepsilon}),\quad \| u - u_h \|_{L_\infty(\Omega_M)} = O(h^{2/3-\varepsilon})$$

2.2 Techniques for Singularities. The error estimates of Section 2.1 indicate the deterioration from the *optimal* state caused by the presence of the singularity. On account of the practical importance of singularaties much effort has been expended in producing special finite element techniques for treating singularities, and a considerable literature now exists. The approaches fall mainly into three classes; augmentation of the trial and test spaces with functions having the form of the dominant part of the singularity, use of singular elements, use of local mesh refinement. These techniques and their effects are now reviewed briefly.

Since for problems of type (2.2) with re-entrant corners the form of the singularity is known, use of this can be made by augmenting the space S^h with functions having the form of the singularity. The solution u of (2.5) is in this case approximated by $\hat{u}_h \in \text{Aug}S^h$. The technique, proposed by Fix [7] and used by Barnhill and Whiteman [8] and Stephan and Whiteman [9], enables estimates as for problems with smooth solutions to be obtained. It does, however, have the disadvantage of producing a system of linear equations in which the coefficient matrix has a more complicated structure than normal.

The technique of employing *singular* elements involves in elements near the singularity the use of local functions which approximate realistically the singular behaviour. Elements of this type have been proposed by Akin [10], Blackburn [11] and Stern and Becker [12], and their use can lead to significant increase in accuracy of u_h. O'Leary [13], specifically for the Stern-Becker element, has proved that use of the element produces no improvement in the rate of convergence in the error estimate. The increase in accuracy must therefore be produced by reduction in size of the constant in the estimate.

Local mesh refinement near a singularity was originally performed on an ad-hoc basis without theoretical backing. In recent years error analysis has been produced which indicates the grading which a mesh should have near a corner in order that the effect of a singularity may be nullified. Examples of such local mesh refinement are given by Schatz and Wahlbin [5] and Babuska and Osborn [14]. Another approach is to use *adaptive* mesh refinement involving a-posteriori error estimation.

With adaptive mesh refinement the region Ω is partitioned initially and the local error in each element is estimated. If, for a particular element, this is greater than a prescribed tolerance, the element is subdivided thus causing the local refinement, see Babuska and Rheinboldt [15],[16]. Hierarchical finite elements have recently been incorporated into the technique, Craig, Zu and Zienkiewicz [17], as have multigrid methods Bank and Sherman [18] and Rivara [19].

2.3 Retrieved Quantities. As has been stated in Section 1, for problems involving boundary singularities the approximation of secondary (retrieved) quantities is most important. Specifically the coefficients a_j in (2.10) of singular terms are of practical significance, so that ways must be found of approximating these accurately. Apart from the obvious approach of using collocation or least squares methods to fit terms to calculated results, it is often possible to exploit the mathematics of the original problem. An important case is that of a problem containing a slit, $\alpha_M = 2\pi$, and here use can be made of the "J-integral" concept to produce an integral expression for the a_M, see Destuynder et al.,[20]. This integral can be approximated using the calculated solution u_h. For piecewise linear test and trial functions on a mesh with local refinement, O(h) estimates are given in [20] for the absolute value in the error in the approximation to the singularity coefficient a_M.

For problem (2.2), when a singularity is present, the integrand of the "J-integral" involves derivatives of the solution u . Thus the accuracy of the approximation to the integral, and hence to the singularity coefficient, depends on the errors in the gradients of u_h. A possibility exists here of exploiting superconvergence properties in the estimation of errors in gradients of u_h, provided local estimates can be obtained. To date the error estimates have depended on the global regularity of the solution u , see Levine [21].

2.4 Three Dimensional Poisson Problems. We consider again problems of the type (2.1), except that now $\Omega \subset \mathbb{R}^3$ is a polyhedral domain. The weak forms and the finite element method for the three-dimensional case can be described similarly, again with $\Omega \subset \mathbb{R}^3$. Singularities can in this case occur on account of re-entrant edges and vertices. The decomposition of the three dimensional weak solution corresponding to (2.10) has been shown, e.g. by Stephan [22] to have the form

$$u = \underset{\text{(vertices)}}{\sum_{j=1}^{M}} a_j \chi_j u_j + \underset{\text{(edges)}}{\sum_{k=1}^{N}} f_k \Xi_k v_k + w \qquad (2.15)$$

where $w \in H^2(\Omega)$, $\chi_j(r_j)$ and $\Xi_k(\rho_k)$ are cut-off functions respectively for the vertices and edges, whilst the u_j and v_k are functions associated also respectively with vertices and edges. For an edge the v_k have the two dimensional form for any plane orthogonal to the edge associated with the appropriate two dimensional problem, whilst the b_k are functions of z_k.

The singular function u_j for each vertex is found by solving a Laplace-Beltrami eigenvalue problem on that part of the surface of the unit ball centred on the vertex cut off by the faces of the vertex. When the vertex is such that the eigenvalue problem is separable (has a single coordinisation), there are special cases when the problem can be solved exactly, see Walden and Kellogg [23]. When this is not so, for example for a vertex made up from three mutually orthogonal planes, Beagles and Whiteman [24], a numerical approximation to the eigenvalue must be obtained with the result that the singular function will not be known exactly.

Clearly this lack of knowledge of the exact singular functions is very important from the finite element point of view, and in particular means that the error analysis of Section 2.1 cannot in general be transferred directly to the three-dimensional singular case. All the singularity methods described in Section 2.1 are affected, although all are used in the three-dimensional context. The augmentation technique is obviously adversely affected, although Beagles and Whiteman [25] have devised the technique of *non-exact augmentation*, whereby the trial and test function spaces in the Galerkin procedure are augmented with the non-exact singular functions. As far as we are aware no method of the "J-integral" type exists for three-dimensional Poisson problems.

The above indicates that the state-of-the-art for treating three-dimensional singularities with finite element methods is far less advanced than that for the two-dimensional case. This arises more from limitations in the theory of three-dimensional Poisson problem than from the finite element methods themselves.

3. LINEAR ELASTIC FRACTURE.

We now consider problems of linear elasticity defined in $\Omega \subset \mathbb{R}^3$, with boundary $\partial\Omega$ such that $\partial\Omega \equiv \partial\Omega_1 \cup \partial\Omega_2$ and $\partial\Omega_1 \cap \partial\Omega_2 = \phi$, where the part $\partial\Omega_1$ is constrained whilst $\partial\Omega_2$ is subject to tractions. The vector function $\underline{u}(\underline{x}) \equiv \{u_1(\underline{x}), u_2(\underline{x}), u_3(\underline{x})\}^T$ is the displacement at the point $\underline{x} \equiv \{x_1,x_2,x_3\}^T \in \Omega$ and satisfies

$$\begin{aligned}
-\mu\, \Delta\underline{u}(\underline{x}) - (\lambda+\mu)\,\mathrm{grad\,div}\,\underline{u}(\underline{x}) &= \underline{f}(\underline{x}) , && \underline{x} \in \Omega , \\
\underline{u}(\underline{x}) &= 0 , && \underline{x} \in \partial\Omega_1 \\
\sum_{j=1}^{3} \sigma_{ij}(\underline{u}(\underline{x}))n_j &= g_i(\underline{x}) , && \underline{x} \in \partial\Omega_2 , \\
& && 1 \leq i \leq 3 ,
\end{aligned} \qquad (3.1)$$

where $\underline{f} \in (L^2(\Omega))^3$, $\underline{g} \equiv (g_1,g_2,g_3)^T \in (L_2(\partial\Omega_2))^3$ are boundary tractions, n_j is the jth component of the unit outward normal $\underline{n} \equiv (n_1,n_2,n_3)^T$ to $\partial\Omega_2$ at $\underline{x}$, the stress tensor $\sigma_{ij}(\underline{u}(\underline{x}))$ is, for $\underline{x} \in \Omega$, defined as

$$\sigma_{ij}(\underline{u}) = \sigma_{ji}(\underline{u}) \equiv \lambda\left(\sum_{k=1}^{3} \varepsilon_{kk}(\underline{u})\right)\delta_{ij} + 2\mu\varepsilon_{ij}(\underline{u}) \quad , \quad 1 \leqq i,j \leqq 3 \quad , \tag{3.2}$$

and in (3.1) and (3.2) $\lambda > 0$, $\mu > 0$ are the Lamé constants which are defined in terms of the Young's modulus E and Poisson's ratio ν as

$$\lambda = \frac{E\nu}{(1+\nu)(1-2\nu)} \quad \text{and} \quad \mu = \frac{E}{2(1+\nu)} \quad .$$

The strain tensor $\varepsilon_{ij}(\underline{u})$ is defined similarly in terms of displacements as

$$\varepsilon_{ij}(\underline{u}) = \varepsilon_{ji}(\underline{u}) = \frac{1}{2}\left(\frac{\partial u_i}{\partial x_j} + \frac{\partial u_j}{\partial x_i}\right) \quad , \quad 1 \leqq i,j < \leqq 3 \ . \tag{3.3}$$

Similarly to (2.2), but now in the space $\tilde{V}$, where

$$\tilde{V} \equiv \left\{\underline{v} : \underline{v} \in (H^1(\Omega))^3 \ , \ v_i\big|_{\partial\Omega_1} = 0 \quad 1 \leqq i \leqq 3\right\} \ , \tag{3.4}$$

a weak form of (3.1) is defined in which $\underline{u} \in \tilde{V}$ satisfies

$$\tilde{a}(\underline{u},\underline{v}) = \tilde{F}(\underline{v}) \quad \forall \ \underline{v} \in \tilde{V} \ , \tag{3.5}$$

where

$$\tilde{a}(\underline{u},\underline{v}) \equiv \int_\Omega \sum_{i,j=1}^{3} \sigma_{ij}(\underline{u}) \ \varepsilon_{ij}(\underline{v}) \ d\underline{x}$$

$$= \int_\Omega \left\{\lambda \operatorname{div}\underline{u} \operatorname{div}\underline{v} + 2\mu \sum_{i,j=1}^{3} \varepsilon_{ij}(\underline{u}) \ \varepsilon_{ij}(v)\right\} d\underline{x} \quad \underline{u},\underline{v} \in \tilde{V} \tag{3.6}$$

and

$$\tilde{F}(v) \equiv \int_\Omega \underline{f}.\underline{v} \ d\underline{x} + \int_{\partial\Omega_2} \underline{g}.\underline{v} \ ds \ , \qquad v \in \tilde{V} \ . \tag{3.7}$$

In a manner similar to that for the Poisson problem of Section 2, problem (3.1) is treated by considering the weak form (3.5). The bilinear form $\tilde{a}(u,v)$ is symmetric and continuous on $\tilde{V}$. It has also been proved by Fichera [26] to be $\tilde{V}$-elliptic. As a result, when the Galerkin method is applied to (3.5) to produce an approximation to the displacement, estimates for the error can again be derived when the regularity of $u(\underline{x})$ is known.

The main interest here in this type of problem is the case of linear elastic fracture, where the region Ω contains a crack so that the re-entrant angle along the crack front is 2π. The mathematical theory for this type of singular problem is much less developed than that for

Poisson problems containing singularities. Expressions for near-tip singular stress and displacement fields for *two-dimensional* problems involving plane-strain and plane-stress have been developed Rice [27] and Williams [28] and involve *stress intensity* factors K . The presence of the singularities again affects the finite element solutions for such problems. The stress intensity factors play a role in linear elastic fracture similar to the singularity coefficients a_j in the Poisson problems of Section 2. The stress intensity factor, which for a problem of linear elastic fracture defined in terms of displacements is a secondary quantity, is used as a fracture criterion. It is thus very important that stress intensity factors be approximated accurately. As they are *retrieved* quantities, usually calculated from approximations to displacements, it is necessary that the approximations to these *primary* variables be accurate.

The similarity to the case of Poisson problems containing boundary singularities is clear, particularly for two dimensions, and motivates in the fracture setting the use of finite element singularity techniques as described for Poisson problems. Indeed many of the special methods and elements for singularities were originally produced to treat fracture problems. Finite element techniques for three-dimensional linear elastic fractures are again less well understood theoretically than are those for two dimensions. This is in the main due to the many open questions which exist in the mathematical theory of three-dimensional linear elastic fracture, for example concerning the form of the singular fields near the point of intersection of a crack front with a stress free surface, see e.g. Whiteman and Thompson [29].

In two dimensions a particularly successful way of calculating the stress intensity factor is to use the path independent J-integral, see Rice [30]. This predates the integrals mentioned in Section 2 for Poisson problems, and is an integral around a contour, which surrounds a crack-tip joining one face of the crack to the other, for which the integrand involves derivatives of the displacements. If approximations to stress intensity factors are calculated using the J-integral based on finite element approximations to the displacements, many of the comments made in Section 2.3 carry over to the linear elasticity context. The J-integral method is used extensively in conjunction with finite element methods and its implementation is discussed at length by Whiteman and Thompson [29]. A mathematical interpretation of the

J-integral is given by Destuynder and Djaoua [31].

In summary it is fair to say that whilst finite element methods are routinely used with success for treating linear elastic fracture, much remains to be done in order that rigorous error estimates comparable to those of Poisson problems may be derived.

REFERENCES

1. Ciarlet P.G., The Finite Element Method for Elliptic Problems. North-Holland, Amsterdam, 1979.
2. Schatz A., An introduction to the analysis of the error in the finite element method for second order elliptic boundary value problems. pp. 94-139 of P.R. Turner (ed.) Numerical Analysis Lancaster 1984. Lecture Notes in Mathematics 129, Springer Verlag, Berlin, 1985.
3. Nitsche J.A., L_∞-convergence of finite element approximations, mathematical aspects of finite element methods. Lecture Notes in Mathematics 606, Springer Verlag, Berlin, 1977.
4. Grisvard P., Behaviour of the solutions of an elliptic boundary value problem in a polygonal or polyhedral domain. pp. 207-274 of B. Hubbard (ed.), Numerical Solution of Partial Differential Equations III, SYNSPADE 1975. Academic Press, New York, 1976.
5. Schaltz A. and Wahlbin L., Maximum norm estimates in the finite element method on plane polygonal domains. Parts I and II. Math. Comp. 32, 73-109, 1978, and Math. Comp. 33, 465-492, 1979.
6. Oden J.T. and O'Leary J., Some remarks on finite element approximations of crack problems and an analysis of hybrid methods. J. Struct. Mech. 64, 415-436, 1978.
7. Fix G., Higher order Rayleigh Ritz approximations. J. Math. Mech. 18, 645-657, 1969.
8. Barnhill R.E. and Whiteman J.R., Error analysis of Galerkin methods for Dirichlet problems containing boundary singularities. J. Inst. Math. Applics. 15, 121-125, 1975.
9. Stephan E. and Whiteman J.R., Singularities of the Laplacian at corners and edges of three dimensional domains and their treatment with finite element methods. Technical Report BICOM 81/1, Institute of Computational Mathematics, Brunel University, 1981.
10. Akin J.E., Generation of elements with singularities. Int. J. Numer. Method. Eng. 10, 1249-1259, 1976.
11. Blackburn W.S., Calculation of stress intensity factors at crack tips using special finite elements. pp. 327-336 of J.R.Whiteman (ed.), The Mathematics of Finite Elements and Applications. Academic Press, London, 1973.
12. Stern M. and Becker E., A conforming crack tip element with quadratic variation in the singular fields. Int. J. Numer. Meth. Eng. 12, 279-288, 1978.
13. O'Leary J.R., An error analysis for singular finite elements. TICOM Report 81-4, Texas Institute of Computational Mechanics, University of Texas at Austin, 1981.

14. Babuska I. and Osborn J., Finite element methods for the solution of problems with rough input data. pp. 1-18 of P. Grisvard, W. Wendland and J.R. Whiteman (eds.), Singularities and Constructive Methods for Their Treatment. Lecture Notes in Mathematics 1121, Springer Verlag, Berlin, 1985.
15. Babuska I. and Rheinboldt W.C., Error estimates for adaptive finite element computations. SIAM J. Num. Anal. 15, 736-754, 1978.
16. Babuska I. and Rheinboldt W.C., Reliable error estimation and mesh adaptation for the finite element method. pp. 67-108 of J.T. Oden (ed.), Computational Methods in Nonlinear Mechanics. North Holland, Amsterdam, 1979.
17. Craig A.W., Zhu J.Z. and Zienkiewicz O.C., A-posteriori error estimation, adaptive mesh refinement and multigrid methods using hierarchical finite element bases. pp. 587-594 of J.R. Whiteman (ed.), The Mathematics of Finite Elements and Applications V, MAFELAP 1984. Academic Press, London, 1985.
18. Bank R.E. and Sherman A.H., The use of adaptive grid refinement for badly behaved elliptic partial differential equations. pp. 18-24 of Computers in Simulation XXII. North Holland, Amsterdam, 1980.
19. Rivara M.C., Dynamic implementation of the h-version of the finite element method. pp. 595-602 of J.R. Whiteman (ed.), The Mathematics of Finite Elements and Applications V, MAFELAP 1984, Academic Press, London, 1985.
20. Destuynder P., Djaoua M. and Lescure S., On numerical methods for fracture mechanics, pp. 69-84 of P. Grisvard, W.L. Wendland and J.R. Whiteman (eds.), Singularities and constructive methods for their treatment. Lecture Notes in Mathematics 1121, Springer Verlag, Berlin, 1985.
21. Levine N., Superconvergence recovery of the gradient from piecewise linear finite element approximations. Technical Report 6/83, Dept. of Mathematics, University of Reading, 1983.
22. Stephan E., A modified Fix method for the mixed boundary value problem of the Laplacian in a polyhedral domain. Preprint Nr. 538, Fachbereich Mathematik, T.H. Darmstadt, 1980.
23. Walden H. and Kellogg R.B., Numerical determination of the fundamental eigenvalue for the Laplace operator on a spherical domain. J. Engineering Mathematics 11, 299-318, 1977.
24. Beagles A.E. and Whiteman J.R., Treatment of a re-entrant vertex in a three dimensional Poisson problem. pp. 19-27 of P. Grisvard, W.H. Wendland and J.R. Whiteman (eds.), Singularities and Constructive Methods for Their Treatment. Lecture Notes in Mathematics 1121, Springer Verlag, Berlin, 1985.
25. Beagles A.E. and Whiteman J.R., Finite element treatment of boundary singularities by augmentation with non-exact singular functions. Technical Report BICOM 85/1, Institute of Computational Mathematics, Brunel University, 1985.

26. Fichera G., Linear Elliptic Differential Systems and Eigenvalue Problems. Lecture Notes in Mathematics 8, Springer Verlag, Berlin 1965.
27. Rice J.R., Mathematical analysis in the mechanics of fractures. pp. 191-311 of H. Liebowitz (ed.). Fracture, Vol.2, Academic Press, New York, 1968.
28. Williams M.L., Stress singularities resulting from various boundary conditions in angular corners of plates in extension. J. Appl. Mech. 24, 526-528, 1952.
29. Whiteman J.R. and Thompson G.M., Finite element calculations of fracture parameters in problems of fractures. pp. 27-47 of J.R. Whiteman (ed.), The Mathematics of Finite Elements and Applications V, MAFELAP 1984. Academic Press, London, 1985.
30. Rice J.R., A path independent integral and the approximate analysis of strain concentration by notches and cracks. J. Appl. Mech. 34, 379-386, 1968.
31. Destuynder P. and Djaoua M., Sur une interpretation mathematique de l'integrale de Rice en theorie de la rupture fragile. Math. Meth. Appl. Sci. 3, 70-87, 1981.

Numerical Approximation of Partial Differential Equations
E.L. Ortiz (Editor)

FINITE ELEMENT SOLUTION OF THE FUNDAMENTAL EQUATIONS OF SEMICONDUCTOR DEVICES

Miloš Zlámal
Technical University of Brno
Brno, Czechoslovakia

We consider the nonstationary equations of the semiconductor device theory consisting of a Poisson equation for the electric potential ψ and of two highly nonlinear continuity equations for carrier densities n and p. We use triangular elements with linear polynomials. There is constructed a finite element solution such that the current densities $\underline{J}_n$ and $\underline{J}_p$ are constant on each element. The scheme is unconditionaly stable and the approximate solution converges strongly to the unique exact solution.

1. We consider a system of three equations in a two-dimensional domain Ω with a polygonal boundary Γ which form the basic model of mobile carrier transport in a semiconductor device:

$$(1) \qquad -\Delta\psi = \frac{q}{\varepsilon}(p - n + N(x))$$

$$(2) \qquad \frac{\partial n}{\partial t} = \nabla.[D_n(x,\|\nabla\psi\|)\nabla n - \mu_n(x,\|\nabla\psi\|)n\nabla\psi] - R_n(n,p)$$

$$(3) \qquad \frac{\partial p}{\partial t} = \nabla.[D_p(x,\|\nabla\psi\|)\nabla p + \mu_p(x,\|\nabla\psi\|)p\nabla\psi] - R_p(n,p) \; .$$

The unknowns are the electrostatic potential ψ and the electron and hole densities n and p. All coefficients appearing in (1) - (3) are positive. q and ε are constants and the diffusion coefficients $D_s(x,\xi)$ (s = n,p) are related to the mobilities $\mu_s(x,\xi)$ by the Einstein relation $D_s(x,\xi) = U_T\mu_s(x,\xi)$ where U_T is the thermal voltage. The doping profile N(x) is a given function of $x = (x_1,x_2)$ and it changes its values extremaly rapidly if x approaches the so called junctions (curves lying in Ω). We simplify the system taking $R_n(n,p) = R_p(n,p) = R(n,p) = \dfrac{np - n_i^2}{\tau_p(n+n_i) + \tau_n(p+n_i)}$ and further $\tau_n = \tau_p = \tau$.

Before formulating the problem completely we scale the system in that we introduce new dependent as well as independent dimensionless variables and the new dimensionless profile and mobilities

$$\psi = U_T\bar{\psi},\ n = n_i\bar{n},\ p = n_i\bar{p},\ x_j = \ell\bar{x}_j\ (j = 1,2),$$
$$t = \tau\bar{t},\ N = n_i\bar{N},\ \mu_s = \mu_0\bar{\mu}_s\ (s = n,p)\ .$$

Here ℓ is a realistic value of diam (Ω) and $n_i = 10^{10}\,cm^{-3}$ is the intrinsic number. Keeping the same notation of all variables, of the mobilities and of the doping profile as before we get

(4) $$-\Delta\psi = \alpha(p - n + N(x))$$

(5) $$\frac{\partial n}{\partial t} = \gamma_n \nabla.\{\mu_n(x,\|\nabla\psi\|)[\nabla n - n\nabla\psi]\} - R(n,p)$$

(6) $$\frac{\partial p}{\partial t} = \gamma_p \nabla.\{\mu_p(x,\|\nabla\psi\|)[\nabla p + p\nabla\psi]\} - R(n,p)$$

(5), (6) in $\Omega \times (0,T)$

$$R(n,p) = \frac{np-1}{n+p+2}$$

where α and γ_s are positive constants. (5) and (6) are equivalent to

(7) $$\frac{\partial n}{\partial t} = \gamma_n \nabla.\underline{J}_n^* - R,\quad \frac{\partial p}{\partial t} = -\gamma_p \nabla.\underline{J}_p^* - R$$

where

(8) $$\underline{J}_n^* = \mu_n(\nabla n - n\nabla\psi),\quad \underline{J}_p^* = -\mu_p(\nabla p + p\nabla\psi)$$

The basic boundary conditions are Dirichlet nonhomogeneous and Neumann homogeneous boundary conditions

(9) $$\psi|_{\Gamma 1} = \psi^*|_{\Gamma 1},\ n|_{\Gamma 1} = n^*|_{\Gamma 1},\ p|_{\Gamma 1} = p^*|_{\Gamma 1},$$
$$n^*(x),\ p^*(x) > 0 \text{ on } \Gamma^1$$

(10) $$\frac{\partial\psi}{\partial\nu}\Big|_{\Gamma 2} = \frac{\partial n}{\partial\nu}\Big|_{\Gamma 2} = \frac{\partial p}{\partial\nu}\Big|_{\Gamma 2} = 0$$

Here, $\Gamma = \Gamma^1 \cup \Gamma^2$, $\Gamma^1 \neq \phi$. In addition, we have the initial condition

(11) $$n = n^0(x),\ p = p^0(x) \text{ in } \Omega,\ n^0(x),\ p^0(x) > 0 \text{ on } \bar{\Omega}$$

and the requirement

(12) $n(x,t) > 0,\ p(x,t) > 0$ on $\bar{\Omega} \times [0,T]$.

We introduce the integral relations which form the basis of a variational formulation. We denote by V the space

$$V = \{v | v \in H^1(\Omega),\ v|_{\Gamma 1} = 0\}$$

Let us remark that we use the usual notations of the Sobolev spaces $H^m(\Omega)$, $m = 1,2,\ldots,$ and $H^{m,p}(\Omega)$, $1 \le p \le \infty$ ($H^{m,2} = H^m$). Also $L^p(\Omega)$ are the Lebesgue spaces and $(.,.)$ will denote the scalar product in $L^2(\Omega)$.

Set $w = e^{-\psi}n$, $z = e^{\psi}p$ which is the so called Boltzmann statistics. Then $\nabla n - n\nabla\psi = e^{\psi}\nabla w$, $\nabla p + p\nabla\psi = e^{-\psi}\nabla z$. Multiplying (4), (5), (6) by a function $v \in V$, integrating over Ω and using Green's theorem and (10) we get

$$\left.\begin{aligned} &(13) \quad d(\psi,v) = \alpha(p - n + N,v) \\ &(14) \quad (\tfrac{\partial}{\partial t}n,v) + \gamma_n \nu^1(\psi;w,v) + (R(n,p),v) = 0 \\ &(15) \quad (\tfrac{\partial}{\partial t}p,v) + \gamma_p \pi^1(\psi;z,v) + (R(n,p),v) = 0 \end{aligned}\right\} \forall v \in V$$

where

$$(16) \quad \begin{cases} d(\psi,v) = \int_\Omega \nabla\psi.\nabla v dx, \quad \nu^1(\psi;w,v) = \int_\Omega \mu_n(x,\|\nabla\psi\|)e^{\psi}\nabla w.\nabla v dx \\ \pi^1(\psi;z,v) = \int_\Omega \mu_p(x,\|\nabla\psi\|)e^{-\psi}\nabla z.\nabla v dx \end{cases}$$

Without Boltzmann statistics the equations (14) and (15) have the form

$$\left.\begin{aligned} &(17) \quad (\tfrac{\partial}{\partial t}n,v) + \gamma_n \nu^2(\psi;n,v) + (R(n,p),v) = 0 \\ &(18) \quad (\tfrac{\partial}{\partial t}p,v) + \gamma_p \pi^2(\psi;p,v) + (R(n,p),v) = 0 \end{aligned}\right\} \forall v \in V$$

$$(19) \quad \begin{cases} \nu^2(\psi;n,v) = \int_\Omega \mu_n(x,\|\nabla\psi\|)[\nabla n - n\nabla\psi].\nabla v dx \\ \pi^2(\psi;p,v) = \int_\Omega \mu_p(x,\|\nabla\psi\|)[\nabla p + p\nabla\psi].\nabla v dx \end{cases}$$

2. We consider a family $\{T_h\}$ of triangulations of Ω. If $K \in T_h$ then h_K denotes the greatest side of K and

$$h = \max_{K \in T_h} h_K \to 0 .$$

We assume that the family $\{T_h\}$ satisfies the minimum angle condition, i.e. if ϑ_h is the minimum angle of all angles of T_h then $\vartheta_h \geq \vartheta_0 > 0$, and is of acute type, i.e. any angle of any triangle from $\bigcup_h T_h$ is not greater than $1/2\pi$. By V_h we denote the space

$$V_h = \{v \mid v \in C(\overline{\Omega}),\ v \text{ is a linear polynomial on each } K \in T_h,\ v|_{\Gamma 1} = 0\}$$

It is known that the discretization of equations (5) and (6) by finite differences does not give good results if it is carried out in the standard way (Scharfetter and Gummel, 1969). This is caused by extremely large values of $\|\nabla\psi\|$ close to junctions. Scharfetter and Gummel proposed for the onedimensional case to treat the equations (8) as differential equations in n and p with $\underline{J}_n$, $\underline{J}_p$, μ_n, μ_p and $\nabla\psi$ assumed constant between mesh points. The scheme which they derived proved to be succesfull. We use the same idea: the above quantities are assumed constant on each element. For more details we refer the reader to (Zlámal, to appear). Here we introduce the result. By W_h we denote the space

$$W_h = \{v \mid v \in C(\overline{\Omega}),\ v \text{ is a linear polynomial on each } K \in T_h\} .$$

The approximations from W_h of the functions ψ, n, p, w, z are denoted by Ψ, N, P, W, Z. W and Z are determined by

$$W_j = e^{-\Psi_j} N_j,\ Z_j = e^{\Psi_j} P_j \quad \forall x^j \in \overline{\Omega}$$

Here x^j is an arbitrary node of T_h and $W_j, \Psi_j, \ldots$ denote the values $W(x^j)$, $\Psi(x^j), \ldots$. We denote by x^j, x^k, x^m the nodes of a triangle $K \in T_h$. The discrete analogues $\nu_h^1(\Psi;W,v)$ and

$\pi_h^1(\Psi;Z,v)$ of the forms $\nu^1(\psi;w,v)$ and $\pi^1(\psi;z,v)$ are

(20) $$\nu_h^1(\Psi;W,v) = \sum_{K\in T_h} \mu_n^K \sum_{r=j,k,m} v_r e^{\Psi_r} \int_K (J^T)^{-1} B^K J^T \nabla W . \nabla v^r dx$$

(21) $$\pi_h^1(\Psi;Z,v) = \sum_{K\in T_h} \mu_p^K \sum_{r=j,k,m} v_r e^{-\Psi_r} \int_K (J^T)^{-1} D^K J^T \nabla Z . \nabla v^r dx$$

Here $\mu_s^K = \mu_s(x^K, \|\nabla\psi\|)$, x^K is the center of gravity of K, J is the mapping which maps K on the unit reference triangle $\hat{K}$ in the ξ-plane with vertices (0,0),(1,0),(0,1) in such a way that x^r is mapped on the vertex (0,0) (this requirement does not determine J uniquely; nevertheless it suffices to a unique determination of ν_h^1 and π_h^1), v_r is the value $v(x^r)$, v^r is the basis function associated to the node x^r and B^K, D^K are the matrices

(22) $$\begin{cases} B^K = \operatorname{diag}(B(\Psi_1 - \Psi_2), B(\Psi_1 - \Psi_3)), \ B(\xi) = \xi(e^\xi - 1)^{-1} \\ D^K = \operatorname{diag}(D(\Psi_1 - \Psi_2), D(\Psi_1 - \Psi_3)), \ D(\xi) = e^\xi B(\xi) = \\ \quad = B(-\xi) \end{cases}$$

Here Ψ_1, Ψ_2, Ψ_3 is the local notation of the values of Ψ at the vertices x^j, x^k, x^m such that $\Psi_1 = \Psi_r$ $(r = j,k,m)$. The forms ν_h^1 and π_h^1 are symmetric with respect to W,v and Z,v, respectively.

The forms ν_h^2 and π_h^2 which are discrete analogs of ν^1 and π^1, respectively, are easy to derive from (20), (21):

(23) $$\nu_h^2(\Psi;N,v) = \sum_{K\in T_h} \mu_n^k \sum_{r=j,k,m} v_r \Big[\int_K (J^T)^{-1} D^K J^T \nabla N . \nabla v^r dx - \\ - \int_K N_r \nabla\Psi . \nabla v^r dx\Big]$$

(24) $$\pi_h^2(\Psi;P,v) = \sum_{K\in T_h} \mu_p^K \sum_{r=j,k,m} v_r \Big[\int_K (J^T)^{-1} B^K J^T \nabla P . \nabla v^r dx + \\ + \int_K P_r \nabla\Psi . \nabla v^r dx\Big]$$

Now we introduce a fully discrete approximate solution. To this end we consider an equally spaced partition of the interval $[0,T]$: $t_i = i\Delta t$, $i = 0,1,\ldots,q$, $q = \frac{T}{\Delta t}$. The value of any function at $t = t_i$ is denoted by the superscript i. The derivatives $\frac{\partial n}{\partial t}$ and $\frac{\partial p}{\partial t}$ are replaced by the difference quocients $\Delta t^{-1}\Delta N^i$ $(\Delta N^i = N^i - N^{i-1})$ and $\Delta t^{-1}\Delta P^i$. The L_2-scalar product $(.,.)$ is computed numerically and replaced by $(.,.)_h$ where

$$(n,v)_h = \sum_K I^K(nv), \quad I^K(f) = \frac{1}{3}\,\overline{\text{meas}}\,(K)\sum_{j=1}^{3} f_j$$

and f_j are values of f at the vertices of K. In this way we get a lumped mass matrix which is diagonal and has positive diagonal elements. By f_I we understand the interpolate of f. The defining equations are

$$\left.\begin{array}{ll}
(25) & d(\Psi^i,v) = \alpha(P^i - N^i + N(x),v)_h \\
(26) & (\Delta N^i,v)_h + \gamma_n\Delta t\nu_h^2(\Psi^i;N^i,v) + (R(N^i,P^i),v)_h = 0 \\
(27) & (\Delta P^i,v)_h + \gamma_p\Delta t\pi_h^2(\Psi^i,P^i,v) + (R(N^i,P^i),v)_h = 0
\end{array}\right\}\forall v \in V_h \quad i=1,\ldots,q$$

$$(28) \qquad \Psi_j^i = \psi^*(x^j),\ N_j^i = n^*(x^j),\ P_j^i = p^*(x^j) \qquad \forall x^j \in \Gamma^1$$

$$(29) \qquad N^0 = n_I^0,\ P^0 = p_I^0 \quad \text{on } \overline{\Omega}$$

$$(30) \qquad N^i > 0,\ P^i > 0 \quad \text{on } \overline{\Omega},\ i = 1,\ldots,q$$

3. All theoretical results concern the case

$\mu_s = \text{const},\ s = n,p$

and we assume that the family $\{\mathcal{T}_h\}$ of triangulations fulfils the minimum angle condition and is of acute type.

We introduce three more assumptions:

A^1: $\psi^*, n^*, p^* \in H^2(\Omega)$, $n^*|_{\Gamma^1} > 0$, $p^*|_{\Gamma^1} > 0$, N, n^0, p^0 are measurable and bounded on $\overline{\Omega}$, n^0 and p^0 are positive on $\overline{\Omega}$.

A2: The angles of the polygonal boundary Γ are smaller than π at corners where two sides of Γ^1 meet and smaller than $\frac{1}{2}\pi$ at corners where a side of Γ^1 and a side of Γ^2 meet.

A3: $n^*, p^* \in H^2(\Omega)$, $\psi^*, n^0, p^0 \in H^{2,q}(\Omega)$, $q = 2 + \varepsilon$ and ε is arbitrarily small but positive, $n^*|_{\Gamma^1} > 0, p^*|_{\Gamma^1} > 0$, n^0, p^0 are positive on $\overline{\Omega}$, $N \in H^2(\Omega)$.

THEOREM 1. *The scheme (25) - (27) is unconditionally stable. More exactly, for case that* $\psi^* = n^* = p^* = N = 0$ *there holds for arbitrary* h *and* Δt

$$\max_i\{\|N^i\|_{L^2(\Omega)}, \|P^i\|_{L^2(\Omega)}, \alpha^{-1}\|\Psi^i\|_{H^1(\Omega)}\} + $$
$$+ \Delta t\{\sum_{i=1}^{q}(\|N^i\|^2_{H^1(\Omega)} + \|P^i\|^2_{H^1(\Omega)})\}^{1/2} \leq$$
$$\leq C(\Omega,T)[N^0\|_{L^2(\Omega)} + \|P^0\|_{L^2(\Omega)} + 1]$$

THEOREM 2. *If* A^1 *is fulfiled there exist* N^i, P^i, Ψ^i, $i = 1,\ldots,q$, *satisfying (25) - (30). If, in addition,* A^3 *is fulfiled and* Δt *is sufficiently small,* $\Delta t \leq \Delta t_0$ *where* Δt_0 *does not depend on* h*, then these solutions are unique.*

Now we extend the approximate solution piecewise linearly on the interval $[0,T]$:

$$N^\delta = N^{i-1} + \frac{t-t_{i-1}}{\Delta t}\Delta N^i, \quad P^\delta = P^{i-1} + \frac{t-t_{i-1}}{\Delta t}\Delta P^i,$$
$$\Psi^\delta = \Psi^{i-1} + \frac{t-t_{i-1}}{\Delta t}\Delta\Psi^i \quad \text{on } [t_{i-1},t_i],\ i = 1,\ldots,q.$$

Here $\delta = (h,\Delta t)$ and Ψ^0 is uniquely defined by $d(\Psi^0,v) = \alpha(P^0 - N^0 + N_I, v)_h \ \forall v \in V_h$, $\Psi^0(x^j) = \psi^*(x^j)\ \forall x^j \in \Gamma^1$.

THEOREM 3. *Let* A^2 *and* A^3 *be fulfiled. If* $\Delta t \geq c_0 h^2$ $(c_0$ *= const* $> 0)$, *there exists a triple* (n,p,ψ), $n - n^*, p - p^* \in L^\infty(0,T;V)$,

$n, p \in C([0,T]; L^p(\Omega))$, $1 \le p < \infty$, $\psi - \psi^* \in C([0,T]; V \cap H^{1,\infty}(\Omega))$ *such that for* $\delta \to 0$ *we have*

$$\|N^\delta - n\|_{C([0,T];L^p(\Omega))} \to 0, \quad \|P^\delta - p\|_{C([0,T];L^p(\Omega))} \to 0$$

$$1 \le p < \infty$$

$$\|\Psi^\delta - \psi\|_{C([0,T];H^{1,\infty}(\Omega))} \to 0$$

The triple (n,p,ψ) *is the unique solution of the following problem: Given* $n^*, p^* \in H^1(\Omega)$, $n^*|_{\Gamma^1} > 0$, $p^*|_{\Gamma^1} > 0$, $\psi^* \in H^{1,\infty}(\Omega)$, $n^0, p^0 \in L^2(\Omega)$, $n^0 > 0$, $p^0 > 0$ *a.e. on* $\bar{\Omega}$, $N \in L^q(\Omega)$, $q = 2 + \varepsilon$ *with* $\varepsilon > 0$, *find* (n,p,ψ) *such that* $n - n^*, p - p^* \in L^2(0,T;V)$, $\psi - \psi^* \in L^\infty(0,T; V \cap H^{1,\infty}(\Omega))$ *and*

$$\forall t \in (0,T) \quad d(\psi, v) = \alpha(p - n + N(x), v) \quad \forall v \in V$$

$$\left.\begin{array}{l} \frac{\partial}{\partial t}(n,v) + \gamma_n \nu^2(\psi; n, v) + (R(n,p),v) = 0 \\ \frac{\partial}{\partial t}(p,v) + \gamma_p \pi^2(\psi; p, v) + (R(n,p),v) = 0 \end{array}\right\} \quad \text{in } \mathcal{D}'((0,T)) \ \forall v \in V$$

$$n = n^0, \; p = p^0 \text{ in } \Omega$$

$$\forall t \in [0,T] \quad n \ge 0, \; p \ge 0 \quad \text{a.e. in } \Omega .$$

REFERENCES

1. Scharfetter D.L. and Gummel H.K., Large Signal Analysis of a Silicon Read Diode Oscillator, IEEE Trans. on Electron Devices, ED-16, 1969, 64-77.
2. Zlámal M., Finite Element Solution of the Fundamental Equations of Semiconductor Devices I, to appear in Math. of Computation.

CONTRIBUTED PAPERS

Part I:
RESULTS ON COMPUTATIONAL LINEAR ALGEBRA

Interpolation and Related Techniques

Computational Linear Algebra

Numerical Approximation of Partial Differential Equations
E.L. Ortiz (Editor)

ON AITKEN-NEVILLE FORMULAE FOR MULTIVARIATE INTERPOLATION

M. Gasca
Depto. Ecuaciones Funcionales
Universidad de Zaragoza , Spain

E. Lebrón
E. U. Arquitectura Técnica
Univ. Politécnica, Madrid, Spain

In a paper in J.S.I.A.M.,8, (1960), p.33-42, Thacher and Milne showed how the solution of certain polynomial interpolation problems in R^s can be constructed from the solutions of s+1 simpler problems. In the present paper we extend the method showing that the number of basic problems to be used depends on the distribution of the points and on the chosen interpolation space, which here is not necessarily a polynomial space.

1. INTRODUCTION.

Very few results are known on the application of Aitken -Neville methods to multivariate interpolation problems. These methods, originally thought for Lagrange polynomial interpolation problems in one variable, have been recently extended to the finite linear interpolation problem (see |1|,|3|,|4|,|7|,|9|-|11|) but in each particular problem adequate conditions, not always easy to check, have to be verified.

In 1960, H.C.Thacher and W.E.Milne |12| showed that the solution $p_m \in P_m$ (space of polynomials of total degree non greater than m in R^s) of an interpolation problem on certain sets S of $\binom{m+s}{s}$ points of R^s can be constructed from the solutions $p_{m-1,i}$ of s+1 interpolation problems (in P_{m-1}) on some subsets S_i $(1 \le i \le s+1)$ of S. p_m is obtained as the quotient of two determinants of order s+1 and the particular case s=1 yields the well-known Aitken-Neville formula for one variable.

The assumtions in |12| being very restrictive, the only example given in the paper is the s-simplicial lattice. In |6| some reflexions were made about similar formulae for other examples. In the present paper we consider less restrictive hypotheses which allow us to extend the result to any distribution of points provided that we may choose the interpolation space freely. Familiar distributions of points yield familiar interpolation spaces. In the last section several examples are given.

2. AN AITKEN-NEVILLE FORMULA FOR MULTIVARIATE INTERPOLATION PROBLEMS.

Let F be the vector space of real valued functions defined on $D \subset R^n$ and P an m-dimensional subspace of F.

Denote by S a set of m points X_r $(1\leq r\leq m)$ of D , and S_j , $1\leq j\leq k$, k subsets of S $(1\leq k\leq m)$ such that

i) $S = \bigcup_{j=1}^{k} S_j$

ii) For each $j=1,2,\ldots,k,$ card. $S_j = q$ $(0<q<m)$

iii) There exist k points in S, which can be numbered $X_1, X_2, \ldots, X_k$, such that for each $j=1,2,\ldots,k$

(2.1) $X_j \in S_j$

(2.2) $X_j \notin \bigcup_{\substack{t=1 \\ t\neq j}}^{k} S_t .$

We are going to give sufficient conditions for the existence and unicity of $p\in P$ satisfying

(2.3) $p(X_i)=f(X_i)$ $1\leq i\leq m$

where $f\in F$, in terms of the existence and unicity of solution of k similar problems respectively associated with the subsets S_j.

Let H be a subspace of dimension k of F such that :

iv) If w is a constant function on D, then $w\in H$

v) For each $f\in F$ there exists a unique function $u\in H$ satisfying

(2.4) $u(X_i)=f(X_i)$ $1\leq i\leq k$

vi) For each $j=1,2,\ldots,k,$ the function $w_j\in H$ defined by

(2.5) $w_j(X_t) = \delta_{jt}$ $1\leq t\leq k$ (δ_{jt} Kronecker delta)

satisfies the condition

(2.6) $r>k$ and $X_r \notin S_j \Rightarrow w_j(X_r)=0.$

Finally, let Q be a subspace of dimension q of P such that

vii) $HQ=\{f=w\cdot g \mid w\in H, \ g\in Q\} \subset P$

viii) For each $f\in F$ and $1\leq j\leq k,$ there exists a unique function $p_j\in Q$ such that

(2.7) $p_j(X_t)=f(X_t)$ $\forall X_t \in S_j$.

Theorem.- Under the above assumptions, for each $f\in F$ there exists a unique function $p\in P$ satisfying (2.3), that can be written

(2.8) $$p = \sum_{j=1}^{k} p_j w_j$$

where $p_j\in Q$ (resp. $w_j\in H$) satisfies (2.7) (resp. (2.5)).

Proof.- Condition vii) implies that p, defined by (2.8), belongs to P, and for $t=1,2,\ldots,k$ (2.3) is a direct consequence of iii), (2.5) and (2.7).

Consider $k<t\leq m$ and denote r ($1\leq r\leq k$) the number of subsets S_j such that $X_t\in S_j$. After renumbering $1,2,..,k$ we may assume

(2.9) $\qquad X_t\in S_j\ , \qquad j=1,2,\ldots,r$

and, if $r<k$,

(2.10) $\qquad X_t\notin S_j, \qquad j=r+1,r+2,\ldots,k.$

From (2.6) (2.7), we have

(2.11) $$\sum_{j=1}^{k} p_j(X_t)w_j(X_t) = f(X_t)\cdot\sum_{j=1}^{r} w_j(X_t).$$

On the other hand, from iv)-vi) we easily get

(2.12) $$1=\sum_{j=1}^{k} w_j$$

and then (2.6),(2.10) imply

(2.13) $$\sum_{j=1}^{r} w_j(X_t) = 1.$$

Therefore, in (2.11)

$$\sum_{j=1}^{k} p_j(X_t)\cdot w_j(X_t)=f(X_t).$$

Since the problem has always a solution, this must be unique.

Remarks.-

1. For q=1 and k=m (2.8) is the Lagrange interpolation formula.

2. Some of the conditions given can be weakened. For example, ii) can be omitted and Q replaced in vii)-viii) by k subspaces Q_j ($1\leq j\leq k$) such that

$$\text{card. } S_j = \text{dim. } Q_j.$$

Condition (2.2) could also be omitted.

However, we have found no example satisfying these weakened conditions and not satisfying those assumed in the theorem. For this reason we have prefered to write it in its present form.

3. Condition vi) can be written in an equivalent form: if $\{u_1,u_2,\ldots,u_k\}$ is a basis of H, then

$$(2.14) \qquad r>k \text{ and } X_r \notin S_j \Rightarrow \begin{vmatrix} u_1(X_1) & u_2(X_1) & \ldots & u_k(X_1) \\ \cdots & \cdots & \cdots & \cdots \\ u_1(X_{j-1}) & u_2(X_{j-1}) & \ldots & u_k(X_{j-1}) \\ u_1(X_r) & u_2(X_r) & \ldots & u_k(X_r) \\ u_1(X_{j+1}) & u_2(X_{j+1}) & \ldots & u_k(X_{j+1}) \\ \cdots & \cdots & \cdots & \cdots \\ u_1(X_k) & u_2(X_k) & \ldots & u_k(X_k) \end{vmatrix} = 0.$$

Similar reasons allow us to write p in the form (see |12|)

$$(2.15) \qquad p(X) = - \frac{\begin{vmatrix} 0 & u_1(X) & u_2(X) & \ldots & u_k(X) \\ p_1(X) & u_1(X_1) & u_2(X_1) & \ldots & u_k(X_1) \\ p_2(X) & u_1(X_2) & u_2(X_2) & \ldots & u_k(X_2) \\ \cdots & \cdots & \cdots & \cdots & \cdots \\ p_k(X) & u_1(X_k) & u_2(X_k) & \ldots & u_k(X_k) \end{vmatrix}}{\begin{vmatrix} u_1(X_1) & u_2(X_1) & \ldots & u_k(X_1) \\ u_1(X_2) & u_2(X_2) & \ldots & u_k(X_2) \\ \cdots & \cdots & \cdots & \cdots \\ u_1(X_k) & u_2(X_k) & \ldots & u_k(X_k) \end{vmatrix}}$$

3.EXAMPLES. Let S be a set of m points of R^2 and $f:D\rightarrow R$ $(S \subset D)$ a function to be interpolated by a polynomial, with f(S) as data.

Through each point of S we take a straight line r_i parallel to an arbitrary, given direction:

$$(3.1) \qquad r_i \equiv ax+by+c_i=0 .$$

We use r_i to denote both, the straight line and the polynomial $ax+by+c_i$. Let us choose (a,b) in such a way that the set $T=\{r_1, r_2,\ldots,r_t\}$, $t\leq m$, of different lines r_i (3.1) contains k $(k\geq 2)$ lines $r_1,r_2,\ldots,r_k$ such that

$$(3.2) \qquad n(1)=n(2)=\ldots=n(k)\geq n(k+1)\geq\ldots\geq n(t)$$

where $n(i)=\text{card.}(r_i \cap S)$.

Note thatthis can be assumed because we do not rule out the

possibility

$$t=m, \quad n(i)=1 \quad i=1,2,\ldots\ldots,m.$$

Denote by X_{ij}, $j=1,2,\ldots,n(i)$, the points of S lying on r_i, $(1\leq i\leq t)$ and through each X_{ij} take a straight line r_{ij} parallel to an arbitrary direction different from that of r_i:

$$(3.3) \qquad r_{ij} \equiv dx+ey+c_{ij}=0.$$

In |5| it was proved that the Lagrange interpolation problem on S has a unique solution in the space P given by

$$(3.4) \qquad P = \text{span}\left\{ \prod_{h=1}^{i-1} r_h \cdot \prod_{s=1}^{j-1} r_{is}, \quad i=1,2,\ldots,t \ ; \ j=1,2,\ldots,n(i) \right\}$$

where each product is taken as 1 if i (resp. j) is 1.

From (3.1) (3.3) we have also

$$(3.5) \qquad P=\text{span} \left\{ r_1^{i-1} r_{11}^{j-1}, \qquad i=1,2,\ldots,t \ ; \ j=1,2,\ldots,n(i) \right\} .$$

To apply the theorem we take for X_i any point of S lying on r_i, $1\leq i\leq k$, renumber $X_{k+1},\ldots,X_m$ the other points of S and denote

$$(3.6) \qquad S_i = S - \bigcup_{\substack{j=1 \\ j\neq i}}^{k} r_j , \qquad H=\text{span}\left\{1,r_1,\ldots,r_1^{k-1}\right\}$$

$$Q = \text{span}\left\{ r_1^{i-k} \cdot r_{11}^{j-1}, \qquad i=k,k+1,\ldots,t, \qquad j=1,2,\ldots,n(i)\right\}.$$

Then all the conditions hold and therefore

$$(3.7) \quad p(X)= - \frac{\begin{vmatrix} 0 & 1 & r_1(X) & \ldots & r_1^{k-1}(X) \\ p_1(X) & 1 & r_1(X_1) & \ldots & r_1^{k-1}(X_1) \\ \ldots & \ldots & \ldots & \ldots & \ldots \\ p_k(X) & 1 & r_1(X_k) & \ldots & r_1^{k-1}(X_k) \end{vmatrix}}{\begin{vmatrix} 1 & r_1(X_1) & \ldots r_1^{k-1}(X_1) \\ 1 & r_1(X_2) & \ldots r_1^{k-1}(X_2) \\ \ldots & \ldots & \ldots \\ 1 & r_1(X_k) & \ldots r_1^{k-1}(X_k) \end{vmatrix}}$$

For k=2

$$(3.8)\quad p(X) = \frac{p_2(X)\,(r_1(X) - r_1(X_1)) - p_1(X)\,(r_1(X) - r_1(X_2))}{r_1(X_2) - r_1(X_1)}$$

recalls inmediately the Aitken-Neville formula for one variable.

In |5| Newton-like formulae based on (3.4) were given.

For certain sets S the reasoning can be repeatedly applied easily. Let S be a rectangular lattice

$$(3.9)\quad S = \left\{(x_i, y_j) \mid i=1,2,\dots,n, \quad j=1,2,\dots,m\right\}$$

and denote

$$(3.10)\quad P = Q_{n-1,m-1} = \text{span}\left\{x^s y^t \mid s=0,1,\dots,n-1, \quad t=0,1,\dots,m-1\right\}.$$

By choosing

$$(3.11)\quad r_i \equiv x - x_i, \qquad r_{ij} \equiv y - y_j$$

we can take, for any $k \in \{2,3,..,n\}$,

$$H = \text{span}\left\{1, x, \dots, x^{k-1}\right\}, \quad Q = Q_{n-k,m-1} \quad \text{(defined as in (3.10))}$$

$$S_i = S - \bigcup_{\substack{t=1\\ t\neq i}}^{k} \left\{(x_t, y_j) \mid j=1,2,\dots,m\right\} \qquad 1 \le i \le k$$

$$X_i = (x_i, y_1) \qquad (1 \le i \le k)$$

and apply (3.7) (and particularly (3.8)) where $p_1, p_2, \dots, p_k$ can be obtained similarly.

A different decomposition of the problem defined by (3.9) (3.10) satisfying all the conditions of the theorem is possible:

$$H = Q_{1,1}, \qquad Q = Q_{n-2,m-2}$$

$$S_1 = \left\{(x_i, y_j) \mid 1 \le i \le n-1, \quad 1 \le j \le m-1\right\}$$

$$S_2 = \left\{(x_i, y_j) \mid 2 \le i \le n, \quad 1 \le j \le m-1\right\}$$

$$S_3 = \left\{(x_i, y_j) \mid 2 \le i \le n, \quad 2 \le j \le m\right\}$$

$$S_4 = \left\{(x_i, y_j) \mid 1 \le i \le n-1, \quad 2 \le j \le m\right\}$$

$$X_1 = (x_1, y_1), \quad X_2 = (x_n, y_1), \quad X_3 = (x_n, y_m), \quad X_4 = (x_1, y_m).$$

Then (2.15) holds with $k=4$, $u_1=1$, $u_2=x$, $u_3=y$, $u_4=xy$.

The problem defined by the simplicial lattice

$$S=\left\{(x_i,y_j) \mid i,j=1,2,\ldots,m; \quad i+j\leq m-1\right\} \tag{3.12}$$
$$P=P_{m-1}$$

is the example considered in |12| giving rise to a formula (2.15) with $k=3$, $Q=P_1$, $H=P_{m-2}$. In this case the sets S_i can be taken similar to S and the points X_i are the 'vertices'

$$X_1=(x_1,y_1), \qquad X_2=(x_m,y_1), \qquad X_3=(x_1,y_m).$$

Other more general examples of this kind of decomposition, like the so called natural lattices (see |2|) are shown in |6|. Note that (3.2) does not hold for this choice of S_i, X_i, and therefore we do not apply (3.7) but (2.15).

For interpolation problems in one variable

$$S=\left\{x_1, x_2,\ldots,x_n\right\}\subset R \qquad P=P_{n-1}$$

(little obvious changes for P space of trigonometrical polynomials of order h if $n=2h-1$) we take, for any $k\in\{2,3,\ldots,n\}$

$$H=P_{k-1} \qquad Q=P_{n-k}$$
$$S_i=S-\left\{x_1,x_2,\ldots,x_{i-1},x_{i+1},\ldots,x_k\right\} \qquad i=1,2,\ldots,k.$$

The resulting formula is a wellknown generalization of the Aitken-Neville formula for one variable |8|.

Finally let us point out that some interpolation problems with derivatives as interpolation data, which can be considered as 'limit' of Lagrange interpolation problems, are also included in the framework of the above theorem. An example is given by the problem defined by the interpolating space $P=Q_{n-1,m-1}$ and the data

$$f(x_i,y_j) \qquad 1\leq i\leq n-1, \quad 1\leq j\leq m$$

$$\frac{\partial f(x_{n-1},y_j)}{\partial x} \qquad j=1,2,\ldots,m.$$

If we take $H=\text{span}\left\{1,x\right\}$, $Q=Q_{n-2,m-1}$, $X_1=(x_1,y_1)$, $X_2=(x_2,y_1)$ and p_1 (resp. p_2) is the solution in Q of the problem

$$p_1(x_i,y_j)=f(x_i,y_j) \qquad 1\leq i\leq n-1, \quad i\neq 2 \quad ; \quad 1\leq j\leq m$$

$$\frac{\partial p_1(x_{n-1},y_j)}{\partial x} = \frac{\partial f(x_{n-1},y_j)}{\partial x} \qquad 1\leq j\leq m$$

(resp.

$$p_2(x_i,y_j)=f(x_i,y_j) \qquad 2\le i\le n-1 \quad , \quad 1\le j\le m$$

$$\frac{\partial p_2(x_{n-1},y_j)}{\partial x} = \frac{\partial f(x_{n-1},y_j)}{\partial x} \qquad 1\le j\le m \quad)$$

we have, as in (3.8)

$$p(X) = \frac{(x-x_1)p_2(X)-(x-x_2)p_1(X)}{x_2-x_1}$$

REFERENCES

1. BREZINSKI C. :The Mühlbach-Neville-Aitken algorithm and some extensions. B.I.T., v.80, 1980, p.444-451.
2. CHUNG K.C. and YAO T.H. : On lattices admitting unique Lagrange interpolation. SIAM J. Num. Anal., v.14, 1977, p.735-743.
3. GASCA M. and LOPEZ CARMONA A.: A general recurrence interpolation formula and its applications to multivariate interpolation. J. of Approx. Theo. v.34, 1982, p.361-374.
4. GASCA M. , LOPEZ CARMONA A. and RAMIREZ V.: A generalization of Sylvester's identity on determinants and its applications to interpolation problems. Multivariate Approximation Theory II, W. Schempp and K.Zeller edit. Birkhäuser V. 1982, p.171-184.
5. GASCA M. and MAEZTU J.I.: On Lagrange and Hermite interpolation problems in R^k. Numer. Mathem. v.39, 1982 , p.1-14.
6. GASCA M. and LEBRON E.:A note on recurrence interpolation formulae for certain sets of points in R^k.Numerical methods of Approximation Theory, vol.7. Collatz, Meinardus and Werner edit. Birkhäuser V. 1983, p. 77-85.
7. HAVIE T.:Remarks on a unified theory for classical and generalized interpolation and extrapolation.B.I.T. v.21, 1981, p.465-474.
8. HENRICI P.: Elements of Numerical Analysis. John Wiley, New York 1964
9. LARKIN F.M.: A class of methods for tabular interpolations. Proc. Cambridge Philo. Socie. v. 63, 1967, p. 1101-1114.
10. MUHLBACH G.: The general Neville-Aitken algorithm and some applications. Nume. Mathem. v.31, 1978 p.97-110.
11. MUHLBACH G.: An algorithmic approach to finite linear interpolation. Approximation Theory III, E.W.Cheney edit. Academic Press. p. 655-660.
12. THACHER H.C. Jr. and MILNE W.E.: Interpolation in several variables. J. S.I.A.M. v.8, 1960, p.33-42.

Numerical Approximation of Partial Differential Equations
E.L. Ortiz (Editor)

ON SOME METHODS OF THE CONSTRUCTION OF SMOOTHING SPLINES

Jiří Hřebíček

František Šik

Institute of Metallurgy, Czechoslovak Academy of Sciences
Brno, Czechoslovakia

Introduction

The algorithm [1] for experimental data smoothing using the cubic spline with an a priori chosen extent of smoothing and its modifications became, at the beginning of the last decade, the most widely used algorithms for curve fitting. The smoothing B-splines [2],[3] currently represent the other often used group of smoothing algorithms, where it is necessary to choose the extent of smoothing explicitely.

In [4] it is proposed to construct interpolating and smoothing splines via reproducing kernel techniques. In [5] - [8] this approach was exploited in constructing smoothing splines in such a way that the extent of smoothing was automatically determined by minimising the general cross-validation.

In our contribution we shall deal with a new method of the smoothing spline construction based on a series of papers [9] - [15]. The new suggested algorithm is faster and more efficient than other ones known from [1] - [8].

1. Smoothing Lg-splines

The smoothing Lg-splines are special cases of smoothing H-splines [10], [15].

Let $H=W^{n,2}[a,b]$ be the Sobolev space of real-valued functions f on $[a,b]$, $f \in AC^{(n-1)}[a,b]$, $f^{(n)} \in L^2[a,b]$, L a differential operator $H \to L^2[a,b]$, $L(f)=\sum_{j=0}^{n} a_j D^j f$, where $a_j(x) \in W^{j,2}[a,b] \cap C[a,b]$, $a_n=1$ and A an operator $H \to E^N$, $A(f)=(\lambda_1 f, \dots, \lambda_N f)^T$, where $\{\lambda_j\}_1^N$ is a linearly independent collection of continuous linear functionals on H and $n \leq N$. Let $\boldsymbol{\varrho}=(\varrho_1,\dots,\varrho_N)^T$ be strictly positive weights and let the vector $\mathbf{r}=(r_1,\dots,r_N)^T$ denote the measurement values.

Definition $s(x) \in H$ is an Lg-spline smoothing $\mathbf{r} \in E^N$ with respect to L, A and $\boldsymbol{\varrho}$ if it is a solution of the extremal problem

$$(1.1) \qquad \min_{f \in H} \left\{ \int_a^b (Lf)^2 dx + \sum_{j=1}^{N} (r_j - \lambda_j f)^2/\varrho_j \right\}.$$

The existence of smoothing Lg-splines is proved in [10]. The uniqueness of s(x) is equivalent to the following condition [16] : The set $\{\lambda_j\}_1^N$ contains n elements linearly independent on N_L (the null space of L). For simplicity we take them to be $\{\lambda_j\}_1^n$.

The construction of the Lg-splines to be given in our contribution is based on the reproducing kernel techniques, the reproducing kernel $K(t,\tau)$ in H being defined as follows:

$$K(t,\tau) = \sum_{j=1}^{n} z_j(t) z_j(\tau) + \int_a^b G(t,\xi) G(\tau,\xi) d\xi \ , t,\tau \in [a,b],$$

where $\{z_j\}_1^n$ is a basis on N_L dual to $\{\lambda_j\}_1^n$, $G(t,\tau)$ the Green function of the operator L satisfying

$$L_{(t)} G(t,\tau) = \delta(\tau - t), \ \lambda_{j(t)} G(t,\tau) = 0, \ j=1,\dots,n, \tau \in [a,b] .$$

The smoothing can be reduced to the interpolation [10], [13] by a suitable modification of the space H. Let $H^+=H \times E^N$, with the inner product in H^+ defined in such a way that the problem (1.1) and the following problem (1.2), (1.3) are equivalent

$$\min_{(f,\boldsymbol{\theta})\in U^+(\mathbf{r})} \|(f,\boldsymbol{\theta})\|^2 , \tag{1.2}$$

where

$$U^+(\mathbf{r})=\{(f,\boldsymbol{\theta})\in H^+ : \lambda_j f+\theta_j=r_j, j\in J\} \text{ with } f\in H, \boldsymbol{\theta}\in E^N \text{ and } J=\{1,\dots,N\} . \tag{1.3}$$

The required inner product in H^+ is defined as

$$\langle(f,\boldsymbol{\theta}),(g,\boldsymbol{\omega})\rangle = \int_a^b (Lf)(Lg) + \sum_{j=1}^{N} \theta_j\omega_j/\rho_j + \sum_{j=1}^{n} (\lambda_j f+\theta_j)(\lambda_j g+\omega_j),\quad \rho_j > 0, j\in J.$$

Then the reproducing kernel $K^+(t,\tau)$ of the Hilbert space H^+ has the form

$$K^+(t,\tau)=\begin{cases} \left.\begin{array}{ll} K_1(t,\tau) & ,t\in \mathbf{I} \\ -\rho_t z_t(\tau) & ,t\in J \end{array}\right\}, & \tau\in\mathbf{I} \\ \left.\begin{array}{ll} -\rho_\tau z_\tau(t) & ,t\in \mathbf{I} \\ \rho_t\delta_{t\tau} & ,t\in J \end{array}\right\}, & \tau\in J, \end{cases}$$

where $K_1(t,\tau) = \sum_{j=1}^{n} (1+\rho_j)z_j(t)z_j(\tau) + \int_a^b G(t,\xi)G(\tau,\xi)d\xi$,

$t,\tau\in\mathbf{I}$ and $\mathbf{I} = [a,b]$, $\mathbf{z}(t)=(z_1(t),\dots,z_n(t),0,\dots,0)^T$ is an N-vector.

The linear functionals λ_j on H will be extended to H^+ as follows: $\lambda_j^+(f,\boldsymbol{\theta}) = \lambda_j f + \theta_j$, $j\in J$.

Thus the representer h_j^+ of λ_j^+ in H^+ is $h_j^+=(q_j,\boldsymbol{\omega}^j)$, where

$$q_j(\tau) = \lambda^+_{j(t)} K^+(t,\tau) = \left\{ \begin{array}{l} z_j(\tau) = h_j(t), j=1,\dots,n \\ \\ \lambda_{j(t)} K_1(t,\tau), j=n+1,\dots,N \end{array} \right\}, \tau \in \mathbf{I},$$

$$(\boldsymbol{\omega}^j)_\tau = \lambda^+_{j(t)} K^+(t,\tau) = \left\{ \begin{array}{l} \mathbf{0}, j=1,\dots,n \\ \\ \varrho_j \mathbf{e}_j, j=n+1,\dots,N \end{array} \right\}, \tau \in J.$$

Now the solution of the smoothing problem (1.1) is reduced to that of the interpolation problem (1.2), (1.3) on the ground of the following

<u>Theorem</u> [13] Let $(\bar{f},\bar{\boldsymbol{\theta}})$ be the solution of the problem (1.2), (1.3). Then the smoothing spline s(t) (i.e. the solution of the problem (1.1)) is $s(t) = \bar{f}(t), \quad t \in [a,b]$.

Consider the differential equation

$$Ly(t,\tau) = G(t,\tau).$$

This equation can be written in the equivalent state variable form given below in (1.4), (1.5). Now, for each fixed $\tau \in \mathbf{I}$, the reproducing kernel $K(t,\tau)$ fulfils the following relations

$$L_{(t)} K(t,\tau) = G(t,\tau), \lambda_{j(t)} K(t,\tau) = h_j(\tau), \ j=1,\dots,N.$$

The following theorem holds:

<u>Theorem</u> [17] The reproducing kernel $K(t,\tau)$ has the dynamical model for each $\tau \in [a,b]$:

(1.4) $\quad K(t,\tau) = \mathbf{c}\mathbf{x}(t,\tau), \qquad t \in [a,b],$

(1.5) $\quad \frac{\partial}{\partial t}\mathbf{x}(t,\tau) = A(t)\mathbf{x}(t,\tau) + \mathbf{b}G(\tau,t)$

where $\mathbf{c} = (1,0,\dots,0)$, $\mathbf{b} = (0,\dots,0,1)^T$,

(1.6) $$A(t) = \begin{pmatrix} 0 & \vdots & I \\ \cdots & \cdots & \cdots \\ -a_0(t) & \vdots & -a_1(t) \ \dots \ -a_{n-1}(t) \end{pmatrix}.$$

The corresponding integral form is

$$K(t,\tau)=\mathbf{c}\Phi(t,t_0)\mathbf{x}(t_0,\tau)+\int_{t_0}^{t}\mathbf{c}\phi(t,\xi)\mathbf{b}G(\tau,\xi)d\xi \ .$$

As it is evident from this integral form, it is sufficient to determine $\mathbf{x}(t_0,\tau)$ to specify $K(t,\tau)$. But in most cases this is difficult to obtain. Thus we must apply what is given, i.e. the set $\{\lambda_{j(t)}K(t,\tau)=z_j(\tau);\ j=1,\dots,n\}$ - a suitable fundamental system of solutions of $Lu = 0$. This leads to $\mathbf{x}(t_0,t) = \sigma^{-1}(\mathbf{Z}(t)-\mathbf{\Omega}(t))$, where

$$\sigma = \begin{pmatrix} \lambda_1\mathbf{c}\phi(t,t_0) \\ \vdots \\ \lambda_n\mathbf{c}\phi(t,t_0) \end{pmatrix}, \tag{1.7}$$

$\phi(t,\tau)$ is the fundamental matrix of the system (1.5) with respect to the condition $\phi(\tau,\tau)=I$, $\mathbf{Z}(t)=(z_1(t),\dots,z_n(t))^T$, $\mathbf{\Omega}(t)=(\omega_1(t),\dots,\omega_n(t))^T$, whose elements are

$$\omega_j(\cdot) = \lambda_{j(t)}\int_{t_0}^{t}\mathbf{c}\phi(t,\xi)\mathbf{b}G(\cdot,\xi)d\xi \ .$$

2. Recursive algorithm for Lg-splines smoothing EHB data

The EHB data are given by the vectors $\mathbf{c}_j=(\alpha_{j1},\dots,\alpha_{jn})$ and the relations

$$\lambda_j f = \mathbf{c}_j(f(t_j),\ f^{(1)}(t_j),\dots,f^{(n-1)}(t_j))^T,\ j\in J.$$

The general recursive algorithms for the interpolating and smoothing splines are described in [18], [13], but the step (52g) in [13] is not correct. For simplicity we describe the corrected algorithm for the special case with $L=D^2$, $\lambda_j f=f(t_j)=r_j$, $j\in J$, $\mathbf{r}=(r_1,\dots,r_N)^T$ thus $n=2$, $\mathbf{c}_j=[1,0]$, $j\in J$.

Then $A=\begin{pmatrix}0 & 1\\ 0 & 0\end{pmatrix}$, $\mathbf{c}=(1,0)$, $\mathbf{b}=(0,1)^T$, $\Phi(t,\tau)=\begin{pmatrix}1 & t-\tau\\ 0 & 1\end{pmatrix}$ and

$\sigma=\begin{pmatrix}1 & t_1-t_2\\ 1 & 0\end{pmatrix}$ for $t_0=t_2$ in (1.7).

Recursive algorithm:

<u>Step A</u> Compute

$$\mathbf{x}_0 := \sigma^{-1}\mathbf{r}_0, \text{where } \mathbf{r}_0^T = (r_1, r_2) ,$$

$$\Pi_0 := \sigma^{-1}(\mathrm{diag}\,(\varrho_1,\varrho_2) + [\omega_i(\cdot).\ \omega_j(\cdot)]_{i,j=1}^{2})\sigma^{-T} .$$

<u>Step B</u> Forward pass for $t_2 \leq t \leq t_N$.

For $3 \leq j \leq N$ compute and store $d_j := r_j - \mathbf{c}_j\mathbf{x}_{j/j-1}, \mathbf{K}_j := P_{j/j-1}\mathbf{c}_j^T$ ($n \times 1$ vector) and $R_j := \mathbf{c}_j\mathbf{K}_j + \varrho_j$ (scalar).

The $n \times 1$ vectors $\mathbf{x}_{j/j-1}$ and the $n \times n$ symmetric matrices $P_{j/j-1}$ are obtained, starting with

$$\mathbf{x}_{2/2} := \mathbf{x}_0, \qquad P_{2/2} := \Pi_0 .$$

(i) Recursive step between knots, $j=2,3,\ldots,N-1$

$$\mathbf{x}_{j+1/j} := \begin{pmatrix}1 & t_{j+1}-t_j\\ 0 & 1\end{pmatrix}\mathbf{x}_{j/j}$$

$P(t_j/j) := P_{j/j}$ (the matrix obtained by the foregoing iteration in (i)),

where $P(t/j)$ has elements $P_{km}(t)$ of the form

$$P_{11}(t)=P_{11}(t_j)+\left\{\left[\tfrac{1}{3}(t-t_j)+P_{22}(t_j)\right](t-t_j)+2P_{12}(t_j)\right\}(t-t_j),$$

$$P_{12}(t)=P_{12}(t_j)+\left[\tfrac{1}{2}(t-t_j)+P_{22}(t_j)\right](t-t_j) ,$$

$$P_{21}(t)=P_{12}(t) , \quad P_{22}(t)=P_{22}(t_j)+t-t_j .$$

Define

$$P_{j+1/j} = P(t_{j+1}/j) .$$

(ii) Recursive step at knots, $j=2,3,\ldots,N-1$

Define

$$\mathbf{x}_{j+1/j+1} := \mathbf{x}_{j+1/j} + \mathbf{K}_{j+1} R_{j+1}^{-1} d_{j+1} \quad ,$$

$$P_{j+1/j+1} := P_{j+1/j} + \mathbf{K}_{j+1} R_{j+1}^{-1} \mathbf{K}_{j+1}^{T} \quad .$$

Note that the original formula for $P_{j+1/j+1}$ in [13] is not correct. The correct formula is given above.

Store $\mathbf{x}(t_N/N) := \mathbf{x}_{N/N}$.

<u>Step C</u> The spline $s(t)$ for $t \geq t_N$ is

$$s(t) = \mathbf{c} \begin{pmatrix} 1 & t-t_N \\ 0 & 1 \end{pmatrix} \cdot \mathbf{x}_{N/N}$$

<u>Step D</u> Backward pass for $t_2 \leq t \leq t_N$.

Compute by parts, for $t_{j-1} \leq t \leq t_j$ and $j=N,N-1,\ldots,3$ the spline $s(t)$

$$s(t) = \mathbf{c}\mathbf{x}(t/N) \quad ,$$

where $\mathbf{x}(t/N) = (x_1(t/N),\ x_2(t/N))^T$, $\mathbf{u}_j = (u_1,u_2)^T$.

$$x_1(t/N) = \left[(-\tfrac{1}{6}u_1(t-t_j) + \tfrac{1}{2}u_2)(t-t_j) + x_2(t_j/N)\right](t-t_j) + x_1(t_j/N) .$$

$$x_2(t/N) = (-\tfrac{1}{2}\,u_1(t-t_j) + u_2)(t-t_j) + x_2(t_j/N) \; .$$

$$\mathbf{u}_j := \Psi_{j+1}^{T}\,\mathbf{u}_{j+1} + \mathbf{c}_j^T R_j^{-1} d_j \quad , \qquad \mathbf{u}_N := \mathbf{c}_N R_N^{-1} d_N \quad ,$$

$$\Psi_{j+1} := \begin{pmatrix} 1 & t_{j+1}-t_j \\ 0 & 1 \end{pmatrix} (I - \mathbf{K}_j R_j^{-1} \mathbf{c}_j) \; .$$

<u>Step E</u> The spline $s(t)$ for $t_1 \leq t \leq t_2$

$$s(t) = \mathbf{c}\mathbf{x}(t/N) \; .$$

where

$$\mathbf{x}(t/N) = \begin{pmatrix} 1 & t-t_2 \\ 0 & 1 \end{pmatrix} \mathbf{x}(t_2/N) + B(t) .\ \sigma^{-T} \Phi^{T}(t_3,t_2) .\ \mathbf{u}_3 \quad .$$

The 2 x 2 matrix B(t) has the elements

$$B_{11}(t) = -\frac{1}{6}(t-t_2)^2\left[(t-t_2) + 3(t_2-t_1)\right], \quad B_{12}(t)=B_{22}(t)=0,$$

$$B_{21}(t) = -\frac{1}{2}(t-t_2)\left[(t-t_2) + 2(t_2-t_1)\right].$$

Step F The spline s(t) for $t \le t_1$

$$s(t) = \mathbf{c}\begin{pmatrix} 1 & t-t_1 \\ 0 & 1 \end{pmatrix}\mathbf{x}(t_1/N) .$$

3. Numerical results

We have compared the cubic smoothing spline algorithms [1],[2],[3],[6] and the one presented above on 5 testing functions, see Table. The used data r_i, $i=1,\dots,N$ $(N=30)$ were generated in such a way that their errors had standard deviation $\sigma = 0.001$, 0.01 and 0.05 respectively. The algorithms have been studied from the following points of view:

a) Maximum error defined by $ER = \max_{i=1}^{N} |f(t_i)-s(t_i)|$, where s(t) denotes the tested smoothing spline.

b) Sum of analytical residuals $RA = \sum_{i=1}^{N} (f(t_i)-s(t_i))^2/N$.

c) Computing costs.

Results of the testing are presented in Table. We can see that the presented algorithm and Reinsch´s one give good results with reasonable computing cost for small errors. Wahba´s algorithm gives results that can be hardly accepted because they ´copy´very closely the data vector $\mathbf{r}$ and badly approximate the function f. For the case of large errors the algorithms [1] - [3], [13] give good results although the

function	σ		Weinert [13]	Wahba [6]	Reinsch [1]	de Boor [2], [3]
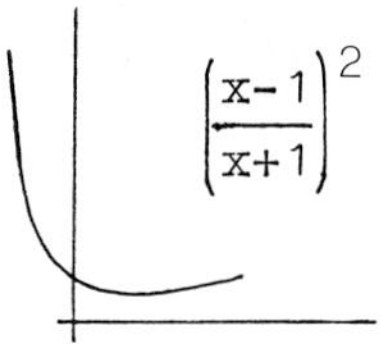 $\left(\frac{x-1}{x+1}\right)^2$	0.001	ER	0.18-2	0.18-2	0.41-2	0.86-1
		RA	0.76-6	0.86-6	0.19-5	0.18-3
		ms	20	7521	59	850
	0.01	ER	0.11	0.18-1	0.42-1	0.91-1
		RA	0.68-3	0.71-4	0.18-3	0.52-3
		ms	19	7505	59	850
	0.05	ER	0.75	0.94-1	0.21	0.11
		RA	0.35-1	0.18-2	0.38-2	0.99-3
		ms	20	7505	75	850
 $\frac{\sin(x)}{x}$	0.001	ER	0.18-2	0.37-2	0.46-2	0.13-2
		RA	0.72-6	0.12-5	0.17-5	0.31-6
		ms	19	7524	59	605
	0.01	ER	0.18-1	0.19-1	0.36-1	0.10-1
		RA	0.66-4	0.12-3	0.98-4	0.21-4
		ms	19	7601	107	480
	0.05	ER	0.69-1	0.82-1	0.90-1	0.49-1
		RA	0.88-3	0.24-2	0.11-2	0.67-3
		ms	19	7466	139	850
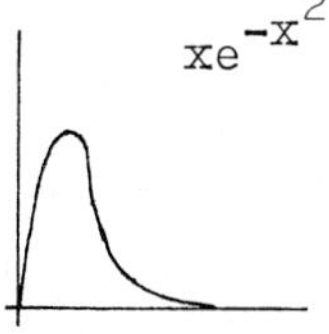 xe^{-x^2}	0.001	ER	0.18-2	0.18-2	0.38-2	0.12-1
		RA	0.72-6	0.72-6	0.13-5	0.22-4
		ms	19	7509	59	843
	0.01	ER	0.18-1	0.71-1	0.35-1	0.17-1
		RA	0.59-4	0.40-3	0.77-4	0.48-4
		ms	19	7603	108	484
	0.05	ER	0.58-1	0.14	0.85-1	0.49-1
		RA	0.65-3	0.18-2	0.1-2	0.70-3
		ms	19	7524	123	850
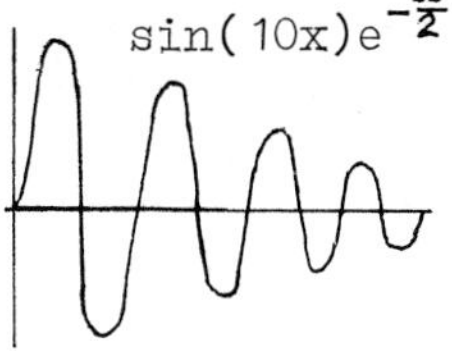 $\sin(10x)e^{-\frac{x}{2}}$	0.001	ER	0.19-2	0.77	0.30-2	0.65
		RA	0.91-6	0.75-5	0.14-5	0.70-1
		ms	19	7301	59	850
	0.01	ER	0.13	0.77	0.29-1	0.66
		RA	0.21-2	0.75-1	0.14-3	0.63-1
		ms	19	7494	59	850
	0.05	ER	0.63	0.77	0.14	0.68
		RA	0.51-1	0.75-1	0.33-2	0.63-1
		ms	19	7621	60	850
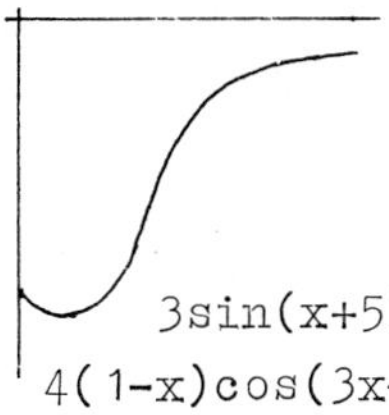 $3\sin(x+5)+4(1-x)\cos(3x+2)$	0.001	ER	0.16-2	0.18-2	0.44-2	0.11-2
		RA	0.61-6	0.71-6	0.18-5	0.27-6
		ms	19	7499	59	850
	0.01	ER	0.42-1	0.55-1	0.44-1	0.99-2
		RA	0.11-3	0.19-3	0.12-3	0.27-4
		ms	19	7525	91	850
	0.05	ER	0.29	0.21	0.12	0.49-1
		RA	0.73-2	0.36-2	0.10-2	0.67-3
		ms	19	7454	123	850

Table

the presented algorithm seems to be more effective with regard to computing cost.

REFERENCES

[1] Reinsch C., Smoothing by spline functions I,II, Num.Math.,10, 1967, 177-183 and 16, 1970, 451-454.
[2] de Boor C., On calculating with B-splines, J.Approx. Theory, 6, 1972, 50-62.
[3] Cox G., The numerical evaluation of B-splines, J. Inst.Maths.Applics., 10, 1972, 134-149.
[4] de Boor C., Lynch R.E., On splines and their minimum properties, J.Math.Mech., 15, 1966, 953-969.
[5] Kimeldorf G., Wahba G., A correspondence between Bayesian estimation of stochastic processes and smoothing by splines, Ann.Math.Stat., 41, 1970,495-502.
[6] Craven O., Wahba G., Smoothing noisy data with spline functions. Estimating the correct degree of smoothing by method of generalized cross-validation, Num.Math., 31, 1979, 377-403.
[7] Utreras Diaz F., Sur le choix du paramètre d`ajustement dans le lissage par fonctions spline, Num.Math., 34, 1980, 15-28.
[8] Silverman B.W., A fast and efficient cross-validation method for smoothing parameter choice in spline regression, J.Am.Statist.Assoc., 79, 1984, 584-589.
[9] Carasso C.,Laurent P.J.,On the numerical construction and practical use of interpolating spline functions, IFIP Congress on Information Processing 68,Edinburgh, North-Holland, Amsterdam, 1969, 86-89.
[10] Jerome J.,Schumaker L.L., On Lg-splines, J.Approx. Theory, 2, 1969, 29-49.
[11] Munteanu M.J.,Schumaker L.L.,On a method of Carasso and Laurent for constructing interpolating splines, Math.Comp.,27, 1973, 317-325.
[12] de Figueiredo R.J.T.,Caprihan A.,An algorithm for the construction of the generalized smoothing splines with applications to system identification,Conf.Scien.Syst., J.Hopkins University,Baltimore,1977, 1-7.
[13] Weinert H.L.,Byrd R.H.,Sidhu G.S.,A stochastic framework for recursive computation of spline functions: Part II, Smoothing splines, JOTA, 30, 1980, 255-268.
[14] Hřebíček J.,Šik F., Comparison of smoothing spline algorithms, NAG Newsletter, 3, 1984, 21-29.
[15] Hřebíček J.,Šik F., On some methods of constructing smoothing splines, V.Sum.School "Software and algorithms of numerical mathematics", Dlouhá Ves, I.Marek, JČMF, Praha, 1984, 99-115.
[16] Weinert H.L.,Sidhu G.S.,On uniqueness conditions for optimal curve fitting,J.Opt.Th.Appl.,23,1977,211-216.
[17] Sidhu G.S.,Weinert H.L.,Dynamical recursive algorithm for Lg-spline interpolation of EHB data,Appl.Math. Commun., 5, 1979, 157-186.
[18] Weinert H.L.,Sidhu G.S.,A stochastic frame work for recursive computation of spline functions:Part I, Interpolating splines.IEEE Trans.Inf.Th.IT 24, 1978, 45-50.

Numerical Approximation of Partial Differential Equations
E.L. Ortiz (Editor)

GENERALIZED L-SPLINES AS A SOLUTION OF N-POINT BOUNDARY VALUE PROBLEM

Carmen Simerská
Czech Technical University
Prague, Czechoslovakia

Little has been known about the Hermite-Birkhoff spline interpolation with mixed conditions i.e., the interpolation conditions involving linear combinations of the values of the function and its derivatives at several points. Some special cases were solved by Karlin, Karon and Pinkus.

The work presented here deals with a generalized L-splines with mixed conditions prescribed at the knots. This interpolation problem can be converted into the N-point boundary value problem with mixed and transition conditions. A theory of this problem for 2n-order selfadjoint operators has been worked out. The easy-to-verify conditions under which the solution exists and is unique are found.

1. Introduction

Many papers e.g. (Jerome, 1970; Lucas, 1970; Karlin, Karon, 1972) have been devoted to the splines and L-splines that satisfy the Hermite-Birkhoff (extended-Hermite-Birkhoff) conditions. Less is so far known about the splines that would satisfy mixed conditions at several points. Some results were obtained by Karlin and Pinkus (Karlin, Pinkus, 1976).

In the present paper the problem of finding the L-spline is converted into N-point boundary value problem (N-BVP). This N-BVP is formulated in a sufficiently general way to cover as general class of L-splines (N-BVP) with mixed conditions as possible. Under mixed conditions we understand a linear combination of the values of the function and its derivatives at N given points.

As the main result of this work, the necessary and sufficient conditions are determined under which the operator generated by N-BVP is selfadjoint and positive definite. These conditions directly imply the existence and uniqueness of the N-BVP solution (L-spline). From practical considerations it is important that these conditions are easy to verify.

2. Problem formulation

Letś suppose real functions p_i, $i=0,\dots,n$ are given such that $1/p_0, p_1, \dots p_n$ are from $L(a,b)$. Let l be a differential operator defined by the expression (1)

$$l \equiv \sum_{i=0}^{n} (-1)^i \frac{d^i}{dt^i}\left(p_{n-i} \frac{d^i}{dt^i}(\ )\right) . \tag{1}$$

Further, y denotes a complex function of real variable.

Definition 1. We say that the function y has quasiderivatives up to 2n-th degree at the point t iff the following terms define functions

$$y^{[0]}(t) = y(t)$$
$$\vdots$$
$$y^{[k]}(t) = y^{(k)}(t) \qquad k=1,\dots,n-1$$
$$- - - - -$$
$$y^{[n]}(t) = p_0 y^{(n)}(t)$$
$$\vdots$$
$$y^{[n+k]}(t) = p_k(t) y^{[n-k]}(t) - \left(y^{(n+k-1)}(t)\right)' \qquad k=1,\dots,n$$

Definition 2. We say that a function f is piecewise absolutely continuous (PAC) on the interval $\langle a,b\rangle$ with respect to the partition $\{x_i\}_{i=1}^{N}$, where $a = x_1 < \dots < x_m = b$, iff there exist functions f_i, $i=1,\dots,N-1$, such that for $i=1,\dots,N-1$ holds:

1. the function f_i is absolutely continuous on $\langle \varkappa_i, \varkappa_{i+1} \rangle$
2. $f(t) = f_i(t)$ for $t \in (\varkappa_i, \varkappa_{i+1})$.

Definition 3. We say that a vector $\vec{x}(t)$ (or matrix $A(t)$) is PAC on $\langle a, b \rangle$ with respect to $\{\varkappa_i\}_1^N$ iff all components of $\vec{x}(t)$ (or $A(t)$) are PAC on $\langle a, b \rangle$ with respect to $\{\varkappa_i\}_1^N$.

Now we are able to formulate the following main problem.

Definition 4. Let´s suppose N points $\{\xi_i\}_{i=1}^N$ are given in the interval $\langle a, b \rangle$ such that

$$a = \xi_1 < \xi_2 < \dots \xi_N = b \quad . \tag{2}$$

Let $\vec{y}(t)$ denotes the vector of quasiderivatives up to (2n-1)-th degree at the point t

$$\vec{y}(t) = \begin{bmatrix} y(t) \\ y'(t) \\ \vdots \\ \\ y^{[2m-1]}(t) \end{bmatrix}$$

Let $y^{[k]}(\xi_i^{+-})$ be one-sided limits at the points ξ_i.

Further, let $\vec{Y}_{2j-1}, \vec{Y}_{2j}, j = 1, \ldots, N-1$, stand for the vectors of lenght $2m$

$$\vec{Y}_{2j-1} = \begin{bmatrix} y(\xi_j^+) \\ \vdots \\ y^{[2m-1]}(\xi_j^+) \end{bmatrix}, \quad \vec{Y}_{2j} = \begin{bmatrix} y(\xi_{j+1}^-) \\ \vdots \\ y^{[2m-1]}(\xi_{j+1}^-) \end{bmatrix}$$

Let D be the set

$$D = \{ y : \vec{y}(t) \text{ is PAC with respect to the partition } \{\xi_i\}_1^N, \ t \in \langle a, b \rangle, \ y^{[2m]} \in L_2(a,b) \}.$$

The problem of finding the function $y \in D$ which is the solution of

$$ly = q \quad \text{a.e. on } (a,b), \text{ where } q \in L_2(a,b) \tag{3}$$

and satisfies at points (2) the boundary conditions (4) given by the matrices $M^{(p)}, p = 1, \ldots, 2(N-1)$ of $J \times 2m$ type,

$$\Lambda y \equiv \sum_{j=1}^{2N-2} M^{(j)} \vec{Y}_j = \vec{m}, \tag{4}$$

where $\vec{m} = [m_1, \ldots, m_J]^T$ is a given vector, is said to be the N-point boundary value problem (N-BVP).

The boundary conditions of type (4) cover all conditions represented as a linear combinations of quasiderivatives (one-sided quasiderivatives) at a point ξ_i .

They include the conditions binding the limit values at points of (2) as well, especially the transition conditions at partition points. Conditions (4) can describe e.g. continuity of quasiderivatives at the point ξ_i of (2).
The condition (4) for two-point boundary value problem looks as follows

$$M^{(1)} \cdot \begin{bmatrix} y(a) \\ y'(a) \\ \vdots \\ y^{[2n-1]}(a) \end{bmatrix} + M^{(2)} \cdot \begin{bmatrix} y(b) \\ y'(b) \\ \vdots \\ y^{[2n-1]}(b) \end{bmatrix} = \vec{m}$$

<u>Definition 5.</u> Under the same propositions as in definition 4, the solution $s \in D$ of the problem

(3´) $\quad \ell s = 0 \quad$ a.e. on (a, b)

that satisfies J conditions (4) i.e. $\Lambda s = [m_1, \ldots, m_J]^T$,

is said to be the <u>generalized L-spline.</u>

3. Solvability

We are concerned with the solvability of N-BVP. The problem beeing linear, it is sufficient to solve the homogeneous N-BVP. We shall require that the operator generated by N-BVP be selfadjoint and, moreover, positive semidefinite.

The following notation will be useful, for $j=1,\dots,N-1$

$$y_{1,2j-1}=\begin{bmatrix} y(\xi_j^+) \\ \vdots \\ y^{(n-1)}(\xi_j^+) \end{bmatrix}, \qquad y_{2,2j-1}=\begin{bmatrix} y^{[n]}(\xi_j^+) \\ \vdots \\ y^{[2n-1]}(\xi_j^+) \end{bmatrix}$$

$$y_{1,2j}=\begin{bmatrix} y(\xi_{j+1}^-) \\ \vdots \\ y^{(n-1)}(\xi_{j+1}^-) \end{bmatrix}, \qquad y_{2,2j}=\begin{bmatrix} y^{[n]}(\xi_{j+1}^-) \\ \vdots \\ y^{[2n-1]}(\xi_{j+1}^-) \end{bmatrix}$$

If $y_1^T \equiv (y_{11}^T,\dots,y_{1,2N-2}^T)$, $y_2^T \equiv (y_{2,1}^T,\dots,y_{2,2N-2}^T)$ then vectors defined by $\vec{y} \equiv \begin{pmatrix} y_1 \\ y_2 \end{pmatrix}$ and analogically $\vec{z} \equiv \begin{pmatrix} z_1 \\ z_2 \end{pmatrix}$ are of lenght $4n(N-1)$.

The matrix $A = \begin{bmatrix} 0 & -B \\ B & 0 \end{bmatrix}$ of the type $4n(N-1) \times 4n(N-1)$

is constructed from the blocks B of type $2n(N-1) \times 2n(N-1)$ such that

$$B \equiv \begin{bmatrix} -T & & & 0 \\ & T & & \\ & & \ddots & \\ 0 & & & -T \\ & & & & T \end{bmatrix},$$

where $T = \begin{bmatrix} 0 & & 1 \\ & \cdot^{\cdot^{\cdot}} & \\ 1 & & 0 \end{bmatrix}$ is a matrix of type $m \times m$.

It can be demonstrated that for every pair of functions y, z from $\mathcal{D}$ the following generalized Lagrange identity holds

$$(5) \qquad (\ell y, z)_{L_2} = (y, \ell z)_{L_2} + (A\vec{y}, \vec{z}) .$$

For our purposes we shall denote the boundary conditions (4) by the matrix $W \equiv (W_1, W_2)$,

$W_1 = (M_1^{(1)}, M_1^{(2)}, \ldots, M_1^{(2N-2)})$,

$W_2 = (M_2^{(1)}, M_2^{(2)}, \ldots, M_2^{(2N-2)})$,

where $M^{(p)} = (M_1^{(p)}, M_2^{(p)})$ and the decomposition matrices $M_i^{(p)}$, $i = 1, 2$, are of type $J \times m$.

Then the following condition (4') are equivalent with (4)

$$(4') \qquad W\vec{y} = W_1\vec{y} + W_2\vec{y} = \vec{m}$$

We have proved these four theorems.

<u>Theorem 1.</u> For every self-adjoint operator L given by the differential expression (1) and by condition (4) there exists a matrix W with the properties

(i) W is of type $J \times 4m(N-1)$, where $J \geq 2m(N-1)$

(ii) rank $(W) = 2m(N-1)$

(iii) $WAW^* = 0$,

such that the domain D_L of L satisfies

(6) $$D_L = \{\psi \in D, W\vec{\psi} = 0\} .$$

Conversely, if the matrix W with the properties (i)-(iii) is given then the self-adjoint operator $L \equiv (D_L, \ell)$ exists such that for its domain (6) holds.

Theorem 2. Let L be the self-adjoint operator, where $p_i \geq 0$ on $\langle a, b\rangle$ for $i = 0, \dots, m$ from the theorem 1. L is positive semidefinite (i.e. $(\ell y, y) \geq 0$ for all $y \in D_L$) iff (iv) $W_1 B W_2^*$, $W_2 B W_1^*$ are positive semidefinite matrices.

For practical purposes the following theorem is useful. It permits to determine generally acceptable matrices W for which the corresponding operators are self-adjoin and positive semidefinite.

Theorem 3. An operator $L \equiv (D_L, \ell)$ is self-adjoint and positive semidefinite iff the corresponding matrix W is of the form

$$W = K(\mathcal{U}, (I - \mathcal{U})B),$$

where K is some regular matrix and $\mathcal{U}$ is a Hermite matrix of type $2m(N-1) \times 2m(N-1)$ for the spectrum of which the following is true

(7) $$\lambda \in \sigma(\mathcal{U}) \Rightarrow \lambda \in \langle 0, 1\rangle .$$

When investigating the uniqueness of (3), (4) it is sufficient, due to the linearity, to study the homogeneous problem ((3´) and $\vec{m}=0$). It holds either

a) in all $\langle \xi_i, \xi_{i+1}\rangle$ is $p_n \neq 0$ in $L(a,b)$ norm. Then it is evident that $(ly, y)>0 \Leftrightarrow y\neq 0$ and the homogeneous problem has only zero solution.

or

b) there exist intervals where

(8) $p_n=0, p_{n-1}=0, \dots, p_{n-l_i+1}=0, p_{n-l_i}\neq 0$ a.e.in $\langle \xi_i, \xi_{i+1}\rangle$,

where l_i is one of the numbers $1, \dots, n$.

Using matrices $S^j_{l_i} \equiv \exp^{R_{l_i} \cdot \xi_j}$, $j=i, i+1$, where

$R_l = \begin{bmatrix} 0 & 1 & & 0 \\ & \ddots & \ddots & \\ & & \ddots & 1 \\ 0 & & & 0 \end{bmatrix}$ of type $l \times l$, we define a matrix $\mathcal{K}$ of type $2n(N-1) \times \sum_1^{N-1} l_i$

$$\mathcal{K} \equiv \begin{bmatrix} S^1_{l_1} & & & & \\ S^2_{l_1} & & & & \\ & S^2_{l_2} & & & \\ & S^3_{l_2} & & & \\ & & \ddots & & \\ & & & & S^{N-1}_{l_{N-1}} \\ & & & & S^N_{l_{N-1}} \end{bmatrix}$$

<u>Theorem 4.</u> Let´s the problem (3),(4) or (3´), (4) respectively be given, where $p_i \geq 0$, (8) holds and the boundary conditions are given by the matrix W satisfying (i)-(iv) then this problem has a unique solution on set D iff for the

matrix $W_1 \cdot \mathcal{K}$ holds

(v) $\operatorname{rank}\,(W_1 \cdot \mathcal{K}) = \sum_{i=1}^{N-1} l_i$

It can be shown that the operator figuring in theorem 4 is even positive definite.

The clasical splines and also L-splines defined in (Ahlberg, Nilson, Walsh, 1967) or in (Schwartz, Varga, 1972) are a special case of our generalized L-splines. Thus, the existence and uniqueness of clasical and L-splines are a direct consequence of our theory. In practice, the conditions (i)-(v) are often fulfilled and so the definitions of several classes of splines acquire now a meaningful sense.

REFERENCES

1. Ahlberg J.H., Nilson E.N., Walsh J.L., The Theory of Splines and Their Applications, Academic Press, New York, 1967.
2. Jerome J.W., Linear Self-Adjoint Multipoint Boundary Value Problem, Numer.Math., 15, 1970, 433-449.
3. Karlin S., Karon J.M., On Hermite-Birkhoff Inderpolation, J.Approximation Theory, 6, 1972, 90-114.
4. Karlin S., Pinkus A., Studies in Spline functions and Approximation Theory, Academic Press, New York, 1976.
5. Lucas T.R., A generalization of L-splines, Numer.Math., 15, 1970, 359-370.
6. Schwartz B.K., Varga R.S., Error Bounds for Spline and L-Spline Interpolation, J.Approximation Theory, 6, 1972, 6-49.

Numerical Approximation of Partial Differential Equations
E.L. Ortiz (Editor)

COMPUTING ERRORBOUNDS FOR EIGENPAIRS USING THE PRECISE SCALARPRODUCT

Götz Alefeld
Institute of Applied Mathematics
University of Karlsruhe
Karlsruhe, Germany

Abstract.
Let there be given a sufficiently accurate approximation to a real eigenpair of a real matrix, where the eigenvalue is a simple zero of the characteristic polynomial. Then we construct bounds for the eigenvalue and for the components of a corresponding eigenvector. Furthermore we introduce an iteration method by which these bounds can be made arbitrarily small. Finally we discuss how this method behaves if all rounding errors are taken into account. For this discussion we assume that the so-called precise scalar product is available which was first introduced by U. Kulisch (U. Kulisch, 1976). The paper closes with two numerical examples.

0. Introduction

We consider the problem of computing errorbounds for a given real approximation λ to a simple real eigenvalue and for the components of an approximation x to the corresponding eigenvector of a real (n,n) matrix A. This problem was considered repeatedly in the past (S. Rump, 1983; H.J. Symm - J. Wilkinson, 1980; T. Yamamoto, 1980).

In this paper we first consider a nonlinear system of equations whose solution delivers the exact eigenpair.

Then we show that with sufficiently accurate approximations λ and x we can enclose the eigenvalue and the components of the eigenvector.

Starting with this inclusion we present an iterative method, which computes a sequence of intervalvectors which theoretically - that is if no rounding errors are involved -is con--vergent to the eigenpair.

Finally we present the result of a careful investigation of the influence of rounding errors if this method is performed on a computer using a floating point system. At the end of the paper we present a numerical example. Proofs of the results of this paper have been given by the author (G. Alefeld, 1985).

1. Computing bounds for an eigenpair and their iterative improvement

Let there be given a real (n,n) matrix A. Assume that λ and x are real approximations to a real simple eigenvalue $\lambda+\mu$ and the corresponding real eigenvector $x+\tilde{y}$ such that

(1) $$A(x+\tilde{y}) = (\lambda+\mu)(x+\tilde{y}).$$

Let $x = (x_i)$ and let an integer s be defined by

(2) $$\|x\|_\infty = |x_s| > 0.$$

s is not necessarily unique. Since the eigenvector $x+\tilde{y}$ is not unique we define

(3) $$\tilde{y}_s = 0$$

where $\tilde{y} = (\tilde{y}_i)$.

Defining the components y_i of a vector $y = (y_i) \in \mathbb{R}^n$ by

(4) $$y_i = \begin{cases} \tilde{y}_i \ , & i \neq 0 \\ \mu \ , & i = s \end{cases}$$

then (1) can be rewritten as

(5) $$By = r + y_s \bullet \tilde{y}$$

where r is the residual vector

(6) $$r = \lambda x - Ax$$

and where B is the matrix which is obtained from $A-\lambda I$ by replacing the s-th column by $-x$:

(7) $B = ((A-\lambda I)_1, \ldots, (A-\lambda I)_{s-1}, -x, (A-\lambda I)_{s+1}, \ldots, (A-\lambda I)_n)$.

The matrix B is nonsingular for sufficiently good approximations λ and x. This follows by continuity arguments from the following theorem which was already proved by Symm and Wilkinson (Symm and Wilkinson, 1980).

Theorem 1. Let (λ, x) be an exact eigenpair of A. Let λ be a simple zero of the characteristic polynomial. Then the matrix B which is obtained from $A-\lambda I$ by replacing the s-th column by $-x$ is nonsingular.

□

(5) is the starting point for the computation of bounds for $\tilde{y}$ and μ and therefore for the eigenpair $(\lambda+\mu, x+\tilde{y})$.

We assume that λ and x are sufficiently good approximations such that B is nonsingular. Assume that L is the inverse of B or some approximation to the inverse of B. Then (5) can be writtes as

(8) $$y = Lr + (I-LB)y + L\bullet(y_s \bullet \tilde{y}).$$

We now determine an interval vector $[y] = ([y]_i)$ such that

(9) $$Lr + (I-LB)y + L\bullet(y_s \bullet \tilde{y}) \in [y]$$

for all $y \in [y]$. Applying Brouwer`s fixed point theorem it then follows that (8) has at least one solution $y^* \in [y]$. We try to determine [y] in the form

(10) $$[y] = [-\beta, \beta]e$$

where $0 < \beta \in \mathbb{R}$, $e = (1, \ldots, 1)^T \in \mathbb{R}^n$.

Defining

(11) $$\varrho = \|Lr\|_\infty$$

(12) $$\kappa = \|I-LB\|_\infty$$

(13) $$\ell = \|L\|_\infty$$

(14) $$\beta_{1/2} = \frac{1-\kappa+\sqrt{(1-\kappa)^2-4\varrho\ell}}{2\varrho}$$

the following result holds.

Theorem 2. *Let* $\kappa < 1$ *and* $(1-\kappa)^2-4\varrho\ell \geq 0$. *If* $\beta \in [\beta_1, \beta_2]$, *where* β_1, β_2 *are defined by* (14) *then there exists at least one solution* y^* *of equation* (8) *in the intervalvector* $[y] = ([y]_i)$ *where* $[y]_i = [-\beta, \beta]$.

□

The proof is performed by using interval arithmetic tools. We do not go into the details of this.

We consider now the *iteration method*

(15) $$\begin{aligned} [y]^0 &= [-\beta, \beta]e \\ [y]^{k+1} &= g([y]^k), \quad k = 0,1,2,\ldots, \end{aligned}$$

where

(16) $$g([y]) = Lr + (I-LB)[y] + L\bullet([y]_s\bullet[\tilde{y}]).$$

The following result holds.

Theorem 3. *Let* $\kappa < 1$, $(\kappa-1)^2-4\varrho\ell > 0$. *Assume that* β_1, β_2 *are defined by* (14). *If* β *in* (15) *is chosen according to*

$$\beta_1 \leq \beta < \frac{\beta_1+\beta_2}{2}$$

then (15) *delivers a sequence of interval vectors* $\{[y]^k\}_{k=0}^{\infty}$ *with*

(17) $$y^* \in [y]^k, \quad k = 0,1,2,\ldots,$$

(18) $$\lim_{k\to\infty} [y]^k = y^*$$

where y^* _is the unique solution of_ (8) _in_ $[y]^0$.

□

Note that compared with the preceding theorem we have the stronger assumption $(1-\kappa)^2-4\varrho\ell > 0$. This together with the restriction on β yields the uniqueness of y^* and the convergence of (15) to y^*.

2. Rounding error analysis

We now choose $L = B^{-1}$ in (16). (15) then reads

$$(19)\qquad \begin{aligned} [y]^0 &= [-\beta,\beta]e \\ [y]^{k+1} &= B^{-1}(r + [y]^k_s \bullet [\tilde{y}]^k), \quad k = 0,1,1,\ldots . \end{aligned}$$

We now investigate how this method behaves if it is performed on a computer using a floating point system.

Let $b > 1$ be the basis of the numbersystem and let t_1 be the mantissa length of a single length floating point number. We make the following assumptions (a)-(b):

(a) For two machine intervals [a] and [b] it holds that

$$(20)\qquad f\ell([a]*[b]) = [(1-\varepsilon_1)([a]*[b])_1,(1+\varepsilon_2)([a]*[b])_2]$$

where

$$* \in \{+,-,\times,/\},$$

$$[a]*[b] = [([a]*[b])_1,([a]*[b])_2],$$

$$|\varepsilon_1|,|\varepsilon_2| \le \varepsilon = b^{1-t_1}.$$

$f\ell(.)$ denotes the result of a machine interval operation. (20) states that the lower bound $([a]*[b])_1$ of the exact result will be rounded downwards to the _next_ machine number (if rounding is necessary at all). The same has to hold for the upper bound.

(b) We assume that the so-called precise scalarproduct which was introduced by U. Kulisch (U. Kulisch, 1976) is available:

For two interval vectors $[x] = ([x]_i)$ and $[y] = ([y]_i)$ which have machineintervals as components it holds that

$$fl\left(\sum_{i=1}^{n} [x]_i \bullet [y]_i\right) = [(1-\varepsilon_1)\sigma_1, (1+\varepsilon_2)\sigma_2] \tag{21}$$

where

$$\sum_{i=1}^{n} [x]_i * [y]_i = [\sigma_1, \sigma_2]$$

denotes the exact scalar product and where again

$$|\varepsilon_1|, |\varepsilon_2| \le b^{1-t_1}.$$

Normally the precision of the precise scalarproduct is comparable to double length accumulation of products of simple length floating point numbers and rounding to single length after completion. If, however, severe cancellation of terms occurs (which is usually the case if residuals are computed) then the precision is much higher. Assume that $[\overline{BI}]$ is an intervalmatrix which has the property that its elements are exactly representable on the machine and for which

$$B^{-1} \in [\overline{BI}] \subseteq B^{-1} + \varepsilon[-1,1] \bullet |B^{-1}| + [-1,1] \bullet \overline{E}$$

holds where $\varepsilon = b^{1-t_1}$. Such an intervalmatrix can be computed as described by Rump (Rump, 1983). Denote by $[\overline{y}]^k$ the iterates computed on the machine. Furthermore let $d[\overline{y}]^k$

be the real vector of the diameters of the components of $[y]^k$. If then

$$\bar{d}_k = \|d[y]^k\|_\infty$$

$$\delta = 2\beta_1 \ell = 1-\sqrt{1-4\varrho\ell} < 1,$$

$$\varepsilon = b^{1-t_1}$$

$$\hat{\varepsilon} = \|\bar{E}\|_\infty$$

$$s = 2(3+3\varepsilon+\varepsilon^2)\bullet\|B^{-1}\|_\infty\bullet\| \; |r+[y]^o_s\bullet[\tilde{\bar{y}}]^k| \; \|_\infty$$

$$\hat{s} = 2(1+\varepsilon)^2\bullet\| \; |r+[\bar{y}]^o_s\bullet[\tilde{\bar{y}}]^k| \; \|_\infty$$

where "$\sim$" has the same meaning as in (3) then the inequality

$$\bar{d}_{k+1} \leq \delta^{k+1}\,\bar{d}_o + \frac{\varepsilon s+\hat{\varepsilon}\hat{s}}{1-\delta}$$

can be proved. The first term on the right hand side tends to zero for $k\to\infty$ such that the final precision on the machine is essentially determined by the second term.

3. Numerical example

We consider the unsymmetric (5,5) matrix (Gregory and Karney, 1969)

$$A = \begin{bmatrix} 15 & 11 & 6 & -9 & -15 \\ 1 & 3 & 9 & -3 & -8 \\ 7 & 6 & 6 & -3 & -11 \\ 7 & 7 & 5 & -3 & -11 \\ 17 & 12 & 5 & -10 & -16 \end{bmatrix}.$$

It has the simple eigenvalue $\lambda = -1$ with

$$(13,\ 22,\ 19,\ 16,\ 28)^T$$

as a corresponding eigenvector. We choose

$$-0.999\ 999\ 99$$

as approximation to this eigenvalue and

$$\begin{bmatrix} 13.000\ 001 \\ 21.999\ 999 \\ 18.999\ 999 \\ 16.000\ 001 \\ 27.999\ 999 \end{bmatrix}$$

as approximation to the eigenvector. After reading in this vector it is normalized to infinity norm one. This gives the vector

$$\begin{bmatrix} 0.464\ 285\ 766\ 582 \\ 0.785\ 714\ 278\ 061 \\ 0.678\ 571\ 417\ 092 \\ 0.571\ 428\ 627\ 551 \\ 1 \end{bmatrix}.$$

Using this vector and the preceding eigenvalue approximation delivers for β_1 defined by (14) the value

$$\beta_1 = 0.561\ 224\ 285\ 947 \bullet 10^{-7}.$$

After one step of the iteration method (19) we have

$$[\ -1.000\ 000\ 000\ 01;\ -0.999\ 999\ 999\ 999\ ]$$

as inclusion interval for the eigenvalue and

$$\begin{array}{l} [\ 0.464\ 285\ 714\ 285;\ 0.464\ 285\ 714\ 286\ ] \\ [\ 0.785\ 714\ 285\ 714;\ 0.785\ 714\ 285\ 715\ ] \\ [\ 0.678\ 571\ 428\ 571;\ 0.678\ 571\ 428\ 572\ ] \\ [\ 0.571\ 428\ 571\ 428;\ 0.571\ 428\ 571\ 429\ ] \\ [\ 1\ ;\ 1\ ] \end{array}$$

as inclusion intervals for the components of the corresponding eigenvector.

This example has been computed using an IBM-PC. Using the programming language PASCAL SC (U. Kulisch and W. Miranker, 1983) this system has available a decimal floating point number system with 12 digits in the mantissa.

REFERENCES

1. Alefeld G. and Herzberger J., Introduction to Interval Computations. Academic Press, New York, 1983.

2. Alefeld G., Berechenbare Fehlerschranken für ein Eigenpaar unter Einschluß von Rundungsfehlern bei Verwendung des genauen Skalarprodukts. Submitted for publication.

3. Gregory R.T. and Karney D.L., A collection of matrices for testing computational algorithms. Wiley Interscience, New York, 1969.

4. Kulisch U., Grundlagen des Numerischen Rechnens, Mathematische Begründung der Rechnerarithmetik. Bibliographisches Institut, Mannheim, 1976.

5. Kulisch U. and Miranker W. (Eds.), A New Approach to Scientific Computation. Academic Press, New York, 1983.

6. Rump S., Solving algebraic problems with high accuracy. In 4., pp. 53-120, 1983.

7. Symm H.J. and Wilkinson J.H., Realistic error bounds for a simple eigenvalue and its associated eigenvector. Numer. Math. 35, 1980, 113-126.

8. Wilkinson J.H. and Reinsch Ch., Handbook for automatic computation, Volume 2: Linear Algebra, Springer Berlin-Heidelber-New York, 1971.

9. Yamamoto T., Error bounds for computed eigenvalues and eigenvectors. Numer. Math. 34, 1980, 189-199.

Numerical Approximation of Partial Differential Equations
E.L. Ortiz (Editor)

SOME RESULTS ABOUT THE AOR ITERATIVE METHOD

M. Madalena Martins
Department of Mathematics
University of Coimbra
Coimbra, Portugal

This work consists of two fundamental sections:

— In the first one, we present a theorem which relates the eigenvalues of the iteration matrix of the AOR method, with those of the iteration matrix of the Jacobi method, when the matrix A of $Ax=b$ is a consistently ordered matrix, using a proof different from the one by Hadjidimos [2].
We also improve the sufficient condition of theorem 2 of [2], when A is a consistently ordered matrix.

— In the second part, by the way of using vectorial norms, we generalize theorem 6 [4], establishing convergence intervals for the AOR method, when A is a block-H matrix.

1. Consistently Ordered matrices. Convergence Conditions.

Let us consider the linear system

$$Ax = b \tag{1.1}$$

where A is a given non singular $n \times n$ matrix, with non vanishing diagonal elements, b is a given column vector and x is a column vector to be determinated. Let us split A, by the following way:

$$A = I - L - U \tag{1.2}$$

with I the identity matrix, $-L$ and $-U$ are, respectively, the strictly lower and upper parts of A.

Hadjidimos [2] defines the AOR method with the iteration matrix

$$L_{r,w} = (I - rL)^{-1} [(1-w)\, I + (w-r)\, L + wU] \tag{1.3}$$

For special values of the parameters w,r, this method becomes respectively, the Gauss-Seidel (GS) $(w=r=1)$, the Successive Overrelaxation (SOR) $(w=r)$, the Jacobi (J) $(r=0, w=1)$ and the Simultaneous Overre laxation (JOR) $(r = 0, w)$ methods.

By means of next lemma we may prove theorem 1 [2] by a different way.

Lemma *1.1 - Let A be a consistently ordered matrix with nonvanishing diagonal elements. Then there exists a nonmigratory permutation matrix P, such that:*

– $A' = P^{-1} A P = \begin{bmatrix} D_1 & H \\ K & D_2 \end{bmatrix}$, *where D_1 and D_2 are diagonal matrices with nonvanishing diagonal elements.*

– $L_{r,w} [P^{-1} A P] = P^{-1} L_{r,w}(A) P$

Theorem *1.1 - Let A be a consistently ordered matrix with nonvanishing diagonal elements, then:*

a) If μ is a nonzero eigenvalue of J and λ satisfies the equality:

$$(\lambda + w - 1)^2 = w \mu^2 [r(\lambda - 1) + w] \qquad (1.4)$$

then λ is an eigenvalue of $L_{r,w}$. If $\mu = 0$ is an eigenvalue of J, then $\lambda = 1 - w$ is an eigenvalue of $L_{r,w}$.

b) If λ is an eigenvalue of $L_{r,w}$, then exists an eigenvalue of J, such that (1.4) holds.

Since the SOR method is a particular case of the AOR method when $w = r$, we achieve theorems 5.2.2 and 6.2.2 [11] as a direct consequence of the last theorem.

Let us consider the stationary iterative method of the first degree:

$$x^{i+1} = T x^i + d \Big|_{i=0,1,2,\ldots} \qquad (1.5)$$

where T is a $n \times n$ matrix, d is a known column vector and x^o is an arbitrary initial approximation for the solution x.

The extrapolated method of (1.5) with the extrapolated parameter β will be then:

$$x^{i+1} = [(1-\beta)\, I + \beta T]\, x^i + \beta d \Big|_{i=0,1,2,\ldots}$$

After some manipulations it is easy to verify that the AOR method is the extrapolated SOR method (ESOR), with the extrapolated parameter $\frac{w}{r}$, provided that $r \neq 0$. So, if we use Theorem of Extrapolation [3], and Theorem 2 [2], we have:

Theorem 1.2

If A is a consistently ordered matrix with nonvanishing diagonal elements, such that the corresponding J has eigenvalues

$$\mu_i \Big|_{i=1(1)n} \quad \text{with} \quad \underline{\mu} = \min_i |\mu_i| \quad \text{and} \quad \bar{\mu} = \max_i |\mu_i|$$

then $\rho(L_{r,w}) < 1$ provided that $\rho(J) < 1$ and the parameters w and r take values in the intervals I_w and I_r, respectivelly defined as follows:

Case A - *a) If $\frac{2r}{1+\rho(L_{r,r})} < 2$ and $\underline{\mu} \neq 0$, we have:*

$$I_w = (-2/(1-\underline{\mu}^2)^{\frac{1}{2}},\ 0) \quad \text{and} \quad I_r = (\beta(\underline{\mu}^2),\ \alpha(\bar{\mu}^2))$$

or

$$I_w = [0,\ 2r/(1+\rho(L_{r,r}))) \quad \text{and} \quad I_r = (\min\{0, \alpha(\bar{\mu}^2)\},\ \max\{2, \beta(\bar{\mu}^2)\})$$

or

$$I_w = [2r/(1+\rho(L_{r,r})),\ 2] \quad \text{and} \quad I_r = (\alpha(\bar{\mu}^2),\ \beta(\bar{\mu}^2))$$

or

$$I_w = \left[2,\ 2/(1-\underline{\mu}^2)^{\frac{1}{2}}\right] \quad \text{and} \quad I_r = (\alpha(\bar{\mu}^2),\ \beta(\underline{\mu}^2))$$

while if

b) $\underline{\mu} = 0$

$$I_w = (0, 2r/(1+\rho(L_{r,r}))) \text{ and } I_r = (\min\{0, \alpha(\bar{\mu}^{-2})\}, \max\{2, \beta(\bar{\mu}^{-2})\})$$

or

$$I_w = (2r/(1+\rho(L_{r,r})), 2) \text{ and } I_r = (\alpha(\bar{\mu}^{-2}), \beta(\bar{\mu}^{-2}))$$

Case B - a) If $2r/(1+\rho(L_{r,r})) \geq 2$ and $\underline{\mu} \neq 0$, we have:

$$I_w = (-2/(1-\underline{\mu}^2)^{\frac{1}{2}}, 0) \text{ and } I_r = (\beta(\underline{\mu}^2), \alpha(\bar{\mu}^{-2}))$$

or

$$I_w = (0, 2] \text{ and } I_r = (\min\{0, \alpha(\bar{\mu}^{-2})\}, \max\{2, \beta(\bar{\mu}^{-2})\})$$

or

$$I_w = (2, 2r/(1+\rho(L_{r,r}))) \text{ and } I_r = (\min\{0, \alpha(\bar{\mu}^{-2})\}, \max\{2, \beta(\underline{\mu}^2)\})$$

or

$$I_w = [2r/(1+\rho(L_{r,r})), 2/(1-\underline{\mu}^2)^{\frac{1}{2}}) \text{ and } I_r = (\alpha(\bar{\mu}^{-2}), \beta(\underline{\mu}^2))$$

while for $\underline{\mu} = 0$, *we have*

b) $I_w = (0, 2)$ and $I_r = (\min\{0, \alpha(\bar{\mu}^{-2})\}, \max\{2, \beta(\bar{\mu}^{-2})\})$

with $\alpha(z) = \frac{1}{wz}\left(\frac{1}{2}w^2 z - \frac{1}{2}w^2 + 2w - 2\right)$ and $\beta(z) = \frac{1}{z}(wz - w + 2)$.

2. Vectorial Norms. Block - H matrices

By using the notions about vectorial regular norms given by F. Robert [4], we may generalize theorem 6 [4].

Theorem 2.1 - *Let* p *be a vectorial regular norm of dimension* k *over*

$\mathbb{R}^n$ and M a vectorial matrix norm generated by p. Let L and U be two $n \times n$ matrices.

Let ρ^* be defined by $\rho^* = \rho(M(L) + M(U))$.

If $\rho^* < 1$, then the iteration matrix of the AOR method

$$L_{r,w} = (I - rL)^{-1} [(1-w) I + (w-r) L + w U]$$

exists and is contracted relatively to p, for the values of w and r, such that $0 \leq r \leq w \leq \frac{2}{1+\rho^*}$. For these values of w and r the matrix

$$\bar{L}_{r,w} = [I - |r| \; M(L)]^{-1} [|1-w| I + |w| \; M(U) + |w-r| \; M(L)]$$

is a convergent majorant of $L_{r,w}$.

Due to the fact that the SOR, G.S and J methods are particular cases of the AOR method, theorem 6 [4] becomes an imediate consequence of this last theorem.

By using the vectorial norm $p(x) = |x|$ and the iterative method by points (all the blocks are (1,1)) we may establish convergence conditions for the AOR method, when A is an H-matrix.

Corollary 2.1 - *If A of (1.1) is an H-matrix, $\rho(L_{r,w}) < 1$, if*

$$0 \leq r \leq w \leq 2/(1+\rho(|J|))$$

Theorem 2.2 - *If A is an H-matrix, the AOR method converges for:*

$$0 < r < 2/(1+\rho(|J|)) \quad \text{and} \quad 0 < w < 2r/(1+\rho(|J|))$$

Proof - This result comes from the last theorem and from the theorem of Extrapolation [3].

***Theorem* 2.3** - *Let A be an H-matrix, the AOR method converges for:*

$$0 \leq r < v \quad \text{and} \quad 0 < w < \max(g(r), v)$$

with $g(r) = 2r/(1+\rho(L_{r,r}))$ *and* $v = 2/(1+\rho(|J|))$

Proof - This result is a consequence of the last theorem and Corollary 2.1.

***Theorem* 2.4** - *For a given H-matrix A, the AOR method converges, if:*

$$\text{(i)} \quad \text{a)} \quad 0 \leq r < \tilde{t} \quad \text{and} \quad 0 < w < \max(\tilde{t}, g(r))$$

or

$$\text{b)} \quad \tilde{t} \leq r < \tilde{q} \quad \text{and} \quad \tilde{f}(r) < w < 1$$

or

$$1 < w < \tilde{m} \quad \text{and} \quad w < r < \tilde{s}(w)$$

if $w < r$ *and* $\tilde{t} \geq v$

$$\text{(ii)} \quad \text{a)} \quad 0 \leq r < v \quad \text{and} \quad 0 < w < \max(v, g(r))$$

$$\text{b)} \quad v \leq r < \tilde{q} \quad \text{and} \quad \tilde{f}(r) < w < 1$$

or

$$1 < w < \tilde{m} \quad \text{and} \quad \tilde{f}(r) < w < 1$$

if $w < r$ *and* $\tilde{t} < v < \tilde{q}$.

$$\text{(iii)} \quad 0 \leq r < v \quad \text{and} \quad 0 < w < \max(g(r), v)$$

if $v \geq \tilde{q}$

where $\tilde{q} = \min_i \left[(1+\tilde{e}_i - \tilde{f}_i)/2\tilde{e}_i\right]$, $\tilde{f}(r) = \max_i \left[2\tilde{e}_i r/(1+\tilde{e}_i - \tilde{f}_i)\right]$, *and*

$$\tilde{s}(w) = \min_i \{\left[2 - w(1-\tilde{e}_i + \tilde{f}_i)\right]/2\tilde{e}_i\} \quad i = 1, 2, \ldots, n$$

Proof - This is a consequence from last theorem, theorem 6 [7] and theorem 9 [8].

Theorem 2.5 - *When* A *is an* H-*matrix, then* $\rho(L_{r,w}) < 1$, *for:*

(i) $0 > r > w > 2r\, \tilde{e}_i/(\tilde{e}_i - \tilde{f}_i - 1)$

or

(ii) $[2r\,\tilde{e}_i/(\tilde{e}_i + \tilde{f}_i - 1)] < w < 1$ *and* $r < (\tilde{e}_i + \tilde{f}_i - 1)/2\,\tilde{e}_i$

if $w < -r$

or

(iii) $1 < w < [(2r\,\tilde{e}_i + 2)/(1 + \tilde{e}_i + \tilde{f}_i)]$ *and* $0 > r > (\tilde{e}_i + \tilde{f}_i - 1)/2\,\tilde{e}_i$

if $-r > w > 0$.

The case $r > -w$, $w < 0$, $r > 0$ might also be considered. However, under this scheme, it is not possible to find intervals for w and r, such that $\rho(L_{r,w}) < 1$.

Theorem 2.6 - *If* A *is an* H-*matrix, then the AOR method converges for:*

(i) a) $0 \leq r < \tilde{t}$ *and* $0 < w < \max(v, g(r))$

or

$\tilde{t} \leq r < q$ *and* $\tilde{f}(r) < w < 1$ *and* $w < r$

or

$1 < w < \tilde{m}$ *and* $w < r < \tilde{s}(w)$

if $\tilde{t} \geq v$

(ii) a) $0 \leq r < v$ *and* $0 < w < \max(v, g(r))$

or

b) $v \leq r < \tilde{q}$ *and* $\tilde{f}(r) < w < 1$

or

$$1 < w < \tilde{m} \quad \text{and} \quad w < r < \tilde{s}(w)$$

if $w < r$ *and* $\tilde{t} < v < \tilde{q}$

(iii) $0 \leq r < v$ and $0 < w < \max(g(r), v)$

if $v \geq \tilde{q}$

(iv) $0 > r > w > 2r\tilde{e}_i/(\tilde{e}_i - \tilde{f}_i - 1)$

(v) $2r\tilde{e}_i/(\tilde{e}_i + \tilde{f}_i - 1) < w < 1$ and $r < (\tilde{e}_i + \tilde{f}_i - 1)/2\tilde{e}_i$

if $w < -r$

(vi) $1 < w < (2r\tilde{e}_i + 2)/(1 + \tilde{e}_i + \tilde{f}_i)$ and $0 > r > (\tilde{e}_i + \tilde{f}_i - 1)/2\tilde{e}_i$

with $-r > w > 0$

Proof - It comes from theorems 2.4 and 2.5

References

1. Avdelas, G. Hadjidimos A. and Yeyios A., Some theoretical and computational results, concerning the accelerated overrelaxation (AOR) method. T.R. No 8, Department of Mathematics, University of Ioannina, Greece, 1978.

2. Hadjidimos A., Accelerated Overrelaxation Method. - Math. Comp., 32 (1978), pp.149-157.

3. Hadjidimos A. and Yeyios A., The principle of extrapolation in connection with the Accelerated Overrelaxation (AOR) method. -Linear Algebra Appl.30 (1980), pp.115-128.

4. Robert F., Bloc H-matrices et convergence des methodes iteratives classiques par blocs. - Linear Algebra and its Appl.2 (1969), pp.223-265.

5. Martins M., On accelerated overrelaxation iterative method for linear systems with strictly diagonally dominant matrix. Math. Comp.,35 (1980), pp.1269-1273.

6. Martins M., Note on irreducible diagonally dominant matrices and the convergence of the AOR iterative method. Math. Comp.37(1981). pp. 101-104.

7. Martins M., Generalized Diagonal Dominance in Connection with the Accelerated Overrelaxation (AOR) Method. Bit.22 (1982), pp.73-78.

8. Martins M., An improvement for the area of convergence of the accelerated overrelaxation iterative method. Revue d'Analyse Numérique et de Théorie de l'Approximation 12 (1983), pp.65-76.

9. Varga R.S., Matrix Iterative Analysis. Englewood Cliffs, N.J.: Prentice--Hall 1962.

10. Varga R.S., On recurring theorems on diagonal dominance. Linear Algebra Appl.13 (1976), 1-9.

11. Young D.M., Iterative Solution of Large Linear Systems. New York-London: Academic Press. 1971.

Numerical Approximation of Partial Differential Equations
E.L. Ortiz (Editor)

ON THE SPEED OF CONVERGENCE OF THE TOTAL STEP METHOD IN INTERVAL COMPUTATIONS

Günter Mayer
Universität Karlsruhe (TH)
Karlsruhe, West Germany

Abstract

We show a possibility to determine the asymptotic convergence factor α_T of the total step method $\mathbf{x}^{m+1} = \mathbf{A}\mathbf{x}^m + \mathbf{b}$ in interval computations. Especially we set up a necessary and sufficient condition for α_T to be less than the spectral radius of the absolute value of $\mathbf{A}$ if this interval matrix fulfils some assumptions.

1. Introduction

In classical numerical analysis the asymptotic convergence factor α_f (Ortega/Rheinboldt, 1970; Ortega, 1972) is known to be a measure for the speed of convergence of an iterative method

$$x^{m+1} := f(x^m), \quad m = 0,1,\ldots , \tag{1}$$

converging for all starting vectors $x^0 \in \mathbb{R}^n$ (set of real vectors with n components) to a unique fixpoint x of the function $f: \mathbb{R}^n \to \mathbb{R}^n$.

In interval analysis we also have the concept of asymptotic convergence factor $\alpha_{\mathbf{f}}$ (Alefeld/Herzberger, 1983) $\mathbf{f}: I\mathbb{R}^n \to I\mathbb{R}^n$ being now an interval function which is defined on the set $I\mathbb{R}^n$ of all real interval vectors with n components consisting of real compact intervals.

Equip $\mathbb{R}^n$ as well as $\mathbb{R}^{n\times n}$ (set of all real $n\times n$ matrices) with the usual componentwise defined partial ordering "$\le$", denote all interval quantities by bold face letters and call the nonnegative real vector

$$q(\mathbf{x},\mathbf{y}) := (\max\{|\underline{x}_i-\underline{y}_i|, |\overline{x}_i-\overline{y}_i|\})$$

the distance of the two interval vectors $\mathbf{x} = ([\underline{x}_i,\overline{x}_i])$, $\mathbf{y} = ([\underline{y}_i,\overline{y}_i])$. Then $\alpha_{\mathbf{f}}$ is defined in the following way:

<u>Definition:</u> (Alefeld/Herzberger, 1983; Mayer, submitted)

Let $\mathbf{f}$: $I\mathbb{R}^n \rightarrow I\mathbb{R}^n$ be an interval function and denote by $\mathbf{C}$ the set of all sequences $\{\mathbf{x}^m\}$ which are defined by the iterative method

$$\begin{cases} \mathbf{x}^o \in I\mathbb{R}^n \\ \mathbf{x}^{m+1} := \mathbf{f}(\mathbf{x}^m), \ m = 0,1,\ldots \end{cases} \tag{2}$$

Let (2) converge for all starting vectors $\mathbf{x}^o$ to a unique fixpoint $\mathbf{x}$ of $\mathbf{f}$.

If $\{\mathbf{x}^m\}$ is an element of $\mathbf{C}$ then

$$R\{\mathbf{x}^m\} := \limsup_{m\to\infty} \|q(\mathbf{x}^m,\mathbf{x})\|^{1/m}$$

is called R-factor of $\{\mathbf{x}^m\}$. Here $\|.\|$ is any vector norm of $\mathbb{R}^n$. ($R\{\mathbf{x}^m\}$ does not depend on some special choice of $\|.\|$!)

The asymptotic convergence factor $\alpha_{\mathbf{f}}$ of (2) is defined by

$$\alpha_{\mathbf{f}} := \sup\{R\{\mathbf{x}^m\} \mid \{\mathbf{x}^m\} \in \mathbf{C}\}.$$

□

Denote by the nonnegative real vectors

$$|\mathbf{x}| := (\max\{|\underline{x}_i|, |\overline{x}_i|\}) \quad \text{and} \quad d(\mathbf{x}) := (\overline{x}_i - \underline{x}_i)$$

the absolute value and the width, respectively, of an interval vector $\mathbf{x} = ([\underline{x}_i, \overline{x}_i])$. Absolute value $|\mathbf{A}|$ and width $d(\mathbf{A})$ of an interval matrix $\mathbf{A} \in I\mathbb{R}^{n\times n}$ (set of all real $n\times n$ interval matrices) are nonnegative real matrices which are defined analogously (Alefeld/Herzberger, 1983). Write int($\mathbf{x}$) for the topological interior of an interval vector $\mathbf{x}$ and $\varrho(M)$ for the spectral radius of a real $n\times n$ matrix M.

In the sequel we consider the total step method

$$\begin{cases} \mathbf{x}^o \in I\mathbb{R}^n \\ \mathbf{x}^{m+1} := \mathbf{f}(\mathbf{x}^m) := \mathbf{A}\mathbf{x}^m + \mathbf{b}, \ m = 0,1,\ldots \end{cases} \tag{3}$$

in interval analysis (Alefeld/Herzberger, 1983) where $\mathbf{A} = ([\underline{a}_{ij}, \overline{a}_{ij}]) \in I\mathbb{R}^{n\times n}$ and $\mathbf{b}, \mathbf{x}^m \in I\mathbb{R}^n$.

It is known (Alefeld/Herzberger, 1983) that (3) converges for every $\mathbf{x}^o \in I\mathbb{R}^n$ to the unique fixpoint of $\mathbf{f}$ if and only if $\varrho(|\mathbf{A}|) < 1$. It is further known that in this case $\varrho(|\mathbf{A}|)$ is an upper bound of the asymptotic convergence factor $\alpha_{\mathbf{f}}$ of (3) which we shall denote by α_T in the sequel. Up to now, however, it was not known whether this bound is sharp for all elements of $I\mathbb{R}^{n\times n}$. Our main theorem states that in some cases α_T can be less than $\varrho(|\mathbf{A}|)$. To this end we remark that under certain circumstances α_T can be computed by considering only two special sequences $\{\mathbf{x}^m\}$ generated by (3).

2. Results

Theorem 1:[*)]

Let $\mathbf{A} \in I\mathbb{R}^{n\times n}$ with $\varrho(|\mathbf{A}|) < 1$. Let $\mathbf{b},\mathbf{x} \in I\mathbb{R}^n$ where $\mathbf{x}$ is the fixpoint of $\mathbf{f}$ in (3). Then the following statements hold:

a) If $d(\mathbf{x}) > 0$ (i.e. $\operatorname{int}(\mathbf{x}) \neq \emptyset$) and if the sequences $\{\mathbf{y}^m\},\{\mathbf{z}^m\}$ are generated by (3) satisfying

$$\mathbf{y}^o \subseteq \operatorname{int}(\mathbf{x}) \subseteq \mathbf{x} \subseteq \operatorname{int}(\mathbf{z}^o)$$

then $\alpha_T = \max\ \{R\{\mathbf{y}^m\},\ R\{\mathbf{z}^m\}\}$.

b) If $d(\mathbf{x}) = 0$ (i.e. if $\mathbf{x}$ is a point interval) and if the sequence $\{\mathbf{z}^m\}$ is generated by (3) satisfying

$$\mathbf{x} \subseteq \operatorname{int}(\mathbf{z}^o)$$

then $\alpha_T = R\{\mathbf{z}^m\}$.

□

In (Mayer, submitted) it is proved that if $|\mathbf{A}|$ is irreducible (Ortega/Rheinboldt, 1970), then either $d(\mathbf{x}) > 0$ or $d(\mathbf{x}) = 0$ holds; thus Theorem 1 is applicable in this case. It forms the basic tool for proving the following main theorem, in which I denotes the real $n\times n$ unit matrix.

---

*) Since the proof of Theorem 1 and of the following results are rather lengthy we must renounce on them. They can be found in a more detailed version of this paper being submitted for publication (Mayer, submitted).

<u>Theorem 2:</u>

Let $\mathbf{b} = [\underline{b},\overline{b}] \in I\mathbb{R}^n$

and let $\mathbf{A} = [\underline{A},\overline{A}] = ([\underline{a}_{ij},\overline{a}_{ij}]) \in I\mathbb{R}^{n\times n}$

have the following properties:

1. $\underline{A} \geq 0$,
2. $\varrho(|\mathbf{A}|) < 1$,
3. $|\mathbf{A}|$ is irreducible.

Let $\mathbf{f}$ be the interval function of (3) and let $S_{\mathbf{A}}$ be the set of all column indices belonging to columns of $\mathbf{A}$ which have at least one entry with positive width.

Then $\alpha_T < \varrho(|\mathbf{A}|)$ if and only if there exist indices $s,t \in S_{\mathbf{A}}$ such that $((I-|\mathbf{A}|)^{-1}\underline{b})_s > 0$ and $((I-|\mathbf{A}|)^{-1}\overline{b})_t < 0$. In all other cases we have $\alpha_T = \varrho(|\mathbf{A}|)$.

If the signs of the bounds $\underline{x}_i,\overline{x}_i, i \in S_{\mathbf{A}}$, of the fixpoint $\mathbf{x} = ([\underline{x}_i,\overline{x}_i])$ of $\mathbf{f}$ are known then

$$\alpha_T = \max\{\varrho(A^*),\varrho(A^{**})\}$$

where the real $n\times n$ matrices $A^* = (a^*_{ij})$, $A^{**} = (a^{**}_{ij})$ are defined by

$$a^*_{ij} := \begin{cases} \overline{a}_{ij} & \text{if } \underline{x}_j \leq 0 \\ \underline{a}_{ij} & \text{if } \underline{x}_j > 0 \end{cases} \qquad a^{**}_{ij} := \begin{cases} \underline{a}_{ij} & \text{if } \overline{x}_j < 0 \\ \overline{a}_{ij} & \text{if } \overline{x}_j \geq 0 \end{cases} \quad \square$$

(Notice that for $j \notin S_{\mathbf{A}}$ we have $\underline{a}_{ij} = \overline{a}_{ij}$, $i = 1,\ldots,n$, thus A^*,A^{**} are uniquely determined by the assumptions.)

$\square$

Corollary:

Let the assumptions of Theorem 2 hold.

a) If $\mathbf{A}$ is a point matrix, i.e. if $S_{\mathbf{A}} = \emptyset$, then we have $\alpha_T = \varrho(|\mathbf{A}|)$.

b) If $S_{\mathbf{A}} = \{1,...,n\}$ then we have $\alpha_T \le \varrho(|\mathbf{A}|)$ with equality if and only if $(I-|\mathbf{A}|)^{-1}\underline{b} \le 0$ or $(I-|\mathbf{A}|)^{-1}\overline{b} \ge 0$. Especially this is true if the assumption "$|\mathbf{A}|$ irreducible" is replaced by "$d(\mathbf{A}) > 0$" or "$d(\mathbf{A})$ irreducible".

□

3. Examples

To illustrate that $|\mathbf{A}|$ is not necessarily the dominant quantity for α_T we present two examples of the total step method. In the first case we use

$$\mathbf{A} = \mathbf{A}_1 := \begin{pmatrix} [0,0] & [0,\frac{1}{2}] \\ [0,\frac{1}{2}] & [0,0] \end{pmatrix} \quad \text{and} \quad \mathbf{b} := \begin{pmatrix} [1,1] \\ [-1,-1] \end{pmatrix},$$

in the second case we use the same vector $\mathbf{b}$, but the matrix

$$\mathbf{A} = \mathbf{A}_2 := \begin{pmatrix} [0,0] & [\frac{1}{2},\frac{1}{2}] \\ [\frac{1}{2},\frac{1}{2}] & [0,0] \end{pmatrix}$$

It is clear that in both cases the absolute value $|\mathbf{A}|$ is the same, namely $|\mathbf{A}| = \begin{pmatrix} 0 & \frac{1}{2} \\ \frac{1}{2} & 0 \end{pmatrix}$. With $\mathbf{A} = \mathbf{A}_1$ the fixpoint $\mathbf{x}$ of $\mathbf{f}(\mathbf{y}) := \mathbf{A}\mathbf{y}+\mathbf{b}$ is

$\mathbf{x} = ([\frac{1}{2},1],\ [-1,-\frac{1}{2}])^T$, with $\mathbf{A} = \mathbf{A}_2$ it is

$\mathbf{x} = ([\frac{2}{3},\frac{2}{3}],\ [-\frac{2}{3},-\frac{2}{3}])^T$. In the first case

$$A^* = \begin{pmatrix} 0 & \frac{1}{2} \\ 0 & 0 \end{pmatrix}, \quad A^{**} = \begin{pmatrix} 0 & 0 \\ \frac{1}{2} & 0 \end{pmatrix},$$ in the second one

$A^* = A^{**} = |\mathbf{A}|$. Denoting by $\alpha_T^{(1)}$, $\alpha_T^{(2)}$ the asymptotic convergence factors of (3) with $\mathbf{A} = \mathbf{A}_1$ and $\mathbf{A} = \mathbf{A}_2$, respectively, we get by Theorem 2

$$\alpha_T^{(1)} = \varrho(A^*) = \varrho(A^{**}) = 0 < \varrho(|\mathbf{A}_1|) = \frac{1}{2} = \varrho(|\mathbf{A}_2|) = \alpha_T^{(2)}. \tag{4}$$

To illustrate this asymptotically faster convergence in the first case, we iterated on a SAM-computer programming in PASCAL-SC (Kulisch/Miranker,1983). In both cases we started with $\mathbf{x}^0 := ([0.6,0.6],[-0.6,-0.6])^T$. We stopped the iteration when

$$\|q(\mathbf{x}^{m+1},\mathbf{x}^m)\| < 10^{-6}$$

was fulfilled for the first time where $\|.\|$ denotes the maximum norm. Although the common starting vector seemed to be more favorable in view of $\mathbf{A}_2$ than in view of $\mathbf{A}_1$ we got the following results confirming (4):

$\mathbf{A} = \mathbf{A}_1$

m	$(\mathbf{x}^m)^T$
0	[0.600 000, 0.600 000] [-0.600 000, -0.600 000]
1	[0.700 000, 1.000 000] [-1.000 000, -0.700 000]
2	[0.500 000, 1.000 000] [-1.000 000, -0.500 000]
3	[0.500 000, 1.000 000] [-1.000 000, -0.500 000]

$\mathbf{A} = \mathbf{A}_2$

m	$(\mathbf{x}^m)^T$
0	[0.600 000, 0.600 000] [-0.600 000, -0.600 000]
1	[0.700 000, 0.700 000] [-0.700 000, -0.700 000]
2	[0.650 000, 0.650 000] [-0.650 000, -0.650 000]
3	[0.675 000, 0.675 000] [-0 675 000, -0.675 000]
4	[0.662 500, 0.662 500] [-0.662 500, -0.662 500]
5	[0.668 750, 0.668 750] [-0.668 750, -0.668 750]
6	[0.665 625, 0.665 625] [-0.665 625, -0.665 625]
7	[0.667 187, 0.667 188] [-0.667 188, -0.667 187]
8	[0.666 406, 0.666 407] [-0.666 407, -0.666 406]
9	[0.666 796, 0.666 797] [-0.666 797, -0.666 796]
10	[0.666 601, 0.666 602] [-0.666 602, -0.666 601]
11	[0.666 699, 0.666 700] [-0.666 700, -0.666 699]
12	[0.666 650, 0.666 651] [-0.666 651, -0.666 650]
13	[0.666 674, 0.666 675] [-0.666 675, -0.666 674]
14	[0.666 662, 0.666 663] [-0.666 663, -0.666 662]
15	[0.666 668, 0.666 669] [-0.666 669, -0.666 668]
16	[0.666 665, 0.666 666] [-0.666 666, -0.666 665]
17	[0.666 667, 0.666 668] [-0.666 668, -0.666 667]
18	[0.666 666, 0.666 667] [-0.666 667, -0.666 666]

References:

1. Alefeld, G. and Herzberger, J., Introduction to Interval Computations, Academic Press, New York, 1983.

2. Kulisch, U.W. and Miranker, W.L., A New Approach to Scientific Computation, Academic Press, New York, 1983.

3. Mayer, G., On the Asymptotic Convergence Factor of the Total Step Method in Interval Computations, submitted to Lin. Alg. Appl. (1985).

4. Ortega, J.M., Numerical Analysis, A Second Course, Academic Press, New York, 1972.

5. Ortega, J.M., and Rheinboldt, W.C., Iterative Solution of Nonlinear Equations in Several Variables, Academic Press, New York, 1970.

Part II:
DISCRETE VARIABLE METHODS

Finite Difference Methods, Finite Element Methods and Related Techniques

Method of Lines

Numerical Approximation of Partial Differential Equations
E.L. Ortiz (Editor)

ON THE STABILITY OF VARIABLE STEPSIZE ADAMS METHODS IN NORDSIECK FORM

Manuel Calvo
Francisco J. Lisbona
Juan I. Montijano
University of Zaragoza
Zaragoza , Spain

The aim of our paper is to show that the stability of Adams methods can be ascertained under weaker assumptions than the ones given in [5] and [13]. In particular it is proved that (k+1)-value Adams methods remain stable if there exists a fixed $p \geq 0$, so that after consecutive arbitrary stepsizes whose number is $\leq p$, there are at least k-1 steps of constant size. On the other hand greater intervals of stability have been obtained.

Introduction. This paper studies the stability of variable stepsize Adams methods that vary the stepsize using the interpolation technique of Nordsieck. It is well known that a number of widely distributed programs as DIFSUB [7] and its successors [9] use this technique for stepsize changing.

In the last years a large number of publications have been devoted to extend some stability results of multistep methods with fixed stepsize to variable steps. Among them we may mention the classical of Piotrowski [11] and Gear-Tu-Watanabe [5],[6], and more recently those of Zlatev [14] ,Grigorieff [8], Crouzeix-Lisbona[4] , Calvo-Lisbona-Montijano [2], Shampine-Gordon [12] and Skeel-Jackson [13] . In particular the stability of Adams -Nordsieck has been studied in [5] and [13] . Jackson and Skeel have improved Gear and Tu results in two directions. First they prove that (k+1)-value Adams methods remain stable if there are at least k-1 steps of constant size between stepsize changes while in [5] k steps of constant size are necessary to insure the stability. On the other hand Skeel and Jackson give some intervals so that if the stepsize ratios belong to them the stability is insured.

In our paper we extend some results of Skeel and Jackson's paper and furthermore we analyze the stability with other strategies of stepsize variation not considered by these authors. Briefly, in section 2 we formulate the variable stepsize Adams methods in Nordsieck form and some stability definitions are introduced. In section 3 we consider the stability of these methods with $k \geq 4$ steps, giving some intervals J_k so that whenever the stepsize ratio belongs to J_k the method remains stable. These intervals are larger than those obtained by a correct application of Skeel and Jackson's technique. Finally it is shown that these intervals can be substantially increased if we assume other strategies of stepsize variation. For the sake of brevity the proofs of our theorems have been omitted and for the interested reader we refer to our Tech. Report [3] .

2. Adams-Nordsieck methods.- Consider the IVP

$$(2.1) \qquad y'(t) = f(t, y(t)) \qquad t \in [0,T] \ , \quad y(0) = y_o$$

whose solution it is assumed to be sufficiently differentiable and let $0=t_o < t_1 < \dots < t_N=T$ be a (possibly) non uniform grid with stepsizes $h_n = t_n - t_{n-1}$, where the solution of (2.1) is to be approximated. Furthermore let r_n be the stepsize ratio h_{n+1}/h_n.

A k-step Adams method that vary the stepsize using the interpolation technique can be written as a Nordsieck method in the form

$$(2.2) \qquad U_{n+1} = S\, C(r_n)\, U_n + h_{n+1}\, d\, f(t_{n+1}, y_{n+1})$$

where $U_n = (y_n, h_n y_n', \dots , h_n^k y_n^{(k)}/k!\ )^t$, $C(r_n)=\mathrm{diag}(1, r_n,\dots,r_n^k)$ d is a fixed (k+1)-vector and S is the so called fixed stepsize propagation matrix. Here y_n and $y_n^{(j)}$ are appoximations to the solution of (2.1) and its derivatives at t_n.

Denoting $S\,C(r_n)$ by S_n or $S(r_n)$, this matrix has the form

$$(2.3) \qquad S(r_n) = \left[\begin{array}{c|c|c} 1 & b_n & \bar{b}_n^t \\ \hline 0 & 0 & 0 \\ \hline 0 & \bar{a}_n & T_n=T(r_n) \end{array}\right]$$

where $b_n = r_n\, b$, $\bar{b}_n^t = \bar{b}^t\, \bar{C}(r_n)$, $\bar{a}_n = r_n\, \bar{a}$ with

$\bar{C}(r_n) = \mathrm{diag}(\ r_n^2\ , \dots , r_n^k\ )$ and $b \in R$, $\bar{b}, \bar{a} \in R^{k-1}$ constants

and $T(r_n) = T\, \bar{C}(r_n)$ with

$$T = \bar{\Omega}^{-1}\, E\, \bar{\Omega}$$

where

$$E = (e_{ij}) \ , \quad e_{ij} = \delta_{i,j+1}$$

$$\bar{\Omega} = (w_{ij}) \ , \quad w_{ij} = (-i)^j\ (j+1)$$

Assuming as usual that $f(t,y)$ is Lipschitz continuous it can be seen [4] that stability is independent of the non linear part of (2.2) and therefore we may give the following

Definition 2.1.- The formula (2.2) is said to be stable if there exist constants $h^* > 0$ and K such that for any grid in $[0,T]$ with $\max h_j \leq h^*$ we have

$$(2.4) \qquad | S(r_n) \dots S(r_j)| \leq K \qquad 0 \leq j \leq n \leq N$$

Note that due to the special form of the propagation matrices, the product in (2.4) depends only on the sequence $\{r_m\}_{m \geq 1}$. This lead us to introduce a new definition.

Definition 2.2.- A sequence of stepsizes $\{r_n\}_{n \geq 1}$ is a stable stepsizes sequence for the method (2.2) if there exists a constant $K < \infty$ such that (2.4) is satisfied.

Clearly it would be desirable to give necessary and sufficient conditions that permit to decide the stability of a stepsize sequence for the Adams methods. A result of this kind has been completely achieved for the three steps Adams methods [1], however it seems to be a difficult problem with more than three steps. Due to this fact let us introduce a new concept by means of which we may state sufficient conditions for the stability of a stepsizes sequence.

Definition 2.3.- Let $J_k \subset R$ be a set of real numbers. J_k is said to be a stability set for the method (2.2), if all sequence $\{r_n\}$ with $r_n \in J_k$ is stable for this method.

It follows at once from the polynomial dependence of $S_k(r)$ on r that the stability sets J_k must be bounded.

3. Stability of Adams-Nordsieck methods. In this section we look for sufficient conditions on a sequence of stepsizes to be stable for the Adams methods with $k \geq 4$. More precisely, let us give some results that permit us to decide whether or not a set $J \subset R$ is a stability set for the methods under consideration. For our purposes we shall state the following theorem whose proof is formally identical to theorem (4.1) [2].

Theorem 3.1.

i) If J_k is a stability set for the k-step Adams-Nordsieck method, then

$$J_k \subset I_k = \{ r \in R^+ \ ; \ \rho(T(r)) \leq 1 \}$$

ii) If there is a constant matrix H and a constant K_1 such that

(3.1) $\| H^{-1} T(r) H \| \leq K_1 < 1$

for all $r \in J_k \subset R^+$, then J_k is a stability set for the k-steps Adams Nordsieck method.

To apply the first part of the above theorem we have computed $p(r) = \rho(T(r))$ for the k-step Adams-Nordsieck methods and we have found that the sets I_k are intervals $[0,r_k^*]$ where the values of r_k with three decimal places are given by

$$r_4^* = 1.439 \quad r_5^* = 1.297 \quad r_6^* = 1.233 \quad r_7^* = 1.187$$

Therefore any stability set of the k-step Adams-Nordsieck methods is contained in the interval $[0,r_k^*]$.

The main difficulty in the application of part ii) of the above theorem is the choice of a suitable matrix H so that the set

(3.2) $J = \{ r \in R^+ \; ; \; \| H^{-1} T(r) H \|_1 < 1 \}$

be as large as posible and contains $r = 1$. It must be remarked that any closed set contained in (3.2) is a stability set for the k-step method.

Next we shall describe the way of obtaining a good H matrix. Given $\bar{r} \in I_k$ such that $T(\bar{r})$ can be diagonalized we calculate the complex matrix $H(\bar{r})$ that reduces $T(\bar{r})$ to the diagonal form scaling its column vectors so that its l_1-norm is unity. We have taken values of $\bar{r} \leq r_k^*$ and computed the sets defined by (3.2). After extensive calculations, that have been carried out in a VAX 11/780 computer of Zaragoza University in double precision using some subroutines from the IMSL Library, the best values of $\bar{r}$ as well as the corresponding J_k sets are given in Table 1.

Note that in the case k=5, the value $\bar{r} = r_5^*$ permits to obtain $J = [0, r_5^*)$ which is an optimal interval. However this situation does not hold in the cases k=4 and k=6. On the other hand, if k=7 we have not been able to get an interval of type $[0,c]$ with $c \geq 1$ and in general we have stability sets $J_7 = [0,r_7') \cup (r_7'',r_7''')$ with $0 < r_7' < r_7'' < 1 < r_7'''$. In particular with the choice $\bar{r} = 1.17$ the set (3.2) becomes

$$J_7 = [0,0.257) \cup (0.9711,1.181)$$

The above stability results can be substantially improved if we make some aditional assumptions on the strategy of stepsize variation.

First, assume that the stepsize remains constant exactly for p steps after changing stepsize. This means that whenever $r_m \neq 1$, we have $r_{m+1} = \dots = r_{m+p-1} = 1$ and $r_{m+p} \neq 1$ and let $A_{k,p}$ the k-step Adams -

Nordsieck method with this strategy of stepsize variation. With these assumptions any product $S(r_r) \ldots S(r_m)$ $(0 < m \leq n)$ of propagation matrices can be written as

$$S^q C(r_{sp+1}) \ldots S^p C(r_{2p+1}) S^p C(r_{p+1}) S^p C(r_1)$$

where without loss of generality we have taken $m=1$, $r_1 \neq 1$ and $n=sp+q$.

Now the stability condition becomes

$$\| S^{(q)}(r_{s+1}) S^{(p)}(r_s) \ldots S^{(p)}(r_1) \| \leq K$$

where the stepsize ratios have been renamed putting r_{i+1} instead of r_{ip+1} and $S^{(p)}(r_{i+1}) = S^p C(r_{i+1})$.

It can be seen that matrices $S^{(p)}(r)$ retain the same form (2.3) as the matrix $S(r)$, therefore theorem 3.1 can also be applied to the $A_{k,p}$ methods. Putting $I_{k,p} = \{ r \in R^+ ; \rho(T^p \bar{C}(r)) \leq 1 \}$, numerical computations show that at least for $3 \leq k \leq 7$ and $p \leq k-2$ the sets $I_{k,p}$ are intervals $I_{k,p} = [0, r^*_{k,p}]$ whose upper bounds are given in Table 2 and furthermore $\rho(T^p \bar{C}(r)) < 1$ for $r \in [0, r^*_{k,p})$.

Next we intend to look for suitable constant matrices $H = H_{k,p}$ so that the sets

$$(3.3) \qquad J_{k,p} = \{ r \in R^+ ; \| H^{-1} T^p \bar{C}(r) H \|_1 < 1 \}$$

contain $r = 1$ and be as large as posible. Clearly any closed set contained in $J_{k,p}$ will be a stability set for the $A_{k,p}$ method.

Taking into account that the transformation $\bar{\Omega} T \bar{\Omega}^{-1}$ reduces the matrix T to its Jordan form, we have taken

$$(3.4) \qquad H_\delta = \bar{\Omega}^{-1} \Delta$$

with $\Delta = \text{diag}(1, \delta, \ldots, \delta^{k-2})$ and δ a parameter that will be chosen appropiately. In the following theorem we state that there exist values of δ so that the sets $J_{k,k-2}$ corresponding to H_δ are optimal.

<u>Theorem 3.2.</u> For all $k \geq 3$, there exist $\delta = \delta(k)$ so that with the matrix H_δ defined by (3.4) we have $J_{k,k-2} = [0, r^*_{k-2})$.

If $4 \leq k \leq 7$ and $2 \leq p \leq k-2$, using the matrix H_δ defined by (3.4) with a suitable value of δ it can be seen numerically that the sets $J_{k,p}$ are intervals of type $[0, \bar{r}_{k,p})$. The greatest values of $\bar{r}_{k,p}$ together with the corresponding values of δ have been collected in the

Table 3.

Consider now another strategy of stepsize variation that can be more interesting from a practical point of view. Suppose that the stepsize remains constant at least p steps after an effective change of it, and let $A'_{k,p}$ the k-step Adams-Nordsieck method with this strategy of stepsize variation. With these assumptions it is clear that if some $r_j \neq 1$, there is an integer $p_j \geq p$ with

$$r_{j+1} = \dots = r_{j+p_j-1} = 1 \quad \text{and} \quad r_{j+p_j} \neq 0.$$

Hence, an arbitrary product of propagation matrices can be written as

$$S^{p'} C(r'_{s+1}) S^{p_s} C(r'_s) \dots \quad S^{p_1} C(r'_1)$$

where $p_i \geq p$, $p' \geq 1$ and $r'_{i+1} = r'_i + p_i$, $r'_1 = r_1$.

Before to study the stability of the $A'_{k,p}$ methods we start giving the following general results:

<u>Theorem 3.3.</u> Let $\{r_m\}_{m\geq 1}$ be a bounded sequence of stepsize ratios. If there exist an index sequence $\{m_j\}$, a constant matrix H and an integer q such that for all $j \geq 1$

i) $1 \leq m_{j+1} - m_j \leq q$

ii) $\| H^{-1} T(r_{m_{j+1}-1}) T(r_{m_{j+1}-2}) \dots \quad T(r_{m_j}) H \|_1 \leq K < 1$

then the sequence $\{r_m\}$ is stable for the method (2.2).

As an inmediate consequence of this theorem we have:

<u>Corollary.</u> If there is a fixed integer $p \geq 0$ so that in any set of $p + (k-1) + (k-2)$ steps we can find at least $(k-1)$ consecutive steps with the same length, the sequence $\{r_j\}$ is stable for the k-step Adams Nordsieck method (2.2).

Proof: From these assumptions it is clear that there is an index sequence $\{m_j\}$ with $1 \leq k-1 \leq m_{j+1} - m_j \leq p+2k-3 = q$ and $r_{m_{j+1}-1} = \dots = r_{m_{j+1}-k+2} = 1$. Therefore

$$\| T\, \bar{C}(r_{m_{j+1}-1}) \dots \quad T\, \bar{C}(r_{m_j}) \| =$$

$$= \| T^{k-1} C(r_{m_{j+1}-k+1}) \dots \quad T\, \bar{C}(r_{m_j}) \| = 0$$

Hence, applying Th. 3.3 it follows that the method (2.2) is stable.

Remarks.

3.1 The hypothesis on the stepsize ratios given in the above Corollary can be made more transparent if the strategy is presented in the following way: With a fixed value of p, we can give p_1 ($\leq p$) arbitrary steps but after that we must have k_1 ($\geq k-1$) equal steps and then $p_2, k_2, p_3, k_3, \ldots$ in the same way.

3.2 The statement ii) of Theorem 3.1 can also be proved applying the Theorem 3.3 with the index sequence $\{m_j\}$ where $m_j = j$. Furthermore the generalization of this result to the methods $A_{k,p}$ can be obtained taking the index sequence so that

$$m_{j+1} - m_j = p$$

$$r_{m_{j+1}-1} = \ldots = r_{m_j+1} = 1.$$

Finally to analyze the stability of the $A'_{k,p}$ methods note that we can find an index sequence $\{m_j\}$ where

$$p \leq m_{j+1} - m_j \leq p + \min\ \{k-2,\ p-1\} = q$$

$$r_{m_j+1} = \ldots = r_{m_{j+1}-1} = 1$$

Then, in order to apply the theorem 3.3 we need again to find suitable H-matrices so that the sets

$$J'_{k,p} = \{ r \in R^+ \ ;\ \| H^{-1}\ T^{p_i}\ C(r)\ H \|_1 < 1,\ \forall p_i\ :\ p \leq p_i \leq q \} \tag{3.5}$$

contain $r = 1$ and be as large as posible.

An interesting property of sets (3.5) when they are defined using the matrices employed in (3.4) is given by

Theorem 3.4. Let $H = H_\delta = \bar{\Omega}\ \Delta^{-1}$ with $\Delta = \text{diag}\ (1, \delta, \ldots, \delta^{k-2})$, $0 < \delta \leq 1$. Then we have for all $p \geq 1$

i) $J_{k,p} \subset J_{k,p+1}$

ii) $J'_{k,p} = J_{k,p}$

Remark: let H_δ, $0 < \delta \leq 1$, δ fixed, be the matrix given by (3.4) and $J_{k,p}$ the sets given by (3.3) with $H = H_\delta$. From Theorems 3.3 and 3.4 it follows that a sequence of stepsizes $\{r_m\}$ such that

$r_j \neq 1, \quad r_j \in \bar{J}_{k,2} - J_{k,1} \implies r_{j+1} = 1$

$r_j \neq 1, \quad r_j \in \bar{J}_{k,3} - J_{k,2} \implies r_{j+1} = r_{j+2} = 1,$

....

$r_j \neq 1, \quad r_j \in \bar{J}_{k,k-2} - J_{k,k-3} \implies r_{j+1} = \dots = r_{j+k-3} = 1,$

$r_j \neq 1, \quad r_j \notin \bar{J}_{k,k-2} \implies r_{j+1} = \dots = r_{j+k-2} = 1,$

where $\bar{J}_{k,i}$ is a closed subset of $J_{k,i}$, is stable.

Tables:

1.

k = 4	$\bar{r}$ = 1.419	J_4 = [0,1.437)
k = 5	$\bar{r}$ = 1.297	J_5 = [0,1.297)
k = 6	$\bar{r}$ = 0.905	J_6 = [0,1.056)

2.

k/p	1	2	3	4	5
3	1.6956				
4	1.4391	2.5747			
5	1.2978	1.8605	3.5121		
6	1.2338	1.6082	2.3280	4.4735	
7	1.1879	1.3157	1.9164	2.8192	5.4470

3.

k/p	2	3	4	5
4	$\delta \leq 0.4$ $\bar{r}_{4,2} = r^*_{4,2}$			
5	$\delta = 0.39$ $\bar{r}_{5,2} = 1.856$	$\delta \leq 0.25$ $\bar{r}_{5,3} = r^*_{5,3}$		
6	$\delta = 0.27$ $\bar{r}_{6,2} = 1.550$	$\delta = 0.31$ $\bar{r}_{6,3} = 2.324$	$\delta \leq 0.15$ $\bar{r}_{6,4} = r^*_{6,4}$	
7	$\delta = 0.38$ $\bar{r}_{7,2} = 1.210$	$\delta = 0.27$ $\bar{r}_{7,3} = 1.910$	$\delta = 0.24$ $\bar{r}_{7,4} = 2.810$	$\delta \leq 0.13$ $\bar{r}_{7,5} = r^*_{7,5}$

REFERENCES

1. M. CALVO, F.J. LISBONA, J.I. MONTIJANO. Estabilidad del método Adams-Nordsieck de tres pasos con paso variable. VIII CEDYA. Santander (1985).

2. M. CALVO, F.J. LISBONA, J.I. MONTIJANO. On the stability of variable stepsize BDF-Nordsieck methods. Peprint submmited to SIAM J. Num. Anal.
3. M. CALVO, F.J. LISBONA, J.I. MONTIJANO. On the stability of Adams-Nordsieck methods. Tech. Report. Dep. Ecuac. Funcionales. Universidad de Zaragoza (1985).
4. M. CROUZEIX, F.J. LISBONA. The convergence of variable-stepsize, variable-formula, multistep methods. SIAM J. Num. Anal. 21, pp. 512-534 (1984).
5. C.W. GEAR, K.W. TU. The effect of variable meshsize on the stability of multistep methods. SIAM J. Num. Anal. 11, pp. 1025-1043 (1974).
6. C.W. GEAR, D.S. WATANABE. Stability and convergence of variable order multistep methods. SIAM J. Num. Anal. 11, pp. 1043-1058 (1974).
7. C.W. GEAR. Algorithm 407, DIFSUB for solution of ordinary differential equations. Comm. ACM, 14, pp. 185-190 (1971).
8. R.D. GRIGORIEFF. Stability of multistep-methods on variable grids. Numer. Math. 42, pp. 359-377 (1983).
9. A.C. HINDMARSH. GEAR: Ordinary differential equations system solver. Lawrence Livermore Laboratory, Report UCID-30001, Revision 3 (1974).
10. A. NORDSIECK. On the numerical integration of ordinary differential equations. Math. Comp. 16, pp. 22-49 (1962).
11. P. PIOTROWSKI. Stability, consistency and convergence of variable k-step methods for numerical integration of large system of ordinary differential equations. Conference on the Numerical Solution of Differential equations. J.L. Morris, ed., Lecture Notes in Math., 109, Springer Verlag, Berlin (1969).
12. L.F. SHAMPINE, M.K. GORDON. Computer solution of ordinary differential equations; the initial value problem. W.H. Freeman and Company. San Francisco (1975).
13. R.D. SKEEK, L.W. JACKSON. The stability of variable stepsize Nordsieck methods.SIAM J. Numer. Anal. 20, pp. 840-853 (1983).
14. Z.ZLATEV. Stability properties of variable stepsize, variable formula methods. Numer. Math. 32, pp. 175-182 (1978).

$$p(r) = \rho(T(r))$$

$$q(r) = \| H(\bar{r})^{-1} T(r) H(\bar{r}) \|_1$$

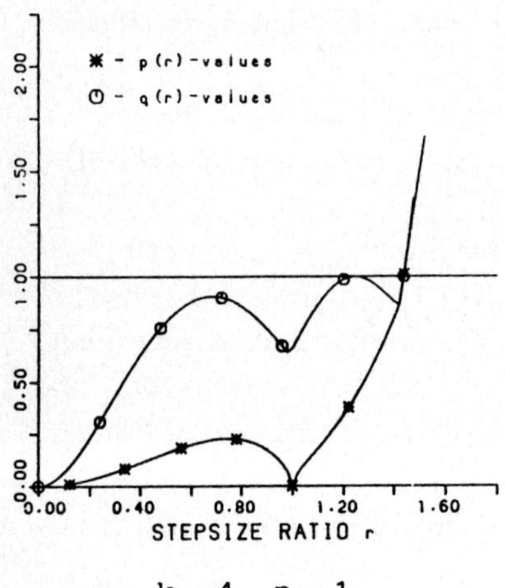

k = 4   p = 1

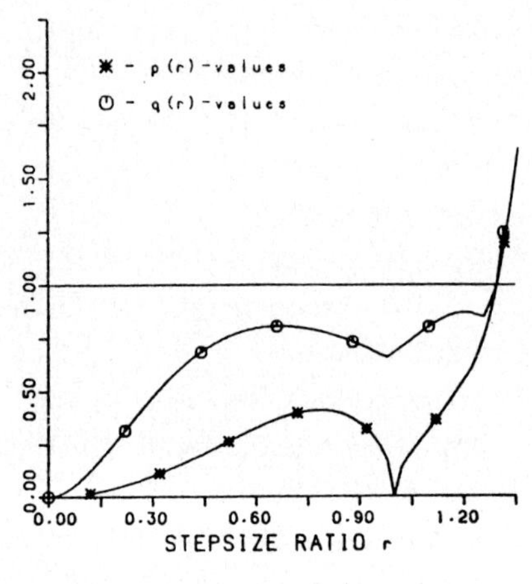

k = 5   p = 1

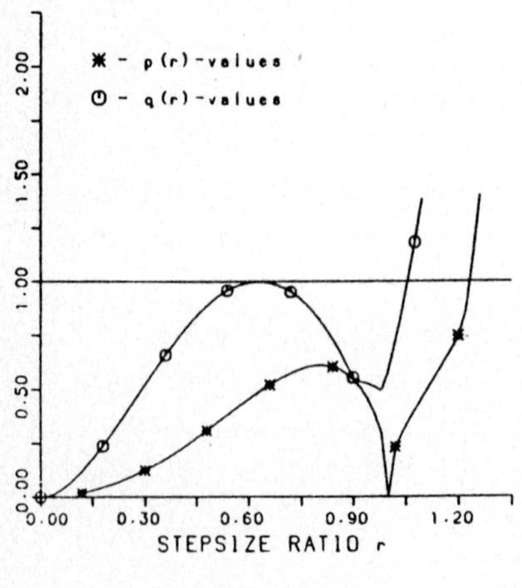

k = 6   p = 1

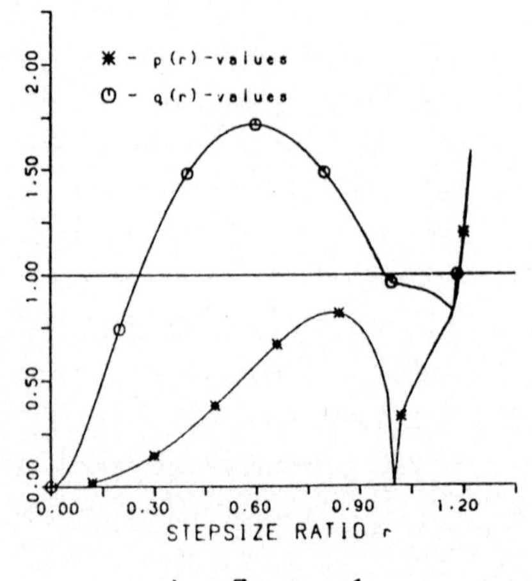

k = 7   p = 1

$$p(r) = \rho(T(r))$$

$$q(r) = \| H_\delta^{-1} T(r) H_\delta \|_1$$

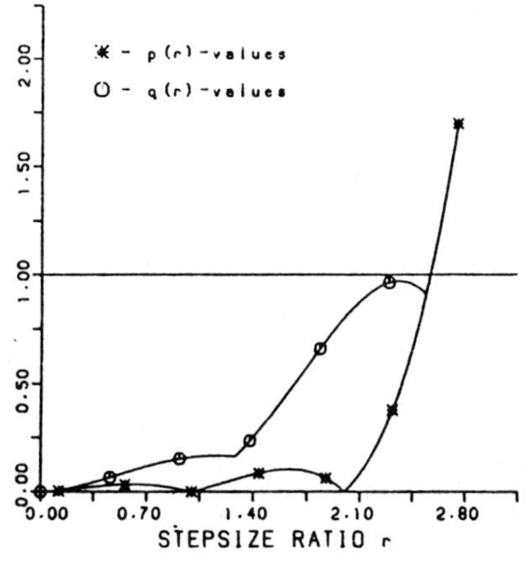

k = 4 p = 2

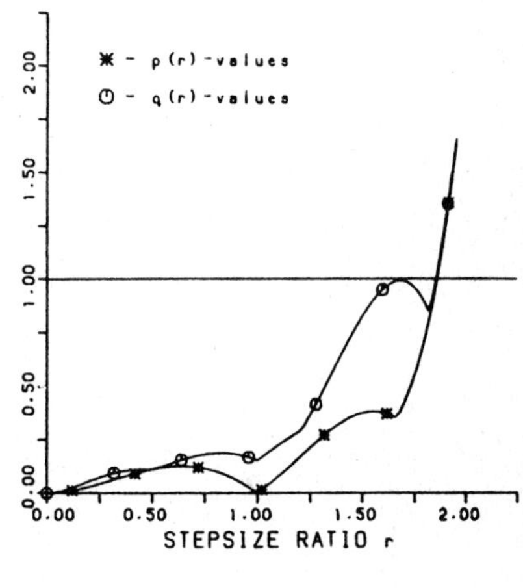

k = 5 p = 2

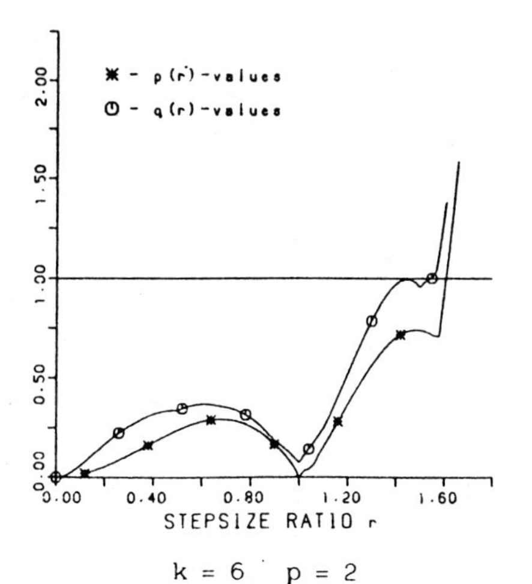

k = 6 p = 2

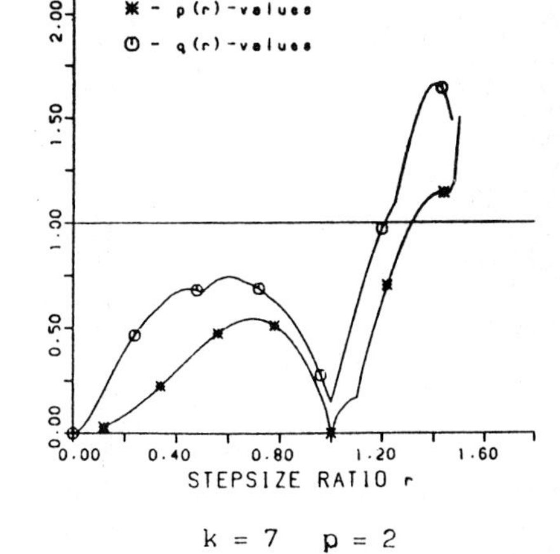

k = 7 p = 2

Numerical Approximation of Partial Differential Equations
E.L. Ortiz (Editor)

THE PRECONDITIONED CONJUGATE GRADIENT METHOD FOR SOLVING ELLIPTIC DIFFERENCE EQUATIONS

D.J. Evans & I.C. Demetriou
Department of Computer Studies
Loughborough University of Technology
Loughborough, Leicestershire,
U.K.

The preconditioned conjugate gradient method for the solution of the large linear systems arising from the discretisation of elliptic partial differential equations requires the pre-determination of an optimal preconditioning iteration parameter to achieve its maximum rate of convergence. However, the determination of this parameter is complex and usually involves the á prìori knowledge of the spectral bounds of some matrices associated with the coefficient matrix A.

In this paper it is shown that the optimal parameter for the model problem can be obtained in $O(N^2)$ sine function evaluations where 1/N is the discretisation mesh size and can be derived in a stable manner, if it is allowed to vary with the iteration index from information derived from the gradient parameters.

1. INTRODUCTION

The optimal use of an algorithm that determines the solution of a linear system of equations arising from the discretization of an elliptic differential operator equation, often relies on a set of acceleration parameters. These parameters usually depend on the spectral properties of the iteration matrix and characterize fast iterative methods. Estimating these optimal parameters may sometimes be more expensive than that of determining the solution of the original problem. However, if one resorts to variational or gradient techniques a self-determining process to the optimal state takes place. For this reason and for its rapid convergence properties a Conjugate Gradient method is one of the most attractive methods for solving linear systems, despite the fact that more work per iteration is required than other similar methods. Although, in general gradient methods do not require the pre-determination of any iteration parameters this is not the case for the Preconditioned Conjugate Gradient (PCG) and the Symmetric Successive Overrelaxation - CG (SSOR-CG) methods where one parameter, i.e. the preconditioning parameter is estimated that usually remains fixed during the course of the iteration process. The calculation of an optimal value of this parameter involves spectral bounds on some matrices which can be obtained in $O(N^2)$ functional evaluations (i.e. sine, cosine, etc.) given that $N=1/h$ where h is the mesh size of the discretization [vd. Section 2]. However it can be shown that this parameter can be chosen in $O(1)$ operations per iteration if it varies with the iteration index. Furthermore the parameters are determined in a stable manner because of the implicit use of the spectral properties of the iteration matrix [vd. Section 3]. Some numerical results are included to illustrate the numerical procedures involved.

2. A GENERALISED NON-STATIONARY PRECONDITIONED ITERATIVE SCHEME

Consider the boundary value problem of a second order self-adjoint equation,

$$L(x) \equiv -\frac{\partial}{\partial x_1}\left(a(x)\frac{\partial u}{\partial x_1}\right) - \frac{\partial}{\partial x_2}\left(b(x)\frac{\partial u}{\partial x_2}\right) = q(x) \quad , \; x \in D, \tag{2.1}$$

$$u(x) = g(x), \; x \in \partial D,$$

where $x=(x_1,x_2)$, D is the interior of a compact region with boundary ∂D, $g(x)$ is a continuous function on the boundary ∂D, $q(x)$ is a continuous function in D. The differential operator $L(x)$ is elliptic, i.e. $a(x)$ and $b(x)$ are strictly positive on D. We seek a function $u(x)$ which is continuous and twice differentiable in D and satisfies both relations (2.1). Assume that D lies in the first quadrant. We construct a net by covering D with the $x_1=ih$, $i=1,2,\ldots,N$ parallel lines to x_1 axis and with the $x_2=ih$, $i=1,2,\ldots,N$ parallel lines to x_2 axis, where h is the net mesh size usually <0.1.

We approximate the operator $L(x)$ with the usual five point difference operator. A system of linear equations,

$$Au = b \; , \tag{2.2}$$

results where A is a symmetric positive definite matrix with positive diagonal elements. Let

$$A=D(I-B), \quad B=L+U, \quad \text{where } D=\text{diag}(A) \ , \tag{2.3}$$

L and U are the lower and upper parts respectively of the symmetric matrix B.

Let B_j, $j\geq 0$, be the easily invertible matrix

$$B_j = D(I-\omega_j L)(I-\omega_j U) \ , \tag{2.4}$$

where we let the preconditioning parameter ω_j be $1\leq\omega_j<2$. For an approximate solution to problem (2.2) we consider a generalized non-stationary preconditioned iterative scheme (Evans [1968]) with an arbitrary initial vector $u^{(0)}$, i.e., the second degree form with varying parameters ρ_j and γ_j,

$$u^{(j+1)} = \rho_{j+1}[u^{(j)}+\gamma_j B_j^{-1}(b-Au^{(j)})]+(1-\rho_{j+1})u^{(j-1)} \ . \tag{2.5}$$

Notice that B_j is a positive definite matrix for any ω_j, since it is similar to the positive definite $B_j=[D^{\frac{1}{2}}(I-\omega_j L)][D^{\frac{1}{2}}(I-\omega_j U)]$. In addition (Evans [1968]) the positive definiteness of A implies the positive definiteness of $B_j^{-1}A$ and vice versa.

The (2.4) form of B_j has received considerable attention recently in relation to the Preconditioned and SSOR iterative methods. Many of the algorithms referring to the schemes (2.5) with (2.4) assume that the spectrum of $B_j^{-1}A$ is known (Evans [1968-1981], Young [1977]). One then has to calculate this spectrum and consequently the choice of the parameter ω_j in (2.4) is very important for the stable and fast application of these algorithms.

We now consider the problem of choosing the parameters ω_j, and hence B_j, in each iteration, by assuming no knowledge of the spectral properties of the matrices A and B_j. Thus, at the jth-step we calculate the parameters τ_j or (ρ_j,γ_j) for a fixed ω_j by using gradient techniques. Then, at the (j+1)th-step we choose ω_{j+1} by requiring the convergence of

(2.5) where now both of these schemes are considered as stationary 1st or 2nd degree methods with fixed iteration parameters $\tau=\tau_j$ or $\rho=\rho_j$, $\gamma=\gamma_j$ respectively, the parameters having been previously calculated in the jth-step. The computational cost for the estimation of ω_j amounts to a total of one addition, subtraction and square root operation per iteration while the requirements for determining ω_j is $O(N^2)$ cosine and sine calculations. Thus, the superiority of the above algorithm for the determination of ω_j for the gradient and variational methods is apparent.

We restrict our attention to only three methods for calculating the preconditioning parameter ω_j in order to show the need for an economical method when gradient techniques accelerate the scheme (2.5).

Let us define the spectral condition number of an invertible matrix T, P(T), say, to be the ratio of the moduli of its two extreme eigenvalues. By the fact that the spectral condition $P(B_j^{-1}A(\omega))$ of the matrix $B_j^{-1}A(\omega)$ is a unimodal function of ω, $1\leq\omega<2$, one way for calculating an optimal ω consists of a simple one dimensional optimization method combined with the power method to determine the optimum value of the spectral number at the optimum $\omega=\omega_0$. However, this procedure may be more expensive than the one for solving the whole problem (2.2) and thus it becomes uneconomical to be implemented for the optimal evaluation of ω. The second technique for estimating the ω relies on adaptive methods that update the acceleration parameters (Benokraitis [1974]). The procedure requires several inner products to be calculated due to the process and seems to suit better the semi-iterative acceleration of (2.5). The third technique permits ω to take an optimal value $\omega=\omega_1$ where ω_1 is the minimizer of $P[B_j^{-1}A(\omega)]=M[B_j^{-1}A(\omega)]/m[B_j^{-1}A(\omega)]$ and where $M[B_j^{-1}A(\omega)]$ and $m[B_j^{-1}A(\omega)]$ are estimated upper and lower respectively bounds to the eigenvalues of the matrix $B_j^{-1}A$. This ω_1 value of ω is calculated once and may be used fixed for all the iterations. To calculate ω_1 one needs an upper bound, M(B) say, on the largest eigenvalue of the matrix B (this matrix has been defined after (2.2)) and an upper bound on the spectral radius S(LU), of the matrix LU (this matrix has also been defined after (2.2). The former can be obtained by Young [1976],

$$M(B)\leq 1-\frac{2\underline{a}\sin^2\pi/2I+2\underline{b}\sin^2\pi/2J}{\frac{1}{2}(\underline{a}+\bar{a})+\frac{1}{2}(\underline{b}+\bar{b})+\frac{1}{2}(\bar{a}-\underline{a})\cos\frac{\pi}{I}+\frac{1}{2}(\bar{b}-\underline{b})\cos\frac{\pi}{J}}=M$$

Here, $0<\underline{a}\leq a(x)\leq\bar{a}$, $0<\underline{b}\leq b(x)\leq\bar{b}$ for the coefficients of (2.1) in the rectangle $0\leq x_1\leq Ih$, $0\leq x_2\leq Jh$, where I and J are integers. Now then the calculation of M(B) requires $O(N^2)$ sine and cosine evaluations plus the additional $O(N^2)$ operations required to calculate the bounds $\underline{a}$, $\bar{a}$, $\underline{b}$, $\bar{b}$ and the bound M, to M(B).

An upper bound on the spectral radius S(LU) can be accurately estimated by the infinity norm $||LU||_\infty$, whose computation is also of order $O(N^2)$ operations (Benokraitis [1974]),

$$S(LU)\leq ||LU||_\infty = b.$$

Then, the estimated ω_1 is given by the formula,

$$\omega_1=\begin{cases}\dfrac{2}{1+\sqrt{1-2M+4b}}, & \text{if } M\leq 4b\\[2ex] \dfrac{2}{1+\sqrt{1-4b}}, & \text{if } M\geq 4b.\end{cases}$$

Since (Young [1977]), $M=1-O(h^2)$ and $b\leq\frac{1}{4}+O(h^2)$, the asymptotic form of ω_1 is $\omega_1=2-O(h)$ whereas $P(B_j^{-1}A)$ asymptotically varies like $h^{-1}O(1)$.

In summary to implement a stationary (i.e. with a fixed parameter $\omega=\omega_1$) Preconditioned or SSOR algorithm, the calculation of ω_1 requires an extra procedure of complexity $O(N^2)$ functional evaluations that is sine, cosine and the ones relevant to the form of a(x) and b(x) in (2.1), and this, to our knowledge, seems to be the cheapest procedure.

Even if one accelerates the scheme (2.5) by the Conjugate Gradient analogue respectively, the determination of the preconditioning parameter ω is still required. Gradient algorithms are adaptive to the optimal as they are implemented and are characterized by fast convergence in the initial iterations. In addition, not only every vector $u^{(j)}$ may be attained as a starting vector for the subsequent iterations but also there is a well known fact that the parameters ω_j and τ_j or (ρ_j,γ_j) are weakly depended functions of the iteration index j (vd. Marchuk, [1971]). This suggests that if one meets some requirements for an optimal ω_j, one can let ω_j change from iteration to iteration and its choice can be based on the derived information from the gradient parameters.

3. ADAPTIVE DETERMINATION OF THE PRECONDITIONING PARAMETER ω IN THE PRECONDITIONED CONJUGATE GRADIENT METHOD (PCG METHOD)

In this section, we shall consider the acceleration of (2.5) by a variant of the well known conjugate gradient techniques, (Evans 1973). Again, no spectral properties of the matrices A and B_j are assumed or required. A CG variant method converges at least as fast as the semi-iteration variant of the preconditioned methods in the sense that the ratio $||\varepsilon^{(j)}||_{A^{\frac{1}{2}}} \div ||\varepsilon^{(0)}||_{A^{\frac{1}{2}}}$ is less for the conjugate gradient method than for the semi-iterative method, where $||\varepsilon||_{A^{\frac{1}{2}}} = ||A^{\frac{1}{2}}\boldsymbol{\varepsilon}||$. Assume again that A is a symmetric positive definite matrix and the Preconditioned Conjugate Gradient method is defined in the form (vd. (2.5)),

$$u^{(j+1)} = \rho_{j+1}(u^{(j)}+\gamma_{j+1}\delta^{(j)})+(1-\rho_{j+1})u^{(j-1)}, \quad j=0,1,\ldots \tag{3.1}$$

where,

$$\delta^{(j)} = B_j^{-1}(b-Au^{(j)}) = B_j^{-1}r^{(j)}, \quad j=0,1,\ldots \tag{3.2}$$

$$\gamma_{j+1} = \frac{\langle\delta^{(j)},B_j\delta^{(j)}\rangle}{\langle\delta^{(j)},A\delta^{(j)}\rangle} = \frac{\langle\delta^{(j)},r^{(j)}\rangle}{\langle\delta^{(j)},A\delta^{(j)}\rangle}, \tag{3.3}$$

with,

$$\begin{cases} \rho_1 = 1 \\ \rho_{j+1} = \left[1-\dfrac{\gamma_{j+1}}{\gamma_j}\cdot\dfrac{\langle r^j,r^j\rangle}{\langle r^{j-1},r^{j-1}\rangle}\cdot\dfrac{1}{\rho_j}\right]^{-1}, \quad j=1,2,\ldots \end{cases} \tag{3.4}$$

and where the matrix B_j has been defined by (2.4). The above scheme of the conjugate gradient is in a slightly different form than that in Young (1975) defining the SSOR-CG method. However, the matrix B_j is allowed to vary in each iteration since the preconditioning parameter ω_j varies with the iteration index j. That is, having calculated ρ_j and γ_j at the jth step of the PCG method and by following the algorithm outlined in the beginning of Section 2, ω_{j+1} is chosen by requiring the convergence of the stationary second degree method (2.5) with $\rho=\rho_j$ and $\gamma=\gamma_j$. Then, the vector $u^{(j)}$ is attained as the starting vector for the next CG iteration with $\omega=\omega_{j+1}$.

Now for a fixed j, set $\rho=\rho_j$, $\gamma=\gamma_j$ and $B=B_j$. Then method (3.1) turns out to be the Preconditioned second order Richardson method (2.5). Let u be the solution of the system (2.2) and let,

$$\varepsilon^{(j)} = u^{(j)}-u = \sum_k \varepsilon_k^{(j)} \phi_k , \tag{3.5}$$

be the expansion of the error in the jth iterate with respect to the basis vectors $\{\phi_k\}$ of the positive definite matrix $B^{-1}A$. Then, method (2.5) by means of (3.5) implies that,

$$\sum_k [\varepsilon_k^{(j)} - \rho(1-\gamma\mu_k)\varepsilon_k^{(j-1)} - (1-\rho)\varepsilon_k^{(j-2)}]\phi_k = 0 , \tag{3.6}$$

where $B^{-1}A\phi_k=\mu_k\phi_k$, μ_k being the eigenvalues of the matrix $B^{-1}A$. The linear independence of ϕ_k yields,

$$\varepsilon_k^{(j)} - \rho(1-\gamma\mu_k)\varepsilon_k^{(j-1)} - (1-\rho)\varepsilon_k^{(j-2)} = 0 ,$$

with characteristic equation,

$$\lambda^2 - \rho(1-\gamma\mu_k)\lambda - (1-\rho) = 0 , \tag{3.7}$$

for every component k of the error vector. By considering (3.7) we derive in the next Theorem (3.8) necessary and sufficient conditions for the convergence of the Preconditioned second order (2.5) method.

Theorem (3.8):

Consider the second order Richardson method (2.5) preconditioned by the matrix $B=B_j$, $1\leq\omega_j<2$, where B is defined by (2.4) for a fixed j.

Then, the method (2.5) converges if and only if,

$$0<\rho<2, \text{ and } 0<\gamma<2/\mu_k \text{ , for every } k.$$

Proof:

The method (2.5) converges if and only if

$$\max\{|\lambda_{1k}|,|\lambda_{2k}|\}<1 \text{ for every } k, \qquad (3.10)$$

where λ_{1k} and λ_{2k} are the roots of the kth characteristic equation (3.7). Condition (3.10) is equivalent to (vd. for example p.171, Young, 1971),

$$|\rho-1|<1 \text{ and } |\rho(1-\gamma\mu_k)|<\rho \text{ for every } k.$$

Thus, the latter inequalities imply (3.9).

Provided now that $1\leq\omega_j<2$, an upper bound is derived in the next Lemma (3.11), on any eigenvalue μ_k of $B^{-1}A$.

Lemma (3.11):

Let A be a symmetric, positive definite matrix. Let $1\leq\omega<2$ and let $\gamma=\gamma_j$ as defined by (3.3) for a fixed j. Then,

$$\gamma \geq \omega(2-\omega).$$

Proof:

Now the matrix B is written as

$$B = \omega(2-\omega)A+N ,$$

where $N=D[(1-\omega)I+\omega L][(1-\omega)I+\omega U]$. Now, since from its definition N is a positive semi-definite matrix and A is a positive definite one, the matrix NA^{-1} is positive semi-definite too. Thus, $<\phi,NA^{-1}\phi>\geq 0$, for every non-zero vector ϕ.

Hence, and by (3.3) the parameter γ is written,

$$\gamma = \frac{<\delta,B\delta>}{<\delta,A\delta>} = \frac{\omega(2-\omega)<\delta,A\delta>+<\delta,N\delta>}{<\delta A,\delta>} \geq \omega(2-\omega) .$$

In view of this Lemma we have an explicit bound on the maximum eigenvalue of the matrix $B^{-1}A$, in terms of the parameter ω, $1\leq\omega<2$,

$$\max_k\{\mu_k\} \leq \frac{1}{\omega(2-\omega)} , \qquad (3.12)$$

This is an important result because it involves spectral properties and will be used in the next Theorem (3.13) in order to proceed to an optimal characterisation of ω_{j+1}, by means of the quantities ρ_j and γ_j.

Theorem (3.13):

Let for some fixed $j \geq 1$ the parameters $\rho = \rho_j \geq 1$ and $\gamma = \gamma_j$ of the PCG method (3.4) satisfy the relations,

$$1 \leq \rho < 2 \text{ and } 0 < \gamma < 2 \ . \tag{3.14}$$

If

$$1 \leq \omega < 1 + \sqrt{1-\gamma/2} \ , \tag{3.15}$$

then the preconditioned second degree method (2.5) converges.

Proof:

By (3.15) and the second equation of (3.14) we have,

$$0 < \gamma < 2\omega(2-\omega) < 2 \ ,$$

since the maximum of the function $f(\omega)-\omega(2-\omega)$ occurs at $\omega=1$. Given (3.12) we conclude that $0 < \gamma < 2/\mu_k$ for any k. Now the first of (3.14) and the latter inequalities satisfy Theorem (3.5). Thus the required result follows.

Relation (3.15) defines the range of the parameters ω depending on the current spectral information as it is expressed by γ_j. Thus, an effective method for choosing ω_{j+1} is given by,

$$\omega_{j+1} = 1 + \sqrt{1-\gamma_j/2}, \ 0 < \gamma_j < 2.$$

However, as noted by Young [1975] the choice of ω does not seem to be very critical for the convergence speed of methods similar to (3.1). This is also the case for the PCG method with respect to the choice of an optimal sequence $\{\omega_j\}$ and the numerical results for the Example I in Table 1 demonstrate this point.

ω	1.00	1.10	1.20	1.30	1.40	1.50	1.55	1.60	1.65	1.70	1.75	1.80	1.85	1.90	1.95
h=1/20	17	16	16	15	14	14	14	14	14	13	14	15	15	15	16
h=1/40	32	29	27	25	23	21	21	20	20	19	18	20	20	21	22

TABLE 1

Example 1: Number of iterations when ω takes values in the range $1 \leq \omega < 2$.

The Preconditioned CG method was applied with a starting guess vector

$u^{(0)}=1$ and the procedure was terminated when the inequality $||r^{(j)}||_\infty \leq 10^{-6}$ was satisfied. The experiments were carried out involving (1.1) with zero boundary conditions, whereas the coefficients $a(x_1,x_2)$ and $b(x_1,x_2)$ used, can be found in Table 2.

Example	
1	$a=b=1$
2	$a=b=\begin{cases}1+x_1, & 0\leq x_1\leq\frac{1}{2}\\ 2-x_1, & \frac{1}{2}\leq x_1\leq 1\end{cases}$
3	$a=1+4\lvert x_1-\frac{1}{2}\rvert^2$ $b=\begin{cases}1, & 0\leq x_1<\frac{1}{2}\\ 9, & \frac{1}{2}\leq x_1\leq 1\end{cases}$

TABLE 2

Now the notation of n_{ω_1} indicates the number of iterations of each example with the preconditioning parameter $\omega=\omega_1$-est computed from (2.8) (vd. Evans and Demetriou, (1983)); n_c, with ω_j to be the average value of the first 4 (resp. 6) initial evaluations of ω, if h=1/20 (resp. 1/40) and then continuing with that value of ω for $j>4$ (resp. 6); and n_a with ω_a being the final value of the sequence produced by

$$\omega_{j+1} = \sum_{\alpha=1}^{j} \omega_\alpha / j\ ,$$

where,

$$\omega_\alpha = \begin{cases} 1+\sqrt{1-\tau_j}/2, & \text{if } 0<\tau_j<2,\\ \omega_j, & \text{otherwise.}\end{cases}$$

the following results in Table 3 were obtained. Now if $1\leq\rho_j<2$ and $0<\gamma_j<2$ then $\{\omega_j\}$, $j>1$, is calculated in $O(1)$ operations by means of the formulae,

$$\omega_1 = 1.6180$$

$$\omega_{j+1} = \begin{cases}\omega_a, & \text{if } \omega_\alpha \in [\omega_0,\omega_M]\\ \omega_j, & \text{otherwise.}\end{cases} \tag{3.16}$$

or

$$\omega_{j+1} = \sum_{\alpha=1}^{j} \omega_\alpha / j\ , \tag{3.17}$$

where,

$$\omega_\alpha = \begin{cases} 1+\sqrt{1-\gamma_j/2}, & \text{if } 0<\gamma_j<2 \\ \omega_j, & \text{otherwise.} \end{cases}$$

One more example 4, was considered with

$$a(x) = \frac{1}{1+2x_1^2+x_2^2} \text{ and } b(x) = \frac{1}{1+x_1^2+2x_2^2} .$$

In Table 3, the smallest (except for the initial value ω_1=1.6180) and the largest value of the sequence $\{\omega_j\}$, j>1 are presented.

We note that the frequencies of the ω observed occur in intervals where the ends are of the form 2-O(h), with ω_f indicating the values of ω varying according to the formulae (3.16) and by ω_a the values of ω varying according to (3.17).

Example	h^{-1}	ω_1	n_{ω_1}	ω_f,formula(3.16)	n_{ω_f}	ω_a,formula(3.17)	n_{ω_a}
1	20	1.7288	14	[1.7207,1.8577]	17	[1.7014,1.7868]	16
	40	1.8544	20	[1.7058,1.9595]	23	[1.7017,1.8003]	22
	60	1.8992	25	[1.7095,1.8604]	28	[1.7017,1.7931]	28
2	20	1.7241	15	[1.7465,1.8626]	17	[1.7019,1.7876]	17
	40	1.8790	20	[1.7087,1.8648]	23	[1.7020,1.7940]	22
	60	1.9126	27	[1.7096,1.8975]	29	[1.7020,1.7982]	28
3	20	1.8782	19	[1.6604,1.8535]	17	[1.7105,1.7897]	17
	40	1.9357	27	[1.6627,1.8805]	23	[1.7058,1.7905]	22
	60	1.9521	30	[1.6724,1.8628]	30	[1.7063,1.7924]	36
4	20	1.8283	17	[1.7323,1.8655]	17	[1.7001,1.8047]	17
	40	1.9105	23	[1.7234,1.8648]	26	[1.7071,1.8031]	25
	60	1.9543	32	[1.7099,1.8780]	35	[1.7065,1.7991]	36

TABLE 3

The numerical results of Table 3 show that the number of iterations for each example for varying ω is quite similar to the number of iterations with constant $\omega=\omega_1$. For these examples we have obtained $O(h^{\frac{1}{2}})$ rate of convergence with the PCG method. A $O(h^{\frac{1}{2}})$ convergence rate has also been obtained even when the coefficient functions were not restricted to the class $C^{(2)}$. As indicated in Table 3, allowing ω to vary is also applicable to problems with certain discontinuities amongst the coefficients as one can see in Examples 2 and 3. The choice of ω_j, $j\geq 1$ does not seem to have been affected by the mesh size h and the rate of convergence of the PCG method appears to be not particularly sensitive to the choice of the preconditioning parameter. Also for both choices of ω from formula (3.16) and (3.17) the results of Table 3 were equivalent.

In Table 4 we see the typical behaviour of the sequence $\{\omega_j\}$ which has been produced by computing the solution of Example 4. One half of the table contains the values of ω with formula (3.16) while the other half by the formula (3.17). In the latter case, the average value of ω's were calculated in each iteration and takes into account the spectral tendencies which occur in γ_j, $j \geq 1$.

4. DISCUSSION

It has been shown in this work that we can implement a PCG or SSOR-CG method where even the parameter ω is self-determined during the course of the iteration, by taking into account the spectral information obtained by the CG parameters. The derivation of ω's is obtained in O(1) operations per iteration and this is incomparably more economical than the $O(N^2)$ sine evaluations necessary to give a good estimation of ω in the normal case.

Example 4, h=1/40, ω_j varying according to formulae (3.16)							
j	ω	j cont.	ω_j cont.	j cont.	ω_j cont.	j cont.	ω_j cont.
1	1.6180	8	1.8613	15	1.8573	22	1.8475
2	1.7949	9	1.7688	16	1.8004	23	1.8433
3	1.7335	10	1.7234	17	1.8461	24	1.7913
4	1.6829	11	1.7706	18	1.7862	25	1.8531
5	1.8648	12	1.8273	19	1.8515	26	1.8004
6	1.8050	13	1.8251	20	1.7938		
7	1.8128	14	1.7674	21	1.7772		
Example 4, h=1/40, $\omega_{j+1} = \sum_{\alpha=1}^{1} \omega_\alpha/j, \omega_\alpha$ according to formulae (3.17)							
j	ω_j	j cont.	ω_j cont.	j cont.	ω_j cont.	j cont.	ω_j cont.
1	1.6180	8	1.7665	15	1.7972	22	1.8000
2	1.7064	9	1.7749	16	1.7950	23	1.7977
3	1.7070	10	1.7810	17	1.7983	24	1.7947
4	1.7070	11	1.7863	18	1.8017	25	1.7918
5	1.7369	12	1.7919	19	1.7981		
6	1.7582	13	1.7966	20	1.8004		
7	1.7728	14	1.8004	21	1.8031		

TABLE 4

REFERENCES

1. Benokraitis, V.J., On the adaptive acceleration of symmetric successive overrelaxation, Ph.D. Thesis, Univ. of Texas at Austin, U.S.A., 1974.
2. Evans, D.J., The use of preconditioning in iterative methods for solving linear equations with symmetric positive definite matrices, J.I.M.A., 4, 295-314, 1968.
3. Evans, D.J., The analysis and application of sparse matrix algorithms in the finite element method, Conf.Proc, 'The Mathematics of Finite Elements and Its Applications', edit. J.R. Whiteman, 427-447, Acad. Press (presented at the MAFELAP Conf., April 1972), 1973.
4. Evans, D.J., On preconditioned iterative methods for elliptic partial differential equations, in 'Elliptic Problem Solvers', ed. M.H. Schultz, 261-269, Acad.Press, 1981.
5. Evans, D.J. and I.C. Demetriou, The preconditioning by direct factorization method for solving self-adjoint partial differential equations, in 'Preconditioning Methods for Analysis and Application', 355-378, ed. D.J. Evans, Gordon & Breach, 1983.
6. Marchuk, G.I., On the theory of the splitting-up method, in 'Numerical Solutions of Partial Differential Equations II', SYNSPADE 170, NY, Acad.Press, 1971.
7. Young, D.M., Iterative solution of large linear systems, Acad. Press, 1971.
8. Young, D.M., Iterative solution of linear systems arising from finite element techniques, in 'Finite Elements and Applications', Acad.Press, 1975.
9. Young, D.M., On the accelerated SSOR method for solving large linear systems, Adv. in Mathematics, 23, 215-271, 1977.

Numerical Approximation of Partial Differential Equations
E.L. Ortiz (Editor)

Accelerated Monotone Scheme for Finite Difference Equations Concerning Steady–State Prey–Predator Interactions

Anthony W. Leung
and
Diego A. Murio

Department of Mathematical Sciences
University of Cincinnati
Cincinnati, Ohio 45221, U.S.A.

In previous work, by adapting a suitable finite difference method to a particular monotone scheme, the authors and A. Lazer have studied the numerical solution of a system of semilinear elliptic partial differential equations which determines the equilibria of the Volterra– Lotka equations describing prey–predator interactions with diffusion. In this paper, in order to improve the efficiency of the method, we show how Newton's method can be succesfully combined with the previous scheme to greatly accelerate the convergence. In some particularly "difficult" problems, the new method reduces the average number of iterations necessary to generate each element of the monotone sequences from 15 to about 3.

1. Introduction.

In this article we present a computationally efficient algorithm for the numerical calculations of the steady states of prey–predator species by means of finite differences, a suitable monotone scheme and Newton's method.

The solutions of the discrete model tend to the solutions of the original reaction–diffusion model, given by

$$\begin{aligned} &u''(x) + u(x)[a - bu(x) - cv(x)]=0 \\ &v''(x) + v(x)[e + fu(x) - gv(x)]=0 \\ &\alpha<x<\beta \\ &u(\alpha)=u(\beta)=v(\alpha)=v(\beta)=0, \end{aligned} \tag{1.1}$$

where a, b, c, e, f and g are positive constants satisfying:

$$cf<gb, \tag{A1}$$

$$e>\lambda_1=\pi^2/(\beta-\alpha)^2, \tag{A2}$$

$$a>gb(\lambda_1 + ce/g)/(gb-cf) \tag{A3}$$

Here $\lambda=\lambda_1$ is the first eigenvalue for the boundary value problem $w''(x) + \lambda w(x)=0$, $\alpha<x<\beta$, $w(\alpha)=w(\beta)=0$. It is shown in [4] that under conditions (A1–A3) there exist functions u^*, u_*, v^* and v_* so that all positive solutions (u,v) of (1.1) will satisfy $u_*\leq u\leq u^*$, $v_*\leq v\leq v^*$. The general uniqueness question remained unanswered.

The discrete version of (1.1) was considered in [3]. With $h=(\beta-\alpha)/N$, $x_k=\alpha+kh$, $k=0, 1,..., N$, the values of $u(x_k),v(x_k)$ are respectively approximated by $u_{N,k}$, $v_{N,k}$, $k=0, 1,..., N$ with $u_{N,0}=u_{N,N}=v_{N,0}=v_{N,N}=0$. The second derivative is replaced by the operator Δ_h where

$$\Delta_h u_{N,k}=[u_{N,k-1} - 2u_{N,k} + u_{N,k+1}]/h_N^2 \text{ for } k=1,..., N-1.$$

The hypotheses (A2), (A3) are respectively replaced by

$$e>2(1 - \cos\pi/N)N^2/(\beta - \alpha)^2=\lambda(N) \tag{A2*}$$

$$a>gb(\lambda(N) + ce/g)/(gb - cf) \tag{A3*}$$

2. Monotone Accelerated Convergence.

We need to solve scalar problems of the following type for constructing our monotone sequences of discrete functions, converging to solutions of the discrete version of (1.1). Consider the problem

$$\begin{aligned}&\Delta_h w_{N,k} + w_{N,k}[p_k - qw_{N,k}]=0, \quad k=1,..., N-1\\ &w_{N,0}=w_{N,N}=0,\end{aligned} \tag{2.1}$$

where $p_k>0$, $k=1,...,N-1$ and $q>0$.

Theorem 2.1 Suppose that $\min_{1\leq k\leq N-1} p_k > \lambda(N) = 2h^{-2}(1 - \cos\pi/N)$, then there exists a unique solution of problem (2.1) with the property that $w_{N,k}>0$ for $k=1,\ldots, N-1$. Furthermore, if K and δ are two constants such that $p_k - qK \leq 0$ and $0<\delta\leq(p_k - \lambda(N))/q$ for $k=1,\ldots,N-1$, then $\delta\sin(k\pi/N)\leq w_{N,k}\leq K$ for $k=1,\ldots N-1$.

The proof of this theorem and the method for computing $w_{N,k}$ by means of Picard succesive approximations is given in Lemmas 2.1 to 2.3 and Theorem 2.1 of [3]. We observe that with obvious modifications, the conclusions of Theorem 2.1 are valid for the corresponding continuous version of problem (2.1), given by

$$\begin{aligned} &w''(x) + w(x)[p(x) - qw(x)]=0, \ \alpha<x<\beta, \\ &w(\alpha)=w(\beta)=0. \end{aligned} \tag{2.2}$$

Here $p(x)>\pi^2/(\beta-\alpha)^2$, with the solution $w(x)=\tilde{w}(x)$ satisfying $\varepsilon\sin((x-\alpha)/(\beta-\alpha))\pi\leq\tilde{w}(x)\leq q^{-1}\max\{p(x): \alpha\leq x\leq\beta\}$, for $\alpha\leq x\leq\beta$, $\varepsilon>0$ sufficiently small. Note that the function $v(x)=\varepsilon\sin((x-\alpha)/(\beta-\alpha))\pi$ satisfies $v''+v[p(x)-qv]\geq0$ in $[\alpha,\beta]$ for $\varepsilon>0$ sufficiently small, and therefore is a lower solution for the problem (2.2).

We will now develop an "accelerated" method for constructing a sequence of positive functions in (α,β) converging to the positive solution of (2.2).

Write

$$f(w,x)=-w[p(x) - qw],$$

$$g(w,x)=\frac{\partial f}{\partial w}(w,x)=-p(x)+2qw,$$

and introduce the sequence of functions $\{w_n(x)\}$, $\alpha \leq x \leq \beta$, $n=1,2,\ldots,$ defined by the recurrence relations

$$\begin{aligned} &w''_{n+1}=f(w_n(x),x) + (w_{n+1} - w_n(x))g(w_n(x),x) \ , \alpha \leq x \leq \beta \\ &w_{n+1}(\alpha)=w_{n+1}(\beta)=0, \end{aligned} \tag{2.3}$$

where $w_1(x)\equiv K$ is a known positive constant, upper solution of (2.2). (i.e. $p(x) - qK \leq 0$ for all $\alpha<x<\beta$). Formula (2.3) is Newton's method of approximation applied to the differential equation (2.2). To see that the functions $w_n(x)$ are uniquely defined positive functions in (α,β) we will use the following important property (see e.g. [2]) of f:

$$f(w,x)= \max_{-\infty<z<\infty} [f(z,x) + (w - z)g(w,x)], \tag{2.4}$$

for each $-\infty<w<\infty$, $\alpha<x<\beta$. Suppose that $w_k(x)$ is uniquely defined by (2.3) with $w_k(x) \geq \varepsilon \sin((x-\alpha)/(\beta-\alpha))\pi$, $\alpha<x<\beta$. If w_{k+1} exists, $u(x)=w_{k+1}(x)$ should satisfy:

$$\begin{aligned} &u''=f(w_k(x),x)+(u-w_k(x))g(w_k(x),x)\equiv h_{k+1}(u,x), \ \alpha<x<\beta \\ &u(\alpha)=u(\beta)=0. \end{aligned} \tag{2.5}$$

For k=1, we have

$$\begin{aligned} h_2(w_1,x)&=f(w_1,x)+(w_1-w_1)g(w_1,x)=f(K,x) \\ &=-K[p(x)-qK] \geq 0=w''_1; \end{aligned} \tag{2.6}$$

and for k>1, we have by (2.3) and (2.4):

$$\begin{aligned} w_k'' &= f(w_{k-1}(x),x)+(w_k-w_{k-1}(x))g(w_{k-1}(x),x) \\ &\leq f(w_k,x)=h_{k+1}(w_k,x). \end{aligned} \tag{2.7}$$

The two inequalities above indicate that $w_k(x)$ is an upper solution for problem (2.5). On the other hand, the function $v(x)=\varepsilon\sin((x-\alpha)/(\beta-\alpha))\pi$ satisfies:

$$v''\geq f(v,x)\geq f(w_k(x),x)+(v-w_k(x))g(w_k(x),x)\equiv h_{k+1}(v,x) \tag{2.8}$$

for $\alpha<x<\beta$. (Note that the last inequality above is again due to (2.4)). The function $v(x)$ is therefore a lower solution for (2.5), and problem (2.5) must have a solution $u=z(x)$ with $v(x)\leq z(x)\leq w_k(x)$, $\alpha\leq x\leq\beta$. To see that $z(x)$ is unique, let $u=z_i(x)$, $i=1,2$ be two different solutions of (2.5) with $v\leq z_i\leq w_k$. The function $y=z_1-z_2$ will satisfy:

$$y''=yg(w_k(x),x), \quad \alpha<x<\beta; \quad y(\alpha)=y(\beta)=0.$$

That is, $\lambda=1$ is an eigenvalue for the problem:

$$\eta''-2qw_k(x)\eta+\lambda p(x)\eta=0 \text{ for } \alpha<x<\beta, \ \eta(\alpha)=\eta(\beta)=0, \tag{2.9}$$

with $\eta=y(x)$ as the corresponding eigenfunction. Referring to (2.6), (2.7) and (2.8) again, we observe that $w_k(x)$ and $v(x)$ are also upper and lower solutions for the problem

$$u''=f(u,x), \ \alpha<x<\beta, \quad u(\alpha)=u(\beta).$$

Therefore (2.2) has a solution $w=\tilde{w}(x)$, with $v(x)\leq\tilde{w}(x)\leq w_k(x)$, $\alpha\leq x\leq\beta$. Clearly $\eta=\tilde{w}(x)$ satisfies

$$\begin{aligned} &\eta''-q\tilde{w}(x)\eta+\lambda p(x)\eta=0, \ \alpha<x<\beta, \\ &\eta(\alpha)=\eta(\beta)=0, \end{aligned} \tag{2.10}$$

with $\lambda=1$ and $\tilde{w}(x)>0$ in (α,β). Consequently $\lambda=1$ is the first eigenvalue for the eigenvalue problem (2.10). Since $2w_k(x)>\tilde{w}(x)$ in (α,β), the first eigenvalue for problem (2.9) must be strictly larger than 1. This is a contradiction, unless $y(x)\equiv 0$ and $\lambda=1$ is not really an eigenvalue for (2.9). Consequently $z(x)$ is uniquely defined, and by letting $w_{k+1}(x)=z(x)$, we have

$$0<v(x)\leq w_{k+1}(x)\leq w_k(x), \quad \alpha<x<\beta. \tag{2.11}$$

By induction, (2.11) is true for $k=1,2,3,\ldots$ The monotone decreasing sequence $\{w_n(x)\}$ must converge to a function $\hat{w}(x)$ $(\geq v(x))$ for $\alpha\leq x\leq\beta$ as $n\to\infty$. Using the integral representation for the solution of (2.3) and the dominated convergence theorem, one sees that $\hat{w}(x)$ is a solution of (2.2), with $\hat{w}(x)\geq v(x)>0$, $\alpha<x<\beta$. (From the uniqueness of the positive solution in (α,β) for (2.2), we actually have $\hat{w}(x)\equiv\tilde{w}(x)$).

We have proved the following:

Theorem 2.2 Suppose that $p(x)$ is continuous on $[\alpha,\beta]$ with $p(x) > (\pi/(\beta-\alpha))^2$ for $\alpha\leq x\leq\beta$. The sequence of functions $\{w_n(x)\}$ defined recursively by (2.3) converges monotonically to the unique positive (for $\alpha<x<\beta$) solution of the problem (2.2).

Remark: A similar result can be found in [2]. However, in [2], the length of the interval has to be small and the corresponding boundary value problem has a unique solution. For our problem (2.2), the trivial solution is another solution other than $\hat{w}$. Theorem 2.2 is therefore different and new.

The next theorem illustrates why our present method is called accelerated.

Theorem 2.3 Under the conditions of Theorem 2.2, the sequence $\{w_n(x)\}$ converges quadratically to the unique positive solution $\hat{w}(x)$ of the problem (2.2), in the sense that:

$$\|\hat{w}(x)-w_n(x)\|\leq L\|\hat{w}(x)-w_{n-1}(x)\|^2, \quad n=2,\ldots \tag{2.12}$$

for some L independent of n. (Here $\|f(x)\|$ is the maximum of $|f(x)|$ for x in $[\alpha,\beta]$).

Proof. From the equation which $\hat{w}$ and w_n satisfy, we find

$$(\hat{w}-w_n)''=f(\hat{w},x)-f(w_{n-1}(x),x)-(\hat{w}-w_{n-1}(x))g(w_{n-1}(x),x).$$

This can be rewritten as:

$$(\hat{w}-w_n)''=f(\hat{w},x)-f(w_{n-1}(x),x)-(\hat{w}-w_{n-1}(x))g(w_{n-1}(x),x)+(\hat{w}-w_n)g(w_{n-1}(x),x).$$

Let $\eta=\hat{w}-w_n$. Then η satisfies, for $n\geq 2$:

$$\begin{aligned}&\eta''-2qw_{n-1}(x)\eta+p(x)\eta=f(\hat{w},x)-f(w_{n-1}(x),x)-(\hat{w}-w_{n-1}(x))g(w_{n-1}(x),x)\\&\eta(\alpha)=\eta(\beta)=0.\end{aligned} \tag{2.13}$$

We have shown at the end of the proof of Theorem 2.2 that the first eigenvalue λ for problem (2.9), $k\geq 1$, must satisfy $\lambda>1$. Conversely, from (2.13), the function $\hat{w}-w_n(x)$ is representable as

$$(\hat{w}-w_n)(x)=\int_\alpha^\beta G_{n-1}(x,\xi)(1/2)\partial^2 f/\partial w^2(\gamma_n(\xi),\xi)[\hat{w}(\xi)-w_{n-1}(\xi)]^2 d\xi \tag{2.14}$$

for $\alpha\leq x\leq\beta$, where G_{n-1} is the Green's function for the operator $(d^2/dx^2)-2qw_{n-1}+p$ on the interval $[\alpha,\beta]$. (Here $\gamma_n(\xi)$ lies between $\hat{w}(\xi)$ and $w_{n-1}(\xi)$, and the mean value theorem is used). For our f, we readily verify that $\partial^2 f/\partial w^2 \equiv 2q$; hence (2.14) gives

$$\|\hat{w}-w_n\| \leq \|\hat{w}-w_{n-1}\|^2 q \int_\alpha^\beta |G_{n-1}(x,\xi)|\, d\xi \tag{2.15}$$

We will now deduce that the functions $G_n(x,\xi)$ are uniformly bounded for $n=1,2,..,\alpha\leq x\leq\beta$, $\alpha\leq\xi\leq\beta$. From the characterization of the Green's function, one readily sees that for $n=2,3,...$ if we let $u_n(x)$, $v_n(x)$ be solutions of

$$\eta''+[p(x)-2qw_{n-1}(x)]\eta=0\ ,\ \alpha<x<\beta \tag{2.16}$$

with $u_n(\alpha)=0$, $u_n'(\alpha)=1$ and $v_n(\beta)=0$, $v_n'(\beta)=-1$, we have:

$$G_{n-1}(x,\xi)=\begin{cases}(-v_n(\xi)/W_n(\xi))u_n(x) & \text{for } x<\xi\\ (-u_n(\xi)/W_n(\xi))v_n(x) & \text{for } \xi<x.\end{cases} \tag{2.17}$$

Here,

$$W_n(\xi)=u_n(\xi)v_n'(\xi)-v_n(\xi)u_n'(\xi) \tag{2.18}$$

which is the Wronskian for the two independent solutions u_n, v_n. For convenience, let $u_0(x)$, $v_0(x)$ be solutions of:

$$\eta''+[p(x)-2q\hat{w}(x)]\eta=0,\ \alpha<x<\beta \tag{2.19}$$

with $u_0(\alpha)=0$, $u_0'(\alpha)=1$ and $v_0(\beta)=v_0'(\beta)=-1$. (Note that the eigenvalue problem: $\eta''-2q\hat{w}(x)\eta+\lambda p(x)\eta=0$, $\alpha<x<\beta$, $\eta(\alpha)=\eta(\beta)=0$, must have its first eigenvalue $\lambda>1$, since $2q\hat{w}(x)>q\hat{w}(x)$ and (2.10) has 1 as its first eigenvalue. Consequently, $v_0(\alpha)\neq 0$).

Let $W_0(x)$ be the Wronskian defined also by (2.18) with $n=0$. From the equations (2.16) and (2.19) which do not have η' term, we find that

$W_n(\xi)=W_n(\alpha)$, for $\alpha\leq\xi\leq\beta$, $n=2,3,...$ and $n=0$.

We will now see that $\lim_{n\to\infty} W_n(\alpha)=W_0(\alpha)=-v_0(\alpha)$. (The difficulty here is that we do not know that $\lim_{n\to\infty} w_n(x)=\hat{w}(x)$ uniformly in $[\alpha,\beta]$).

From the uniform boundedness of $w_n(x)$ in $[\alpha,\beta]$, $n=1,2,...$ and the equations (2.16), we must have $v_n(x)$, $v_n'(x)$ uniformly bounded for x in $[\alpha,\beta]$, $n=2,3,...$(cf. Lemma 4.1, p. 54, [1]). Fom (2.16) again, we have uniform boundedness of $v_n''(x)$ in $[\alpha,\beta]$. Consequently, we can extract a subsequence of $v_n(x)$ convergent uniformly in $[\alpha,\beta]$ to $v_0(x)$, which is uniquely defined by an initial value problem at $x=\beta$. From the uniqueness of the initial value problem defining $v_0(x)$, we conclude that $\lim_{n\to\infty} v_n(\alpha)=v_0(\alpha)$. (cf. Theorem 2.4, p. 4, [1]). Therefore, we have $\lim_{n\to\infty} W_n(\alpha)=\lim_{n\to\infty}-v_n(\alpha)=-v_0(\alpha)\neq0$, and $|W_n(\xi)|=|W_n(\alpha)|\geq c>0$ for some positive constant c, $n=2,3,....$ Again, using Lemma 4.1, p. 54, in [1] and equations (2.16) we find that u_n, v_n are uniformly bounded in $[\alpha,\beta]$, $n=2,3,...$ Using (2.17), we conclude that the functions $G_n(x,\xi)$ are uniformly bounded for $n=1,2,...$, $\alpha\leq x\leq\beta$, $\alpha\leq\xi\leq\beta$.

From (2.15), we arrive at (2.12). This concludes the proof of Theorem 2.3.

<u>Remark</u>: In [2], results similar to Theorem 2.3 can be found. The conditions in [2] would be satisfied if $p(x)$ is less than about $2/(\beta-\alpha)^2$. However, in Theorems 2.2 and 2.3 we assume $p(x)>(\pi/(\beta-\alpha))^2$. Consequently [2] is

not applicable to our problem and the new proof in Theorem 2.3 is needed.

In actual computations, we solve (2.1) by restricting (2.3) to the corresponding grid functions, i.e., we solve

$$\begin{aligned} &\Delta_h w_{N,k}^{(n+1)} + p_k w_{N,k}^{(n+1)} - 2q w_{N,k}^{(n)} w_{N,k}^{(n+1)} = -q(w_{N,k}^{(n)})^2 \\ &w_{N,0}^{(n+1)} = w_{N,N}^{(n+1)} = 0, \quad k=1,\dots,N-1 \end{aligned} \tag{2.20}$$

for $n=1,2,\dots$, where $w_{N,k}^{(1)}=K$, $k=1,\dots,N-1$, is a known positive constant, upper solution of (2.1) with $w_{N,0}^{(1)}=w_{N,N}^{(1)}=0$. One can proceed to prove the existence of a positive solution $w_{N,k}^{(n)}$, $n=2,3,\dots$ for (2.20), where $w_{N,k}^{(n)}$ converges monotonically and quadratically as $n\to\infty$ to $w_{N,k}$, the solution of (2.1), as it is done in Theorem 2.2 and Theorem 2.3 for the continuous case. However, the details are too lengthy for our present purpose.

3. Construction of Monotone Sequences.

For completeness, we state some results which had been proved in [3], and are used to generate functions $u_{N,k}^{(2i)}$, $u_{N,k}^{(2i+1)}$, $v_{N,k}^{(2i)}$, $v_{N,k}^{(2i+1)}$, $i=1,2,\dots$; $k=1,\dots,N-1$. These discrete functions will converge to the solutions of (1.1) or their bounds, as $N \to \infty$. We recall that we assume (A2*) and (A3*) for all N under consideration.

First let $u_{N,k}^{(1)}>0$, $k=1,\dots,N-1$ satisfy

$$\begin{aligned} &\Delta_h u_{N,k}^{(1)} + u_{N,k}^{(1)}(a-bu_{N,k}^{(1)})=0, \\ &u_{N,0}^{(1)}=u_{N,N}^{(1)}=0. \end{aligned} \tag{3.1}$$

Such solution exists by Theorem 2.1. Similarly, let $v_{N,k}^{(1)}>0$, $k=1,\dots,N-1$ be the solution of

$$\Delta_h v_{N,k}^{(1)} + v_{N,k}^{(1)}(e + fu_{N,k}^{(1)} - gv_{N,k}^{(1)})=0,$$
$$v_{N,0}^{(1)}=v_{N,N}^{(1)}=0. \tag{3.2}$$

For i=2,3,... define $u_{N,k}^{(i)}>0$, $v_{N,k}^{(i)}>0$, for $k=1,\dots,N-1$, inductively as follows:

$$\Delta_h u_{N,k}^{(i)} + u_{N,k}^{(i)}(a - bu_{N,k}^{(i)} - cv_{N,k}^{(i-1)})=0, \tag{3.3}$$

$$\Delta_h v_{N,k}^{(i)} + v_{N,k}^{(i)}(e + bu_{N,k}^{(i)} - gv_{N,k}^{(i)})=0, \tag{3.4}$$

$$u_{N,0}^{(i)}=u_{N,N}^{(i)}=v_{N,0}^{(i)}=v_{N,N}^{(i)}=0.$$

We notice that by Theorem 2.1 and (3.1), we have $u_{N,k}^{(1)}\leqq a/b$, $k=0,\dots,N$. Hence, by (3.2) and Theorem 2.1 again, we have $v_{N,k}^{(1)}\leqq(e + fa/b)/g$, $k=0,\dots,N$. We solve (3.3) and (3.4) at each step by the algorithm indicated in (2.20).

For the purpose of ordering these sequences of functions the following Lemma will be used:

<u>Lemma</u> 3.1 (<u>Comparison</u>). <u>Let</u> $l_k^{(1)}\geqq l_k^{(2)}>\lambda(N)$ <u>for</u> <u>each</u> $k=1,\dots,N-1$. <u>For</u> i=1,2,<u>let</u> $w_{N,0}^{(i)}=w_{N,N}^{(i)}=0$, $w_{N,k}^{(i)}$, $k=1,\dots,N-1$ <u>be</u> <u>positive</u> <u>numbers</u> <u>satisfying</u>

$$\Delta_h w_{N,k}^{(i)} + w_{N,k}^{(i)}[l_k^{(i)} - pw_{N,k}^{(i)}]=0$$

<u>for</u> $k=1,\dots,N-1$. (<u>Here</u> p <u>is</u> <u>a</u> <u>positive</u> <u>constant</u>). <u>Then</u> $w_{N,k}^{(1)}\geqq w_{N,k}^{(2)}$ <u>for</u> <u>each</u> $k=0,\dots,N$.

Lemma 3.1 leads to the following important properties:

<u>Lemma</u> 3.2 <u>The</u> <u>functions</u> $u_{N,k}^{(i)}$, $v_{N,k}^{(i)}$ <u>are</u> <u>uniquely</u> <u>defined</u> <u>positive</u> <u>functions</u> <u>for</u> $k=1,\dots,N-1$; $i=1,2,\dots$, <u>such</u> <u>that</u> $0<u_{N,k}^{(i)}\leqq u_{N,k}^{(1)}$ <u>and</u> $0<v_{N,k}^{(i)}\leqq v_{N,k}^{(1)}$. <u>Furthermore,</u> <u>for</u> <u>each</u> <u>nonnegative</u> <u>integer</u> i, <u>the</u> <u>following</u> <u>are</u> <u>true</u>:

$$u_{N,k}^{(2i+2)} \leq u_{N,k}^{(2i+4)} \leq u_{N,k}^{(2i+3)} \leq u_{N,k}^{(2i+1)},$$
$$v_{N,k}^{(2i+2)} \leq v_{N,k}^{(2i+4)} \leq v_{N,k}^{(2i+3)} \leq v_{N,k}^{(2i+1)}, \qquad (3.5)$$

for each $k=0,\ldots,N$.

The proof of Lemmas 3.1 and 3.2 can be found in Lemma 3.1 and Lemmas 3.3–3.4 of [3]. Lemma 3.2 implies that

$$0 \leq u_{N,k}^{(2)} \leq u_{N,k}^{(4)} \leq u_{N,k}^{(6)} \leq \ldots \leq u_{N,k}^{(5)} \leq u_{N,k}^{(3)} \leq u_{N,k}^{(1)},$$

and

$$0 \leq v_{N,k}^{(2)} \leq v_{N,k}^{(4)} \leq v_{N,k}^{(6)} \leq \ldots \leq v_{N,k}^{(5)} \leq v_{N,k}^{(3)} \leq v_{N,k}^{(1)},$$

for $k=0,\ldots,N$.

It is also possible to show (cf. Section 4 of [3]) that the piecewise linear extensions of $u_{N,k}^{(2i)}$, $u_{N,k}^{(2i-1)}$, $v_{N,k}^{(2i)}$, $v_{N,k}^{(2i-1)}$, $i=1,2,\ldots$ will tend respectively to $u(x)^{(2i)}$, $u(x)^{(2i-1)}$, $v(x)^{(2i)}$, $v(x)^{(2i-1)}$ as $N \to \infty$. The functions $u(x)^{(2i)}$, $u(x)^{(2i-1)}$, $v(x)^{(2i)}$ and $v(x)^{(2i-1)}$ are the solutions of the corresponding continuous versions of (3.1)–(3.4) respectively for $\alpha \leq x \leq \beta$, with homogeneous boundary data. These functions satisfy

$$0 \leq u(x)^{(2)} \leq u(x)^{(4)} \leq u(x)^{(6)} \leq \ldots \leq u(x)^{(5)} \leq u(x)^{(3)} \leq u(x)^{(1)},$$
$$0 \leq v(x)^{(2)} \leq v(x)^{(4)} \leq v(x)^{(6)} \leq \ldots \leq v(x)^{(5)} \leq v(x)^{(3)} \leq v(x)^{(1)}.$$

Moreover, if we let $\lim_{i \to \infty} u(x)^{(2i)} = u_*(x)$, $\lim_{i \to \infty} u(x)^{(2i-1)} = u^*(x)$, $\lim_{i \to \infty} v(x)^{(2i)} = v_*(x)$ and $\lim_{i \to \infty} v(x)^{(2i-1)} = v^*(x)$, then the positive solutions (u,v) of the boundary value problem (1.1) satisfy $u_* \leq u \leq u^*$, $v_* \leq v \leq v^*$ for all $\alpha \leq x \leq \beta$ (cf. [4]).

<u>Computational results.</u>

In this section we show some numerical results obtained by

applying the accelerated monotone scheme described in the previous sections to the boundary value problem (1.1). Numerical experience shows that when condition (A1) is barely fulfilled and the coefficient a is large relative to e, the computations are more difficult, in the sense that a larger number of iterations are needed for the numerical sequences to satisfy a particular convergence criterion. In our example we let $\alpha=0$, $\beta=\pi$ and discretize the problem by choosing N=64, $h=\pi/64$. Figures 1, 2, 3 and 4 show some elements (i=1,11,19,29,40) of the monotone decreasing sequences $u_{N,k}^{(2i-1)}$, $v_{N,k}^{(2i-1)}$ and the monotone increasing sequences $u_{N,k}^{(2i)}$, $v_{N,k}^{(2i)}$ for problem (1.1) with a=80,b=2,c=1 and e=1.5,f=3,g=2. Conditions (A1) to (A3) are all fulfilled, but with the condition (A1), cf=3.9<4=bg, barely satisfied. Note also that the ratio a/e is large. The algorithm stops when the relative error in maximum norm of two consecutives iterates in all four monotone sequences is less than 10^{-8}; i.e., when

$$\left(\|z_{N,k}^{(j+2)} - z_{N,k}^{(j)}\| \,/\, \|z_{N,k}^{(j+2)}\|\right) \leq 10^{-8}, \quad \text{with } \|z_{N,k}^{(j)}\| = \max_{0\leq k\leq N} |z_{N,k}^{(j)}|.$$

Here $z_{N,k}^{(j)}=u_{N,k}^{(2j-1)}$ or $v_{N,k}^{(2j)}$.

The original monotone scheme in [3], needs 18,003 Picard's iterations to solve the proposed problem. The accelerated monotone scheme, in contrast, needs only 3,509 iterations to produce the same 1,236 elements of the monotone sequences. The average number of iterations necessary to generate each element of the monotone sequences has been reduced from 15 to about 3.

In all the figures, the limiting values are plotted with full lines and dashed lines are used for any other elements in the sequences.

Note that the vertical scale for the decreasing sequence for u is different from that for the increasing sequence. A similar situation is true for v. The computations suggest that we have uniqueness although the proof of uniqueness in [4] does not include this case since we definitely do not have cf<<bg. The general question of uniqueness remains open.

References

1. Hartman,P.: Ordinary Differential Equations, Wiley, New York, (1964).
2. Kalaba R.: *On Nonlinear Differential Equations, the Maximum Operation, and Monotone Convergence*, Journal of Mathematics and Mechanics, Vol. 8, No. 4 (1959), pp. 519–574.
3. Lazer A., Leung A. and Murio D.: *Monotone scheme for finite difference equations concerning steady–state prey–predator interactions*, Journal of Computational and Applied Mathematics, Vol. 8, No. 4 (1982), pp. 243–251.
4. Leung A.: *Monotone schemes for semilinear elliptic systems related to ecology*, in Math. Methods in Applied Sciences, Vol. 4 (1982), pp. 272–285.

Captions for the Figures

Fig 1. Monotone decreasing sequence for u, for the problem

$$u'' + u(80 - 2u - v) = 0$$
$$v'' + v(1.5 + 3.9u - 2v) = 0$$

Fig 2. Monotone decreasing sequence for v, for the problem

$$u'' + u(80 - 2u - v) = 0$$
$$v'' + v(1.5 + 3.9u - 2v) = 0$$

Fig 3. Monotone increasing sequence for u, for the problem

$$u'' + u(80 - 2u - v) = 0$$
$$v'' + v(1.5 + 3.9u - 2v) = 0$$

Fig 4. Monotone increasing sequence for v, for the problem

$$u'' + u(80 - 2u - v) = 0$$
$$v'' + v(1.5 + 3.9u - 2v) = 0$$

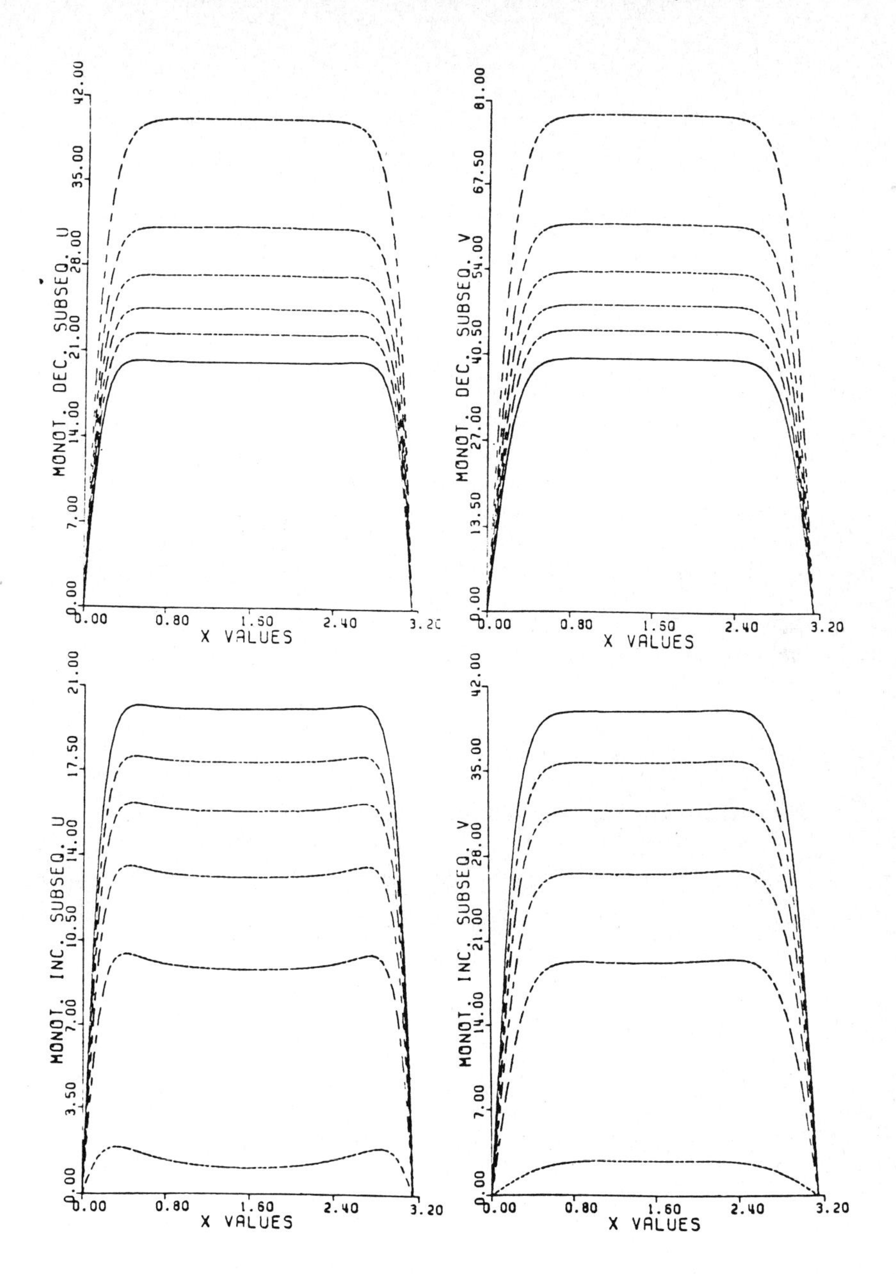
MONOT. DEC. SUBSEQ. U
X VALUES
MONOT. DEC. SUBSEQ. V
X VALUES
MONOT. INC. SUBSEQ. U
X VALUES
MONOT. INC. SUBSEQ. V
X VALUES

Numerical Approximation of Partial Differential Equations
E.L. Ortiz (Editor)

A NOTE ON THREE-TERM RECURRENCES AND THEIR NUMERICAL TREATMENT

Ivo Marek
Karel Žitný

Charles University of Prague
Prague, Czechoslovakia,
Institute of Thermomechanics
of the Czechoslovak Academy of Sciences
Prague, Czechoslovakia
and
Technical University of Tunis
Ras Tabia, Tunis, Tunisia

Three-term recurrences appear in a natural way in the study of group representations, (Vilenkin 1965) discretization methods (Babuška et al. 1969) and further topics of mathematics. Since three-term recurrences form a suitable basis for creating numerical algorithms of computation values of special functions, or their derivatives and of constructing solutions to discretized problems a need of numerical analysis of such recurrences has already angaged many specialists to consider three-term recurrences from point of view of numerical analysis (see Babuška et al. 1969 and Babušková 1964). In our contribution we present a general numerically stable method of solving three-term recurrences.

Let $\mathbb{N}$ denote the set of all positive integers. Let $(a_j)_{j \in \mathbb{N}}$, $(b_j)_{j \in \mathbb{N}}$ and $(c_j)_{j \in \mathbb{N}}$ be given sequences complex functions of a real variable x.

Let us consider three-term recurrence

$$b_{k+1}(x)S_{k+1}(x) + a_k(x)S_k(x) + c_{k-1}S_{k-1}(x) = 0 \tag{1}$$

$k = 1,2,\ldots$, with $S_0(x)$ and $S_1(x)$ given.

The matrix

$$T(x) = \begin{pmatrix} a_1(x) & b_2(x) & 0 & \\ c_1(x) & a_2(x) & b_3(x) & \\ & \ddots & \ddots & \ddots \end{pmatrix}$$

is called the recurrence matrix. The square diagonal block of $T(x)$ containing the (p+1)-th to (p+N)-th rows, where $p = 0,1,\ldots$ and N is a positive integer, is denoted by $T_{p,N}$, i.e.

$$T_{p,N}(x) = \begin{pmatrix} a_{p+1}(x) & b_{p+2}(x) & 0 & & \\ c_{p+1}(x) & a_{p+2}(x) & b_{p+3}(x) & & \\ \cdots & \cdots & \cdots & \cdots & \cdots \\ & & c_{p+N-2}(x) & a_{p+N-1}(x) & b_{p+N}(x) \\ & & 0 \quad 0 & c_{p+N-1}(x) & a_{p+N}(x) \end{pmatrix}$$

Further, we let

$$S = (S_{p+1},\ldots,S_{p+N})$$

so that (1) becomes

$$(2) \qquad -T_{p,N}S = \begin{pmatrix} c_p S_p \\ 0 \\ \vdots \\ 0 \\ b_{p+N+1}S_{p+N+1} \end{pmatrix}$$

Our analysis is based on the following assumptions:

(i) The coefficients a_j, b_j, c_j, $j = 1,2,\ldots$, in the recurrences considered do not vanish, at least in some open interval J;

(ii) To each $\alpha > 0$ there are positive integers $p > \alpha$ and $N > \alpha$ such that the matrix $T_{p,N}$ is regular;

(iii) The exact solutions $S_N = S_N(x)$ posseses the properties that

$$\lim_{N\to\infty} S_N(x) = 0 \quad \text{for all} \quad x \in J$$

and

$$\lim_{N\to\infty} b_N(x)S_N(x) = 0 \quad \text{for all} \quad x \in J\ ;$$

(iv) There is a fixed index $p_0 \geqq 0$ such $S_{p_0}(x) \neq 0$ is known exactly.

REMARK. In some cases instead of (iv) we may assume

(iv)′ There exist functions $\alpha_0,\dots,\alpha_k,\dots$ and γ such that

$$\sum_{k=0}^{\infty} \alpha_k(x) S_N(x) = \gamma(x) \ , \quad x \in J \ ,$$

where explicite formulas of $\alpha_1(x),\dots,\alpha_k(x),\dots$ and $\gamma(x)$ are known in a closed form.

AN ALGORITHM

Let $\tau > 0$ be a chosen tolerance of the computation e.g. the smallest positive number represented by the computer used. Instead of considering an initial value problem for (1) we consider a boundary value problem for (1) i.e. we assume that $S_{p_0}(x)$ and $\lim_{N\to\infty} S_N(x)$ are given.

We assume also that $\tilde{N}$ is such that

$$|b_N(x) S_N(x)| < \tau \quad \text{for} \quad N > \tilde{N} \ .$$

In other words, within the tolerance given by τ

$$b_N(x) S_N(x) = 0 \quad \text{for} \quad N > \tilde{N} \ .$$

Then up to the tolerance τ the right hand side of (2) contains a single nonzero component. Instead of that system we consider the modified system

$$(3) \qquad - T_{p,N}(x)\tilde{S}(x) = \begin{pmatrix} c_p(x)\tilde{S}_p(x) \\ 0 \\ \vdots \\ 0 \end{pmatrix} ,$$

where $\tilde{S}(x) = (\tilde{S}_{p+1}(x),\dots,\tilde{S}_{p+N}(x))$ and $\tilde{S}_p(x) \neq 0$ is arbitrary.

We easily see that $\tilde{S}$ is related to S by

$$\tilde{S}(x) = \rho(x) S(x)$$

with some nonzero function ρ which is to be determined in order to get $S(x)$.

Knowing $\tilde{S}(x)$ we can compute $\tilde{S}_{p-1}(x),\dots,\tilde{S}_0(x)$ using

$$\tilde{S}_{k-1}(x) = \frac{a_k(x)}{c_{k-1}(x)} \tilde{S}_k(x) - \frac{b_{k+1}(x)}{c_{k-1}(x)} \tilde{S}_{k+1}(x)$$

$k = p, p-1, \dots$.

The quantity $\rho(x)$ is obtained from relation

$$S_{p_0}(x) = \rho(x)\tilde{S}_{p_0}(x) \quad \text{in case (iv)}$$

or from

$$\sum_{k=0}^{N} \alpha_k(x)\tilde{S}_k(x) = \gamma(x)\rho(x) \quad \text{in case (iv)'}.$$

MAIN RESULT. The process shown above has the following features:

1° It can be realized by use of fast algorithms of solving linear systems. (This means that the determining $\hat{S}(x)$ requires at most $0(N \log N)$ arithmetic operations.)

2° It is numerically stable.

EXAMPLE 1 (Bessel functions). In this section we show a typical example of applying the previous algorithm (Babušková, 1964).

It is well known (Vilenkin, p. 213) that the Bessel functions of the first kind J_N of the integer index N are solutions to the following three-term recurrence

$$J_{N-1}(x) + J_{N+1}(x) - \frac{2N}{x} J_N(x) = 0 \;, \quad N = 1,2,\ldots \;, \tag{4}$$

with $J_0(x)$ given and $\lim_{N\to\infty} J_N(x) = 0$.

Let $x \neq 0$. Then

$$T_{p,N}(x) = \begin{pmatrix} \frac{-2(p+1)}{x} & 1 & 0 \ldots & & \\ 1 & \frac{-2(p+2)}{x} & 1 \ldots & & \\ \cdots & \cdots & \cdots & \cdots & \cdots \\ & & & \ldots 0 & 1 \\ & & & 1 & \frac{-2(p+N)}{x} \end{pmatrix}$$

Thus $T_{p,N}$ is regular if $0 < x \leq 1$ for every $p \geq 0$. Moreover, $-T_{p,N}$ is an M-matrix (Berman, Plemmons 1979) so that $-T^{-1}_{p,N}$ has all its elements positive. This property has some rather interesting consequences concerning the signs of $J_k(x)$ depending upon J_p, $k = p+1,\ldots,N$. For $x \geq 1$ there exists a positive number p and a positive integer N such that

$$\det T_{p,N}(x) \neq 0 \;.$$

For a fixed x we choose p such that $2(p+1) > x$. Then $-T_{p,N}$ is again an M-matrix. So far the fulfilement of hypotheses (i) and (ii) is checked. That also (iii) holds follows from the fact (Vilenkin 1965, p. 217)

$$e^{ix \sin \theta} = \sum_{k=0}^{\infty} J_k(x)\, e^{ik\theta}, \quad \theta \in R, \quad x \in R. \tag{5}$$

(We also see that (iv)' takes place since according to (5) on obtains that

$$1 = J_0(x) + \sum_{k=1}^{\infty} J_{2k}(x) \,.)$$

EXAMPLE 2 (Legendre polynomials). It is well known that Legendre polynomials P_k fulfil (Vilenkin 1965, p. 164)

$$(1 + k)P_{k+1}(x) - 2kxP_k(x) + kP_{k-1}(x) = 0$$

with $P_0(x) = 1$ and $P_1(x) = x$.

Moreover, we have that

$$(1 - hx + h^2)^{-\frac{1}{2}} = \sum_{k=0}^{\infty} P_k(x)h^k, \quad |h| < 1, \quad x \in [-1,1], \tag{6}$$

and therefore

$$\frac{\frac{1}{2}x - h}{(1 - hx + h^2)^{3/2}} = \sum_{k=1}^{\infty} kP_k(x)h^{k-1}, \quad x \in [-1,1].$$

We see that the hypotheses (i) and (iii) hold. It is easy to see that the corresponding reduced matrix

$$T_{p,N}(x) = \begin{pmatrix} -2px & p+1 & 0 & \dots & & 0 \\ p+1 & -2(p+1)x & p+2 & \dots & & 0 \\ \vdots & & & & & \vdots \\ \dots & \dots & \dots & \dots & p+N & -2(p+N)x \end{pmatrix}$$

is symmetric and positive definite. This implies the regularity of $T_{p,N}$ and thus (ii) holds. Setting $h = \frac{1}{2}$ in (6) we get

$$(1 - \frac{1}{2}x + \frac{1}{4})^{-1/2} = \sum_{k=0}^{\infty} 2^{-k} P_k(x)$$

and thus, (iv)' holds.

References

[1] BABUŠKA, I., PRÁGER, M., VITÁSEK, E.: *Numerical Processes in Differential Equations.* SNTL - Publishers of Technical Literature, Prague 1966.

[2] BABUŠKOVÁ, R.: *Über numerische Stabilität einiger Rekursions formeln.* Apl. mat. 9, 186-193 (1964).

[3] BERMAN, A., PLEMMONS, R.: *Nonnegative Matrices in the Mathematical Sciences.* Academia Press, New York 1979.

[4] VILENKIN, N. Ya.: *Special Functions and the Theory of Group Representations.* Nauka, Moscow 1965 (in Russian).

Numerical Approximation of Partial Differential Equations
E.L. Ortiz (Editor)
© Elsevier Science Publishers B.V. (North-Holland), 1987

MIXED FINITE ELEMENTS METHODS FOR CONVECTION-DIFFUSION PROBLEMS

Jean-Marie THOMAS

Université de Pau et des Pays de l'Adour

PAU, France

Mixed finite elements are proposed and analysed for the n-dimensional problem

$$\begin{cases} -\,div(\mathcal{K}\ grad\ u) + \Phi.\ grad\ u = f & in\ \Omega\ , \\ \mathcal{K}\ \dfrac{\partial u}{\partial \nu} + \varphi u = g & on\ \Gamma = \partial\Omega \end{cases}$$

with $div\ \Phi = 0$ *in* Ω *. We will study with particular attention the methods well adapted to the convection-dominated flows.*

1. THE STEADY CONVECTION-DIFFUSION PROBLEMS. Let Ω be a bounded domain in $\mathbb{R}^n$, $n \leq 3$, with a piecewise smooth boundary Γ . We will consider numerical approximations of the solution of the n-dimensional steady convection-diffusion problem. This problem is defined as follows :

$$\text{(1.1a)} \qquad -\operatorname{div}(\mathcal{K}\ \operatorname{grad} u) + \Phi\cdot\operatorname{grad} u = f \quad \text{in} \quad \Omega \ ,$$

$$\text{(1.1b)} \qquad \mathcal{K}\frac{\partial u}{\partial \nu} + \varphi u = g \quad \text{on} \quad \Gamma \ ,$$

f and g given functions in $L^2(\Omega)$ and $L^2(\Gamma)$, $\mathcal{K}$ is the diffusivity, Φ the convective velocity and φ the Robin boundary condition coefficient ; for concision it will be assumed that φ and φ^{-1} (notation : $\varphi^{-1} = \frac{1}{\varphi}$) are bounded on the whole boundary Γ . The extensions to the case $\varphi = +\infty$ on a part Γ_0 of Γ , which corresponds to an homogeneous Dirichlet boundary condition on Γ_0 and to the case $\varphi = 0$ on an another part Γ_1 of Γ , which corresponds to a Neumann boundary condition, will be considered in remark. We assume the following regularities $\mathcal{K} \in L^\infty(\Omega)$, $\Phi \in (L^\infty(\Omega))^n$, $\varphi \in L^\infty(\Gamma)$ as well as φ^{-1} . Φ is a solenoïdal vector-field, that is :

$$\text{(1.2)} \qquad \operatorname{div} \Phi = 0 \quad \text{in} \quad \Omega \ .$$

At last, we assume the existence of a positive constant α such that :

(1.3a) $\mathcal{K} \geq \alpha$ in Ω ,

(1.3b) $\varphi + \frac{1}{2}\,\Phi\cdot\nu \geq \alpha$ on Γ ,

where ν is the unit outward normal along $\Gamma = \partial\Omega$.

The classical "variational" formulation of this problem consists in finding $u \in H^1(\Omega)$ (that is the Sobolev space of order 1) solution of

(1.4) $A(u,v) = F(v)$ for all $v \in H^1(\Omega)$,

with

(1.5a) $A(u,v) = \int_\Omega \mathcal{K}\,\mathrm{grad}\,u.\,\mathrm{grad}\,v\,dx + \int_\Gamma \varphi\,uv\,d\sigma + \int_\Omega \Phi\cdot\mathrm{grad}\,u\,v dx$,

(1.5b) $F(v) = \int_\Omega fv\,dx + \int_\Gamma gv\,d\sigma$.

With the previous assumptions, the bilinear form $A(.,.)$ is continuous and elliptic on $H^1(\Omega)$ but not symmetric unless $\Phi = 0$. A very standard Galerkin finite element method (Raviart and Thomas, 1983) for the approximation of the solution of (1.4) is : Let V_h be a finite-dimentional space of continuous and piecewise polynomial functions. Let $u_h \in V_h$ the solution of

(1.6) $A(u_h,v_h) = F(v_h)$ for all $v_h \in V_h$

Problem (1.6) has a unique solution u_h and optimal asymptotic estimates of the error $u-u_h$ in $H^1(\Omega)$-norm and $L^2(\Omega)$-norm can be obtained. In fact, it is also well-known that this finite element method applied to convection-dominated flow problems gives unphysical oscillatory solutions with a reasonable mesh size. The same difficulties arise in finite difference methods when using centered difference approximations for the convective term. Even for a one-dimensional problem with constant coefficients oscillations appear unless if the grid Peclet number $\frac{\Phi h}{\mathcal{K}}$ is inferior or egal to 2 when we discretize with linear trial functions. The first remedy for the difference methods was to introduce upwind by differencing for the convective term. The same upwind concept was adapted to the finite element methods by skewing the trial functions or by shifting the numerical integration points. All these procedures generate spurious crosswind diffusion. In the streamline diffusion method (Hughes and Brooks, 1979 ; Johnson and Nävert, 1981) the artificial added diffusion act only in the direction of the streamlines ; these last methods may be now obtained with a Petrov-Galerkin formulation or with the use of a generalized equation which is discretized by a standard Galerkin formulation. In the most recent litterature on this subject see in particular (Mizukami and Hughes, 1985 ; Donéa, Belytschko and Smolinski, 1985) and the schemes comparisons (Patel and co., 1985 ; Huang and co., 1985).

2. THE MIXED FORMULATION ; FIRST APPROACH. In the case of pure diffusion, i.e. $\Phi = 0$, mixed methods have been discussed and analyzed in detail for the Dirichlet problem (Raviart and Thomas, 1977 ; Falk and Osborn, 1980). It is easy to extend the analyse for the Robin problem. For this purpose, we introduce W and M the following functional spaces :

(2.1) $$W = \{q \in (L^2(\Omega))^n \ ; \ \mathrm{div}\, q \in L^2(\Omega) \ ; \ q.\nu \in L^2(\Gamma)\}$$

and

(2.2) $$M = L^2(\Omega) \quad .$$

Then the mixed formulation of the pure diffusion problem consist in finding the pair $(p,u) \in W \times M$ solution of

$$\begin{cases} \displaystyle\int_\Omega \mathcal{K}^{-1} p.q \, dx + \int_\Gamma \varphi^{-1} p.\nu \, q.\nu \, d\sigma + \int_\Omega u \, \mathrm{div}\, p \, dx = \\ \qquad\qquad = \displaystyle\int_\Gamma \varphi^{-1} g.q\nu \, d\sigma \quad \text{for all} \quad q \in W , & (2.3a) \\ \displaystyle\int_\Omega v \, \mathrm{div}\, p \, dx = - \int_\Omega fv \, dx \qquad \text{for all} \quad v \in M . & (2.3b) \end{cases}$$

The equation (2.3a) is a weak form of

(2.4a) $$p = \mathcal{K} \,\mathrm{grad}\, u \ \text{ in } \ \Omega \ , \ p.\nu + \varphi u = g \ \text{ on } \ \Gamma \quad ,$$

while (2.3b) corresponds to the equilibrium relation :

(2.4b) $$\mathrm{div}\, p + f = 0 \ \text{ in } \ \Omega \ .$$

Remark. If $\varphi = +\infty$ on a part Γ_o of Γ (Dirichlet condition), it suffices to make formally $\varphi^{-1} = 0$ on Γ_o in (2.3a). If $\varphi = 0$ on a part Γ_1 of Γ , the Neumann condition $p.\nu = g$ on Γ_1 is imposed a priori and the trial functions $q \in W$ must satisfy the constraint $q.\nu = 0$ on Γ_1 .

We come back to the study of the convection-diffusion problem. The relation (2.4b) is changed into

(2.5b) $\quad \mathrm{div}\, p - \mathcal{K}^{-1} p.\Phi + f = 0$ in Ω

while (2.4a) is unchanged : (2.5a) $\equiv$ (2.4a) . So a straight-forward generalization of the mixed method is here to find a pair $(p,u) \in W \times M$- solution of

$$\begin{cases} a_o(p,q) + b_o(q,u) = G_o(q) & \text{for all } q \in W , \quad (2.6a) \\ b_\Phi(p,v) = F_o(v) & \text{for all } v \in M . \quad (2.6b) \end{cases}$$

with the notations

(2.7a) $$a_o(p,q) = \int_\Omega \mathcal{K}^{-1}\, p.q + \int_\Gamma \varphi^{-1} p.\nu\; q.\nu \;d\sigma \quad ,$$

(2.7b) $$b_o(q,v) = \int_\Omega v \;\mathrm{div}\, q \;dx \quad ,$$

(2.7c) $$b_\Phi(q,v) = \int_\Omega v \;\mathrm{div}\, q \;dx - \int_\Omega \mathcal{K}^{-1} p.\Phi \; v \;dx$$

and

(2.8a) $$G_o(q) = \int_\Omega \varphi^{-1}\, g\; q.\nu \;d\sigma \quad ,$$

(2.8b) $$F_o(v) = - \int_\Omega fv \;dx \quad .$$

For the construction of numerical solutions, let $\overline{\Omega} = \bigcup_{T \in \mathcal{T}_h} T$ be a triangulation of $\overline{\Omega}$ with n-simplexes (triangles for $n = 2$) of diameter $\leq h$ and let k be a given nonnegative integer number. To this decomposition of $\overline{\Omega}$ we associate the spaces (Thomas, 1977 ; Nedelec, 1980)

(2.9) $$W_h = \{q_h \in W \; ; \; \forall T \in \mathcal{T}_h \; , \; q_h|_T \in (P_k)^n \oplus x\, P_k\}$$

and

(2.10) $$M_h = \{v_h \in M \; ; \; \forall\, T \in \mathcal{T}_h \; , \; v_h|_T \in P_k\}$$

where P_k is the space of the polynoms of degree $\leq k$. Let us remark that a function $v_h \in M_h$ is fully discontinuous from one simplex to another, while a function $q_h \in W_h$ has its normal trace continuous at an interface. The mixed finite elements method associated with (2.6) is defined as follows : find $(p_h,u_h) \in W_h \times M_h$ such that

$$\begin{cases} a_o(p_h,q_h) + b_o(q_h,u_h) = G_o(q_h) & \text{for all } q_h \in W_h \; , \quad (2.11a) \\ b_\Phi(p_h,v_h) = F_o(v_h) & \text{for all } v_h \in M_h \; . \quad (2.11b) \end{cases}$$

It is proved (Douglas and Roberts, 1985) that for "h sufficiently small" problem (2.11) has a unique solution and optimal order estimates are obtained.

But when the Peclet number is not small, it is not sure that problem (2.11) with a reasonable mesh size admits a solution. So, this formulation seems ill adapted to convection-dominated flow problems.

3. THE MIXED FINITE ELEMENTS METHODS FOR THE CONVECTION-DIFFUSION PROBLEMS.

Let $\overline{\Omega} = \bigcup_{T \in \mathcal{T}_h} T$ be a triangulation of $\overline{\Omega}$ as in §.2. To such a triangulation, we associate the space

(3.1) $$\mathcal{M}^h = \{v \in L^2(\Omega) \; ; \; \forall \; T \in \mathcal{T}_h \, , \; v|_T \in H^1(T)\} \; .$$

On an interface $T' = \partial T_1 \cap \partial T_2$, a function $v \in \mathcal{M}^h$ is discontinuous ; we note $[v]$ its mean value on this interface and δv its jump :

$$[v] = \frac{1}{2} \sum_{i=1}^{2} v|_{\partial T_i} \quad \text{on} \quad T' = \partial T_1 \cap \partial T_2 \; ,$$

$$\delta v = - \sum_{i=1}^{2} \text{sgn}(\Phi \cdot \nu_{T_i}) \; v|_{\partial T_i} \quad \text{on} \quad T'$$

where ν_{T_i} is the unit outward normal on ∂T_i. The above notations will be extended to a boundary face with the convention $v = 0$ outward of Ω .

The solution (p,u) of the system (2.6) belongs to $W \times \mathcal{M}^h$ and is such that

(3.2a) $$p = \mathcal{K} \text{ grad } u \quad \text{in} \quad \Omega \, , \; p.\nu + \varphi \, u = g \quad \text{on} \quad \Gamma \; ,$$

(3.2b) $$\text{div } p - \Phi \cdot \text{grad } u + f = 0 \quad \text{in} \quad \Omega \; .$$

So it can be characterized as the unique solution of

(3.3a) $$a(p,q) + b(q,u) = G(q) \quad \text{for all} \quad q \in W \; ,$$

(3.3b) $$b(p,v) - c^{\circ}(u,v) = F(v) \quad \text{for all} \quad v \in \mathcal{M}^h$$

with

(3.4a) $$a(p,q) = \int_{\Omega} \mathcal{K}^{-1} p.q \; dx + \int_{\Gamma} \varphi^{-2} (\varphi + \tfrac{1}{2} \, \Phi \cdot \nu) \; p.\nu \; q.\nu \; d\sigma \; ,$$

(3.4b) $$b(q,v) = \int_{\Omega} v \text{ div } q \; dx + \frac{1}{2} \int_{\Gamma} \varphi^{-1} \, \Phi \cdot \nu \; v \; q.\nu \; d\sigma \; ,$$

(3.4c) $$c^{\circ}(u,v) = \frac{1}{2} \sum_{T\in\mathcal{T}_h} \int_T \Phi\cdot(v \text{ grad } u - u \text{ grad } v)dx + \sum_{T\in\mathcal{T}_h} \int_{\partial T} \Phi\cdot\nu_T ([u] - \frac{1}{2} u)v \, d\sigma$$

and

(3.5a) $$G(q) = \int_\Gamma \varphi^{-2}(\varphi + \frac{1}{2}\Phi\cdot\nu) g \, q.\nu \, d\sigma ,$$

(3.5b) $$F(v) = -\int_\Omega fv \, dx + \frac{1}{2}\int_\Gamma \varphi^{-1} \Phi\cdot\nu \, gv \, d\sigma .$$

Remark that in this formulation, it is the same bilinear form $b(.,.)$ on $W \times \mathcal{M}^h$ which appears in the two equations (3.3a) and (3.3b). The bilinear form on $\mathcal{M}^h \times \mathcal{M}^h$ is antisymmetric, in particular we have

(3.6) $$c^{\circ}(v,v) = 0 \quad \text{for all } v \in \mathcal{M}^h .$$

Let W_h and M_h be the spaces given by (2.9) and (2.10) ; M_h is a subspace of $\mathcal{M}^h$ and we deduce from (3.3) the following mixed finite elements method : find$(p_h,u_h) \in W_h \times M_h$ such that

(3.7a) $$a(p_h,q_h) + b(q_h,u_h) = G(q_h) \quad \text{for all } q_h \in W_h ,$$

(3.7b) $$b(p_h,v_h) - c^{\circ}(u_h,v_h) = F(v_h) \quad \text{for all } v_h \in M_h .$$

Theorem 1 : Let $\mathcal{T}_h$ be any triangulation of $\overline{\Omega}$; let k be any non-negative integer ; then with the choices (2.9) for the space W_h and (2.10) for the space M_h , the problem (3.7) has a unique solution (p_h,u_h) .

Proof : Let (p_h,u_h) a solution of (3.7) in the homogeneous case ($f = 0$ and $g = 0$) ; it is sufficient to prove that $p_h = 0$ and $u_h = 0$. In this way, let us take $q_h = p_h$ in (3.7a) and $v_h = -u_h$ in (3.7b) and let us sum the equations ; we obtain

(3.8) $$a(p_h,p_h) + c^{\circ}(u_h,u_h) = 0 .$$

Using (3.6), we conclude that $a(p_h,p_h) = 0$, $p_h = 0$ and therefore $b(q_h,u_h) = 0$ for all $q_h \in W_h$. With the choices (2.9) and (2.10), we deduce $u_h = 0$ as for the pure diffusion problem. □

The scheme which results from (3.7) corresponds to centered difference, no artificial diffusion has been added. As for standard finite elements method, the above method gives solutions with spurious oscillations when the convection is preponderant. An artificial diffusion can be easily introduced by changing the bilinear form $c^\circ(.,.)$ into $c(.,.)$:

$$c(u,v) = c^\circ(u,v) + d(u,v) \tag{3.9}$$

where $d(.,.)$ is a bilinear form on $\mathcal{M}^h \times \mathcal{M}^h$ such that

$$d(v,v) \geq 0 \quad \text{for all} \quad v \in \mathcal{M}^h \ . \tag{3.10}$$

The associated mixed finite elements method is : to find $(p_h,u_h) \in W_h \times M^h$ the solution of

$$\begin{cases} a(p_h,q_h) + b(q_h,u_h) = G(q_h) & \text{for all} \quad q_h \in W_h \ , \quad (3.11a) \\ b(p_h,v_h) - c(u_h,v_h) = F(v_h) & \text{for all} \quad v_h \in M_h \ . \quad (3.11b) \end{cases}$$

<u>Theorem 2</u> : If $c(.,.)$ is a bilinear form on $\mathcal{M}^h \times \mathcal{M}^h$ such that (3.9) and (3.10) hold, then the problem (3.11) has a unique solution (p_h,u_h) .

<u>Proof</u> : In the proof of the theorem 1, we have only to change (3.8) into

$$a(p_h,p_h) + c(u_h,u_h) = 0 \ . \tag{3.12}$$

With (3.6), (3.9) and (3.10), we have $a(p_h,p_h) = - c(u_h,u_h) \leq 0$ and since $a(.,.)$ is positive : $a(p_h,p_h) = 0$.

<u>Example</u> : (Jaffré, 1984). We can choose

$$d(u,v) = -\frac{1}{2} \sum_{T \in \mathcal{T}_h} \int_{\partial T} \Phi \cdot \nu_T \ (\delta u) v \ d\sigma \ . \tag{3.13}$$

In this case, $d(.,.)$ is a symmetric bilinear form such that

$$d(v,v) = \frac{1}{2} \sum_{T' \in \partial \mathcal{T}_h} \int_{T'} |\Phi \cdot \nu_{T'}| \ (\delta v)^2 \ d\sigma$$

where $\partial \mathcal{T}_h$ is the set of the faces of the finite elements $T \in \mathcal{T}_h$. This choice (3.13) leads to an upwind discretization of the convective term which coïncides with the Lesaint-Raviart upwinding method for first order systems (Lesaint and Raviart, 1974).

A variant of this example is to choose

$$(3.14) \qquad d(u,v) = - \sum_{T \in \mathcal{T}_h} \int_{\partial T} \theta_{\partial T} \, (\delta u) v \, d\sigma$$

with a function $\theta_{\partial T}$ defined on each element boundary ∂T such that $\operatorname{sgn}(\theta_{\partial T}) = \operatorname{sgn}(\Phi \cdot \nu_T)$ on ∂T and $\theta_{\partial T_1} + \theta_{\partial T_2} = 0$ on all $\partial T_1 \cap \partial T_2$.

4. ERROR ESTIMATES. Henceforth, (p,u) designs the solution of problem (3.3) and (p_h,u_h) the solution of problem (3.11). Any constant independent of h will be noted C ; the triangulation $\mathcal{T}_h$ is assumed uniformly regular (but this uniformity is used only for a simpler presentation of the results).

It is convenient to work with the following norms :

$$(4.1) \qquad \|q\|_W = \left\{ \|q\|^2_{0,\Omega} + \|q.\nu\|^2_{0,\Gamma} \right\}^{1/2}$$

and

$$(4.2) \qquad \|v\|_{\mathcal{M}} = \left\{ \|v\|^2_{0,\Omega} + h^2 \sum_{T \in \mathcal{T}_h} \|\operatorname{grad} v\|^2_{0,T} \right\}^{1/2}$$

where $\|\cdot\|_0$ are the usual L^2-norms.

We assume moreover that the bilinear form $d(.,.)$ satisfies

$$(4.3) \qquad |d(v,w)| \le C\, h^{-1} \|v\|_{\mathcal{M}} \, \|w\|_{\mathcal{M}} \quad \text{for all } v,w \in \mathcal{M}^h$$

and

$$(4.4) \qquad d(u,v) = 0 \qquad \text{for all } v \in \mathcal{M}^h .$$

Then (4.4) implies that both the equations (3.11a) and (3.11b) hold for the solution (p,u).

First, using (3.3a) and (3.11a) we obtain

$$(4.5) \qquad b(q_h,u_h-v_h) = a(p-p_h,q_h) + b(q_h,u-v_h) \quad \text{for all } (q_h,v_h) \in W_h \times M_h ,$$

and then by the inf-sup property (Brezzi, 1974)

$$(4.6) \qquad \|u-u_h\|_{\mathcal{M}} \le C \left\{ \inf_{v_h \in M_h} \|u-v_h\|_{\mathcal{M}} + \|p-p_h\|_W \right\} .$$

Secondly, we have for all (q_h,v_h) in $W_h \times M_h$:

$$(4.7) \qquad a(p-p_h,p-p_h) + c(u-u_h,u-u_h) = a(p-p_h,p-q_h) + b(p_h-q_h,u-v_h) + b(p-q_h,v_h-u_h) + c(u-u_h,u-v_h) \quad .$$

Let $\rho_u u$ be the $L^2(\Omega)$- projection of u onto M_h ; we have

$$(4.8) \qquad b(p_h-q_h,u-\rho_h u) = 0 \quad \text{for all} \quad q_h \in W_h \quad .$$

Let $\pi_h p$ be the W_h-interpolate of p (Thomas, 1977) ; then we have for all v_h in M_h :

$$(4.9) \qquad b(p-\pi_h p,v_h-u_h) = \frac{1}{2}\int_\Gamma \varphi^{-1}\Phi\cdot\nu\,(v_h-u_h)(p-\pi_h p).\nu\,d\sigma \le C\,h^{-1/2}\|v_h-u_h\|_{\mathcal{M}}\,\|p-\pi_h p\|_W \quad .$$

Now let us take in (4.7) $q_h = \pi_h p$ and $v_h = \rho_h u$; by using (3.10), (4.3), (4.6), (4.8) and (4.9) we obtain

$$(4.10) \qquad \|p-p_h\|_W \le C\left\{\|p-\pi_h p\|_W + h^{-1}\|u-\rho_h u\|_{\mathcal{M}}\right\}$$

As conclusion we have the

Theorem 3 : Let us assume that the solution (p,u) be smooth ; if the artificial diffusion $d(.,.)$ satisfies (3.10), (4.3) and (4.4), then the solution (p_h,u_h) of the mixed finite elements methods (3.11) is such that

$$(4.11) \qquad \|p-p_h\|_{0,\Omega} + \|u-u_h\|_{0,\Omega} \le C\,h^k \quad .$$

As a consequence, the convergence is proved only for $k \ge 1$.

Remark : In the case of the example (3.13), all the conditions (3.10), (4.3) and (4.4) are satisfied.

Numerical experiments on an academic Dirichlet problem in a square (Joly, 1982) confirm these theorical results, for an application of the method to a more physical problem, see for example (Chavent and cø, 1984).

REFERENCES

BREZZI F., On the existence, uniqueness and approximation of saddle-point arising from Lagrangian multipliers, R.A.I.R.O. Anal. Numer. 2, 1974, pp.129-151.

CHAVENT G., COHEN G., DUPUY M., JAFFRE J., DIESTE I., Simulation of two dimensional waterflooding using mixed finite elements, 6th SPE Symposium on Reservoir Simulation, New Orléans, 1984, pp.382-390.

DONEA J., BELYTSCHKO T., SMOLINSKI P., A generalized Galerkin method for steady convection-diffusion problems with application to quadratic shape function elements, Comp. Methods in Applied Mech. and Eng. 48, 1985, pp.25-43.

DOUGLAS J., ROBERTS J.E., Global estimates for mixed methods for second order elliptic equations, Math. Comp., 44, 1985, pp.38-52.

FALK R., OSBORN J., Error estimates for mixed methods, R.A.I.R.O. Anal. Numer., 14, 1980, pp.249-277.

HUANG P.G., LAUNDER B.E., LESCHZINER M.A., Discretization of nonlinear convection processes : a broad-range comparison of four shemes, Comp. Methods in Applied Mech. and Eng., 48, 1985, pp.1-24.

HUGHES T.J.R., BROOKS A., A multidimensional upwind scheme with no crosswind diffusion, in : Finite Element, Methods for Convection Dominated Flows, T.J.R. Hughes ed., ASME, New York, 1979.

JAFFRE J., Eléments finis mixtes et décentrage pour les équations de diffusion-convection, Calcolo, 23, 1984, pp.171-197.

JOHNSON C., NÄVERT U., An analysis of some finite element methods for advection-diffusion problems, in : Analytical and numerical approaches for asymptotic problems in Analysis, Axelsson O., Frank L.S. and Van der Sluis A. (eds), North Holland, 1981.

JOLY P., La méthode des éléments finis mixtes appliquée au problème de diffusion-convection, Thèse 3ème cycle, Univ. Paris VI, 1982.

LESAINT P., RAVIART P.A., On a finite element method for solving the neutron transport equation, in : Mathematical Aspects of Finite Elements in Partial Differential Equations, Carl de Boor ed., Academic Press, 1974.

MIZUKAMI A., HUGHES T.J.R., A Petrov-Galerkin finite element method for convection-dominated flows : an accurate upwinding technique for satisfying the maximum principle, Comp. Methods in Applied Mech. and Eng., 50, 1985, pp.181-193.

NEDELEC J.C., Mixed finite elements in $\mathbb{R}^3$, Numer. Math., 35, 1980, pp.315-341.

PATEL M.K., MARKATOS N.C., CROSS M., A critical evaluation of seven discretization schemes for convection-diffusion equations, Int.J. for Numer. Methods in Fluids, 5, 1985, pp.225-244.

RAVIART P.A., THOMAS J.M. A mixed finite element method for 2nd order elliptic problems, in : Mathematical Aspects of the Finite Element Method, Lecture Notes in Mathematics 506, Springer-Verlag, 1977.

RAVIART P.A., THOMAS J.M., Introduction à l'analyse numérique des équations aux dérivées partielles, Collection Mathématiques Appliquées pour la maîtrise, sous la direction de P.G. CIARLET et J.L. LIONS, Masson, 1983.

THOMAS J.M., Sur l'analyse numérique des méthodes d'éléments finis hybrides et mixtes, Thèse d'Etat, Univ. Paris VI, 1977.

Numerical Approximation of Partial Differential Equations
E.L. Ortiz (Editor)

NUMERICAL METHODS OF SOLUTION OF EVOLUTION EQUATIONS OF HYPERBOLIC TYPE

Milan Pultar
Technical University in Prague
Czechoslovakia

We shall deal with the hyperbolic equations of type

$$\frac{d^2u}{dt^2} + A(t)u(t) = f(t) , \quad t \in (0,T) , \quad T < \infty \qquad /1/$$

with the initial conditions

$$u(0) = u_o , \quad \frac{du}{dt}(0) = u_1 , \qquad /2/$$

where $A(t)$ is supposed to be a symmetric elliptic operator which depends on t.

The above mentioned equations are solved by the method of discretization in time which means that we solve the equations in the following way:

a/ for a uniform partition of $\langle 0,T\rangle$, $h = T/n$ and $t_j = j \cdot h$ we construct an approximate solution

$$u(t) = z_{j-1} + (t - t_{j-1}) h^{-1} (z_j - z_{j-1}) \qquad /3/$$

$$\text{for } t_{j-1} \le t \le t_j , \quad j = 1,\dots,n,$$

where z_j , $j = 1,\dots,n$ are the solutions of the system of equations

$$z_{j+1} - 2 z_j + z_{j-1} + \frac{h^2}{2}\left(A_{j+1} z_{j+1} + A_{j-1} z_{j-1}\right) = h^2 \cdot f_j , \qquad /4/$$

$$z_o = u_o , z_1 - u_o + \frac{h^2}{2} \cdot A_1 z_1 = \frac{h^2}{2} \cdot f_o + h \cdot u_1 , \qquad /5/$$

where $A_j = A(t_j)$, $f_j = f(t_j)$, $j = 1,\dots,n$. In the equations /4/ we compute from the unknown values z_{j-1}, z_j the new value z_{j+1};

b/ under certain assumptions on $A(t)$, $f(t)$ and the initial conditions u_o and u_1 we can prove that the sequence $u_n(t)$ converges for $n \to \infty$ to the unique solution $u(t)$ of /1/, /2/;

c/ under certain assumptions we can prove that the rate of convergence of this method is $O(h^2)$;

d/ the described method is convergent even if we compute the solutions of equations /4/, /5/ only approximately.

The above described method is also called the Rothe method or the method of lines. This method has been used by many authors for solution of parabolic equations (Rektorys, 1971; Kačur, 1978). Hyperbolic equations have been solved by the Rothe method first in (Streiblová, 1980). Questions concerning hyperbolic equations equations have been solved also, for instance (Lions, Magenes, 1968), by using other methods.

Notation

Let V be a reflexive Banach space which is contained in a Hilbert space H. We assume that V is dense in H and V is continuously imbedded in H. We identify the space H with its dual space H^* and denote by V^* the dual space of V. In this way we can identify H with the subspace of V^*:

$$V \subset H \subset V^*. \qquad /6/$$

We denote by $|.|$ and $\|.\|$ the norms in H and V, respectively. The inner product in H as well as the pairing between V and V^* are both denoted by $(.,.)$ which is possible due to /6/ and density V in H .

$((.,.))$ denotes a continuous symmetric bilinear form defined on VxV. Such form can be identified with a symmetric operator $A \in \mathcal{L}(V,V^*)$, where for $u \in V$ we define $Au \in V^*$ in the following way

$$((u,v)) = (Au,v) \quad \text{for all } v \in V .$$

The space of all such operators is denoted by $\mathcal{A}$. Further $((.,.))_t$ denotes the set of continuous bilinear forms defined on VxV for which there exists $\alpha>0$ such that

$$((u,u))_t \geq \alpha \|u\|^2 \quad , \quad u \in V \ , \quad t \in \langle 0,T \rangle . \qquad /7/$$

A_t denotes the set of corresponding continuous symmetric operators. For brevity we denote $((u,u))_t = \|u\|_t^2$. Lastly we denote

$$\delta z_j = \frac{z_j - z_{j-1}}{h} \ , \quad \delta^{n+1} z_j = \frac{\delta^n z_j - \delta^n z_{j-1}}{h} .$$

<u>A priori estimates</u>

The equations /4/ are fulfilled if and only if the equality

$$(\delta^2 z_{j+1},v) + \frac{1}{2}((z_{j+1},v))_{j+1} + \frac{1}{2}((z_{j-1},v))_{j-1} = (f_j,v) \qquad /8/$$

holds for all $v \in V. \left(((.,.))_j = ((.,.))_{t_j} \right)$.

If we put $v = z_{j+1} - z_{j-1}$ we receive the recurrent relation

$$|\delta z_{j+1}|^2 + \frac{1}{2}\|z_{j+1}\|_{j+1}^2 = |\delta z_j|^2 + \frac{1}{2}\|z_{j-1}\|_{j-1}^2 +$$

$$+ h\left(f_j, \delta z_{j+1} + \delta z_j\right) + ((z_{j+1}, z_{j-1}))_{j+1} - ((z_{j+1}, z_{j-1}))_{j-1} . \quad /9/$$

If the operator function A_t is supposed to have the first derivative bounded by the constant K_1 then we obtain from /9/ the following inequality

$$|\delta z_{j+1}|^2 + \frac{1}{2}\left(\|z_{j+1}\|^2 + \|z_j\|^2\right) \leq |\delta z_1|^2 + \frac{1}{2}\left(\|z_1\|^2 + \|z_o\|^2\right) +$$

$$+ h\left(f_j, \delta z_{j+1}\right) + h\sum_{i=2}^{j}\left(f_i + f_{i-1}, \delta z_i\right) + h\left(f_1, \delta z_1\right) +$$

$$+ K_1 \, h \sum_{i=1}^{j} ((z_{i+1}, z_{i-1}))_i \; . \quad /10/$$

Similarly we can derive from /5/

$$|\delta z_1|^2 + \frac{1}{2}\|z_1\|^2 = \frac{1}{2}((z_1, z_o))_1 + \frac{1}{2}\left(f_o, \delta z_1\right) h + \quad /11/$$

$$+ \left(u_1, \delta z_1\right) .$$

From /10/, /11/ Gronwall Lemma yields

<u>Lemma 1.</u> Let $A_t \in W^{1,\infty}(0,T;\mathcal{A})$. Then there exists a constan stant M independent on h such that

$$\|z_j\|_j^2 + |\delta z_j|^2 \leq M\left(\|z_o\|^2 + |\delta z_1|^2 + \max_{i=1,\dots,n} |f_i|^2 \cdot T\right),$$

holds for $j = 1,\dots,n$.

Using the same process gives us next a priori estimates as the following Lemma:

<u>Lemma 2.</u> Let $A_t \in W^{2,\infty}(0,T;\mathcal{A})$, $f \in W^{1,\infty}(0,T;H)$. Then

$$\|z_j\|_j^2 + |\delta^2 z_j|^2 \le C_1\left(\|u_0\| , \|u_1\| , |f_0 - A_0 u_0| , \|f\|_{1,\infty}\right)$$

holds where C_1 is constant independent on h.

The main results

From the a priori estimates we can prove that the sequence $u_n(t)$ is convergent, the limit function exists and has certain properties. The following theorms can be formulated:

Theorem 1. Let $A_0 u_0 \in H$, $u_1 \in V$, $A_t \in W^{2,\infty}(0,T;\mathcal{A})$, $f \in W^{1,\infty}(0,T;H)$. Then there exists a unique function u with the following properties:

a/ $A_t\, u(t) \in H$ for a.a. $t \in \langle 0,T\rangle$,

b/ $u' \in L_\infty(0,T;V)$, $u'' \in L_\infty(0,T;H)$,

c/ $A_t\, u(t) + u''(t) = f(t)$ holds for a.a. $t \in \langle 0,T\rangle$,

d/ $u(0) = u_0$, $u'(0) = u_1$,

e/ if two functions u and v have the above defined properties, where u is a solution corresponding to initial conditions u_0, u_1 and the right side f, v is a solution corresponding to the initial conditions v_0, v_1 and the right hand side g, then the estimate

$$\|u(t) - v(t)\|_t^2 + |u'(t) - v'(t)|^2 \le$$

$$\le \left(K_0 \|u_0 - v_0\|_0^2 + |u_1 - v_1|^2 + \|f - g\|^2_{L_2 0,T;H}\right) e^{T\max(1,K_1)}$$

holds for $t \in \langle 0,T\rangle$, where K_0 is constant such that $\|A_t\| \le K_0$,

f/ the estimate

$$\alpha \|u(t) - u_n(t)\|^2 + |u'(t) - U_n(t)|^2 \le TMh\, e^{K_1 T}$$

holds for $t \in \langle 0,T\rangle$ where M is a certain constant and $U_n(t)$

is a function constructed from the values $(z_{j+1} - z_{j-1})/2h$ in the same way as the function u_n from the values z_j (see /3/).

<u>Theorem 2.</u> Let $A_t \in W^{3,\infty}(0,T;\mathcal{A})$, $f \in W^{2,\infty}(0,T;H)$, $f(0) - A_o u_o \in V$. Let there exist a derivative in H of the function $A_t u_o$ at the point $t = 0$ and let there exist $\delta > 0$ such that the function $A_t u_1$ is bounded in H for $0 \leq t < \delta$. Then, in addition to the assertions of Theorem 2., the following assertions take place:

a/ $u'' \in L_\infty(0,T;V)$, $u''' \in L_\infty(0,T;H)$,

b/ an estimate similar to the estimate f/ of the previous theorem holds for the first and second derivatives of solutions,

c/ there exist constants M_1 and M_2 such that

$$\|u_n(t) - u(t)\| \leq M_1 h \quad , \quad |U_n(t) - u'(t)| \leq M_2 h$$

takes place for $t \in \langle 0,T \rangle$.

<u>Theorem 3.</u> Let $A_t \in W^{4,\infty}(0,T;\mathcal{A})$, $f \in W^{3,\infty}(0,T;H)$, $f'(0) - A_o u_1 \in V$, $f''(0) - A_o f_o + A_o^2 u_o \in H$. Then there exists a unique function u such that

a/ $u''' \in L_\infty(0,T;V)$, $u'''' \in L_\infty(0,T;H)$,

b/ there exist constants M_2 and M_3 such that

$$\|u_n(t) - u(t)\| \leq M_2 h^2 \quad , \quad |U_n(t) - u'(t)| \leq M_3 h^2$$

takes place for $t \in \langle 0,T \rangle$.

We suggest shortly the proof of the assertion b/ of the last theorem. Let us consider two systems of partition

$h = T/n$ with the approximate solutions z_j, $j = 0,\dots,n$,
$h' = T/2n$ with the approximate solutions z'_j, $j = 0,\dots,2n$.

We can write

$$z_{j+1} - 2\,z_j + z_{j-1} + \frac{h^2}{2}\left(A_{j+1}\,z_{j+1} + A_{j-1}\,z_{j-1}\right) = h^2 \cdot f_j \,,$$

$$z'_{2j+1} - 2\,z'_{2j} + z'_{2j-1} + \frac{h^2}{8}\left(A'_{2j+1}\,z'_{2j+1} + A'_{2j-1}\,z'_{2j-1}\right) = $$

$$= \frac{h^2}{4} \cdot f_j \,,$$

where $A'_j = A\,(jh/2)$. After denoting $p_j = z'_{2j} - z_j$, multiplying the second equation by 4 and subtracting these two equations we obtain

$$p_{j+1} - 2\,p_j + p_{j-1} + \frac{h^2}{2}\left(A_{j+1}\,p_{j+1} + A_{j-1}\,p_{j-1}\right) =$$

$$= h^4\left[\,\delta^4 z'_{2j+2} + \frac{1}{2}\,\delta^2\left(A'_{2j+2}\,z'_{2j+2} + A'_{2j+1}\,z'_{2j+1} + A'_{2j}\,z'_{2j}\right)\right]$$

where the expression in square brackets is bounded independently on h (a priori estimate) and from Lemma 1. we can prove the existence of constant N such that $\|z'_{2j} - z_j\| \le$ $\le N\,h^2$. Repeating this process for the double h' and $h'' = h/4$ gives us

$$\|z''_{4j} - z'_{2j}\| \le N \cdot \frac{h^2}{4}.$$

Further we consider the double h'' and $h''' = h/8$, etc. From the triangular inequality and by the limiting process then we obtain

$$\|u(t_j) - z_j\| \le \frac{4}{3}\,N \cdot h^2 \,.$$

The existence of the constant M_2 is a consequnce of Lemma 2. The proof ofthe second estimate is analogous.

In the end we mention the problem of convergence of the Rothe method in the case when the equations /4/,/5/

are solved only approximately, i.e. we consider the system of equations

$$\delta^2 z_{j+1} + \frac{1}{2}\left(A_{j+1} z_{j+1} + A_{j-1} z_{j-1}\right) = f_j + R_j ,$$

$$z_o = u_o , (z_1 - u_o)/h^2 + \frac{1}{2} A_1 z_1 = \frac{1}{2} \cdot f_o + u_1/h + R_o ,$$

where R_j , j = 0,...,n, express the error of the solution of each equation of system /4/, /5/. Using Lemma 1. yields the existence of constant K such that

$$\| \bar{z}_j - z_j \|^2 + | \delta \bar{z}_j - \delta z_j |^2 \leq K \max_{i=0,\dots,n-1} | R_i |^2$$

holds for j = 1,...,n.

REFERENCES

1. Dahlquist G., A special stability problem for linear multistep methods, BIT 2/1963, 27-43
2. Kačur J., Method of Rothe and nonlinear parabolic boundary value problems of arbitrary order. Czech. Math. J., 28/1978/, 507-524
3. Lions J., Magenes L., Problèmes aux limites non homogènes et applications. Dunod, Paris, 1968
4. Pultar M., Solution of abstract hyperbolic equations by Rothe method, Aplikace matematiky, vol.29, 1984, 23/39
5. Rektorys K., The method of discretization in time and partial differential equations, D. Reidel publ. comp., Dordrecht/Boston/London, 1982
6. Streiblová J., Solution of hyperbolic problems by the Rothe method, Bull. of the Faculty of Civil Engineering in Prague /To appear/
7. Vitásek E., A-stability and numerical solution of evolution equations, IAC Istituto per le applicazioni del calcolo, serie III-N.186, 5/42

Numerical Approximation of Partial Differential Equations
E.L. Ortiz (Editor)

SOME ASPECTS OF NUMERICAL SOLUTION OF EVOLUTION EQUATIONS BY THE METHOD OF DISCRETIZATION IN TIME

Karel Rektorys
Technical University Prague
Prague, Czechoslovakia

The method of discretization in time (the Rothe method, or horizontal method of lines) has been shown to be a powerful tool of solution of a broad scale of evolution problems, as concerns both theoretical and numerical questions. The technics, suggested by the author (Rektorys, 1971), made it possible to treat these questions in a particularly simple way, while, at the same time, new results have been obtained. Some of them, interesting especially from the point of view of numerical analysis (convergence theorems, including convergence of the "Ritz-Rothe method", error estimates, etc.), are presented below.

Many years ago, E. Rothe suggested an approximate method of solution of a parabolic problem of the second order in two dimensions. Let us show his idea on the following trivial example:

$$\frac{\partial u}{\partial t} - \frac{\partial^2 u}{\partial x^2} = 0 \quad \text{in} \quad (0,\ell) \times (0,T) \ ,$$

$$u(x,0) = u_0(x) \ ,$$

$$u(0,t) = u(\ell,t) = 0 \ .$$

Let us divide the interval $I = [0,T]$ into p subintervals I_j , $j = 1,\dots,p$, of the lengh $h = T/p$, with points of division t_j . For every $t = t_j$ let us approximate the (not yet known) function $u(x,t_j)$ by a function $z_j(x)$ and the derivative $\partial u/\partial t$ in (1) by the difference quotient

$$Z_j(x) = \frac{z_j(x) - z_{j-1}(x)}{h} .$$

Starting by the function $z_0(x) = u_0(x)$, the functions $z_1(x),\ldots,z_p(x)$ are determined subsequently as solutions of ordinary boundary value problems

$$\frac{z_1 - z_0}{h} + z_1'' = 0 , \quad z_1(0) = 0 , \quad z_1(\ell) = 0 ,$$

$$\frac{z_2 - z_1}{h} + z_2'' = 0 , \quad z_2(0) = 0 , \quad z_2(\ell) = 0 ,$$

$$\cdots\cdots\cdots\cdots$$

Having obtained the functions $z_1(x),\ldots,z_p(x)$, the so-called *Rothe function* $u_1(x,t)$ is defined in the whole region $(0,\ell) \times (0,T)$ by

$$u_1(x,t) = z_{j-1}(x) + \frac{z_j(x) - z_{j-1}(x)}{h}(t - t_{j-1})$$

in I_j , $j = 1,\ldots,p$, thus piecewise linear in t for every x fixed and assuming the values $z_j(x)$ for every $t = t_j$. By refining the original division, thus constructing divisions with the steps $h/2$, $h/4$, $h/8,\ldots$, we obtain a sequence $\{u_n(x,t)\}$ of corresponding Rothe functions which can be expected to converge (in an appropriate space) to the solution u (in an appropriate sence) of the given problem.

The problem solved by E. Rothe was a simple one, and for years, the method was forgotten. It was revived by O. A. Ladyženskaja, and further applied by other renowned mathematicians to the solution of substantially more complicated problems (parabolic problems, linear and quasilinear, in N dimensions, then problems of arbitrary order, nonlinear problems, the Stephan problem, hyperbolic problems, mixed parabolic-hyperbolic problems, etc.). Many of the obtained results were obtained, as well, by other methods (method of semigroups, method of monotone oprators, the Fourier method, etc.). As concerns numerical point of view, related methods (methods of space-, or time-space discretization) were applied as well. Each of these methods, including the method of time-discretization (= the Rothe method) has its preferences and its drawbacks. However, the Rothe method has its significance both as a numerical method and theoretical tool. Existence theorems are proved in a constructive way. Thus no other methods are needed to give preliminary information on existence, or regularity of the solution as required in many other numerical methods when questions on convergence, or order of convergence, etc. are to be answered. The Rothe method is a stable method. To the solution of elliptic problems generated by this method, current methods, especially the variational ones, can be applied. As concerns theoretical results, they are obtained in a relatively simple way, as usual. Moreover, the Rothe method, being a very natural one, makes it possible to get a particularly good insight into the structure of the solutions. This is why I prefer it.

In one of my works (Rektorys, 1971) a slightly different technics was given than that applied currently in this method, making it possible to treat corresponding elliptic problems in a particularly simple way. This technics was followed by other authors and became a base for the work of my seminar at the Technical University in Prague. Results obtained in this seminar were summarized in a monograph

(Rektorys, 1982). I would like to present some of them here, namely those, interesting from the numerical point of view, pointing out the very simple way in which they have been obtained.

1. *A convergence theorem.* To begin with, consider a relatively simple parabolic problem

$$\frac{\partial u}{\partial t} + Au = f \quad \text{in} \quad G \times (0,T) \ , \tag{1}$$

$$u(x,0) = 0 \ , \tag{2}$$

$$B_i u = 0 \quad \text{on} \quad \Gamma \times (0,T) \ , \quad i = 1,\ldots,\mu \ , \tag{3}$$

$$C_i u = 0 \quad \text{on} \quad \Gamma \times (0,T) \ , \quad i = 1,\ldots,k-\mu \ . \tag{4}$$

Here, G is a bounded region in E_N with a Lipschitz boundary Γ , A is a linear operator of order 2k, of the form

$$A = \sum_{|i|,|j| \leq k} (-1)^{|i|} D^i (a_{ij}(x) D^j) \ , \tag{5}$$

with a_{ij} bounded and measurable in G , $f \in L_2(G)$; (3), or (4) are (linear) boundary conditions, stable (thus containing derivatives of orders $\leq k - 1$), or unstable with respect to the operator A , respectively. Denote

$$V = \{v;\ v \in W_2^{(k)}(G) \ , B_i v = 0 \ \text{on} \ \Gamma \ \text{in the sence of traces}, \ i = 1,\ldots,\mu\} \ . \tag{6}$$

Let the Rothe method be applied, thus let, for each $t = t_j$, the derivative $\partial u/\partial t$ be replaced by

$$Z_i(x) = \frac{z_i(x) - z_{i-1}(x)}{h} \ .$$

In the weak formulation, we thus have to find, subsequently, such functions

$$z_j \in V \ , \quad j = 1,\ldots,p \ , \tag{7}$$

for which the integral identities

$$((v,z_j)) + (v,Z_j) = (v,f) \quad \forall v \in V \tag{8}$$

are satisfied, with $z_0(x) = u(x,0) \equiv 0$. Here, $(.,.)$ is the scalar product in $L_2(G)$, $((.,.))$ is the bilinear form corresponding to the operator A and to the boundary conditions (3), (4), familiar from the theory of variational methods, see, e.g. (Rektorys, 1979). (For example, for $A = -\Delta$, $u = 0$ on Γ , we have

$$V = \overset{o}{W}{}_2^{(1)}(G) \ , \quad ((v,u)) = \sum_{i=1}^{N} \int_G \frac{\partial v}{\partial x_i} \frac{\partial u}{\partial x_i} \, dx \ .)$$

Let positive numbers K , α (independent of v and u) exist such that the inequalities

$$|((v,u))| \leqq K||v||_V||u||_V \ , \tag{9}$$

$$((v,v)) \geqq \alpha||v||_V^2 \tag{10}$$

hold for all $v, u \in V$. (V-boundedness and V-ellipticity of the form $((.,.))$.) Under the assumptions (9), (10) each of the problems (7), (8) is uniquely solvable and thus the above mentioned Rothe function can be constructed. To obtain information about the sequence $\{u_n(x,t)\}$ of these Rothe functions, some *a priori estimates* are needed. Let us draw the attention to the fact in what a simple way they are obtained: Put $v = Z_1 = (z_1 - z_0)/h = z_1/h$ into (8) written for $j = 1$. We get

$$h((Z_1,Z_1)) + (Z_1,Z_1) = (Z_1,f) \ . \tag{11}$$

Because of (10) and $|(Z_1,f)| \leqq ||Z_1|| \ ||f||$, (11) yields

$$||Z_1||^2 \leqq ||Z_1|| \ ||f|| \Rightarrow ||Z_1|| \leqq ||f|| \ . \tag{12}$$

Subtracting (8), written for $j - 1$, from (8), gives

$$h((v, Z_j)) + (v, Z_j - Z_{j-1}) = 0 .$$

Putting $v = Z_j$, we obtain, in a similar way as before,

$$||Z_j|| \leqq ||Z_{j-1}|| , \quad j = 2, \ldots, p ,$$

what gives, together with (12),

$$||Z_j|| \leqq ||f|| = c_1 . \tag{13}$$

Let us refine our division, considering the divisions d_n with the steps $h_n = h_1/2^{n-1}$, $n = 1, 2, \ldots$, $h_1 = h$. Denote the corresponding functions

$$z_j^n , \quad Z_j^n = \frac{z_j^n - z_{j-1}^n}{h_n} .$$

The estimate (13), having been obtained independently of the lengh of the step h , remains valid as well for the division d_n ,

$$||Z_j^n|| \leqq c_1 . \tag{14}$$

Because $z_j^n = (z_j^n - z_{j-1}^n) + \ldots + (z_1^n - z_0^n)$, it follows

$$||z_j^n|| \leqq jh_n(||Z_j|| + \ldots + ||Z_1||) \leqq Tc_1 = c_2 . \tag{15}$$

Putting then $v = z_j^n$ into (8) written for the functions z_j^n and Z_j^n and using (10), we get

$$||z_j^n||_V \leqq c_3 . \tag{16}$$

(14), (15) and (16) are the basic needed a priori estimates. They have actually been obtained in a very, very simple way. Let

$$u_n(t) = z_{j-1}^n + (z_j^n - z_{j-1}^n) \frac{t - t_{j-1}^n}{h} \qquad \text{for } t_{j-1}^n \leqq t \leqq t_j^n , \quad j = 1, \ldots, p \cdot 2^{n-1} , \tag{17}$$

$n = 1,2,\ldots$ (the Rothe functions), or

$$U_n(t) = \begin{cases} z_1^n & \text{for} \quad t = 0 \,, \\ z_j^n & \text{for} \quad t_{j-1}^n < t \leqq t_j^n \,, \quad j = 1,\ldots,p\cdot 2^{n-1} \end{cases} \tag{18}$$

$n = 1,2,\ldots$ be abstract functions, considered as functions from $I = [0,T]$ into V, or $L_2(G)$, respectively. In consequence of their form and of (16) and (14), they are uniformly bouded (with respect to n) in $L_2(I,V)$, or $L_2(I,L_2(G))$, respectively (even in $C(I,V)$, or $L_\infty(I,L_2(G))$. The spaces $L_2(I,V)$, and $L_2(I,L_2(G))$ being Hilbert spaces, a subsequence $\{u_{n_k}\}$, or $\{U_{n_k}\}$ can be found such that

$$u_{n_k} \rightharpoonup u \quad \text{in} \quad L_2(I,V) \,, \quad U_{n_k} \rightharpoonup U \quad \text{in} \quad L_2(I,L_2(G)) \,. \tag{19}$$

It easily follows (Rektorys, 1982, Chap. 11) that

$$u \in L_2(I,V) \cap AC(I,L_2(G)) \,, \tag{20}$$

$$u' = U \in L_2(I,L_2(G)) \,, \tag{21}$$

$$u(0) = 0 \quad \text{in} \quad C(I,L_2(G)) \,, \tag{22}$$

$$\int_0^T ((v,u))\, dt + \int_0^T (v,u')\, dt = \int_0^T (v,f)\, dt \quad \forall v \in L_2(I,V) \,. \tag{23}$$

A function with the properties (20) - (23) is called the *weak solution* of the problem (1) - (4). Uniqueness is then proved implying convergence of the whole sequence $\{u_n\}$ to u :

Theorem 1. Let (9), (10) be satisfied, let $f \in L_2(G)$. Then there exists exactly one weak solution of the problem (1) - (4) and

$$u_n \rightharpoonup u \quad \text{in} \quad L_2(I,V) \;, \quad u_n \Rightarrow u \quad \text{in} \quad L_2(I,L_2(G)) \;. \qquad (24)$$

Remark 1. By a more detailed treatment it can be proved that even $u_n \Rightarrow u$ in $C(I,V)$, $u' \in L_\infty(I,L_2(G))$, further that the assumption (10) can be weakened, etc. We do not go into details here. See (Rektorys, 1982, Chap. 21).

2. *An error Estimate.* As can be expected, to get an efficient error estimate, some supplementary assumptions are needed: Let the assumptions of Theorem 1 be fulfilled. Let, moreover,

$$f \in V \;, \quad Af \in L_2(G) \;, \quad ((v,f)) = (v,Af) \quad \forall\, v \in V. \qquad (25)$$

Then

$$||u(x,t_j) - z_j(x)|| \leqq \frac{Mjh^2}{2} \;, \quad j = 1,\dots,p \;, \qquad (26)$$

where $M = ||Af||$.

Using the technics applied above, the estimate (26) is easily derived, estimating, for $t = t_j$ fixed, the difference between $z_j(x)$ and corresponding functions, obtained by the Rothe method when refining the division of the interval I and coming to the limit (Rektorys, 1982, Chap. 12). For these estimations, uniform boundedness, in $L_2(G)$, of higher difference quotients is needed, what is ensured when (25) is fulfilled.

From among the assumptions (25), only the first is restrictive in applications. The estimate (26) is very sharp, as seen from a lot of numerical examples given in (Rektorys, 1982), and cannot be improved substantially.

3. *Convergence of the Ritz-Rothe method.* Let the problems (7), (8) be solved approximately, to be concrete, by the Ritz method (or by a method with similar properties). We speak briefly of the "Ritz-Rothe method. So let $v_1, \ldots$ $\ldots, v_n$ be the first n terms of a base in V and let z_1^* be the Ritz approximation of the function z_1. Put z_1^* instead of z_1 into the second of the identities (8),

$$((v, \tilde{z}_2)) + \frac{1}{h}(v, \tilde{z}_2 - z_1^*) = (v, f) , \tag{27}$$

and let z_2^* be the Ritz approximation of the function $\tilde{z}_2$, etc. We thus can construct the function

$$u_1^*(t) = z_{j-1}^* + \frac{z_j^* - z_{j-1}^*}{h}(t - t_{j-1}) \quad \text{for } t_{j-1} \leqq t \leqq t_j , \quad j = 1, \ldots, p , \tag{28}$$

which is an analogue of the Rothe function $u_1(t)$. (27) announces that using the Ritz method, the errors become cumulated with increasing j. However, according to a very simple law: Substract (27) from the second of the identities (8). We obtain

$$((v, z_2 - \tilde{z}_2)) + \frac{1}{h}(v, (z_2 - \tilde{z}_2) - (z_1 - z_1^*)) = 0 \quad \forall v \in V .$$

Putting $v = z_2 - \tilde{z}_2$, we get

$$||z_2 - \tilde{z}_2|| \leqq ||z_1 - z_1^*|| .$$

Etc. Using this result, in (Rektorys, 1982, Chap. 14) the convergence of the Ritz-Rothe method is then proved in the following sense: $\varepsilon > 0$ being given arbitrarily,

$$||u - u_1^*||_{C(I, L_2(G))} < \varepsilon$$

can be achieved if h is sufficiently small and the number

of basic elements in the Ritz method is sufficiently large.

Remark 2. Basic ideas how to obtain results of the type 1,2,3 were shown on the rather simple example (1) - (4) with time-independent operator A and homogeneous initial and boundary conditions. In (Rektorys, 1982), the case of nonhomogeneous initial and boundary conditions is discussed, then the case $A = A(t)$, A nonlinear, further integrodifferential parabolic problems are treated, problems with an integral condition, etc. Existence and convergence theorems (including convergence of the Ritz-Rothe method) are proved. Hyperbolic problems are then treated in a rather nontraditional way and an error estimate, similar to the estimate (26), is derived. See also Pultar, 1984.

Remark 3. In (Rektorys, 1982), the technics shown above is applied as well to obtain results of purely theoretical character, for example regularity results. In particular, the "smoothing effect" for a homogeneous parabolic equation with a nonhomogeneous initial condition $u(x,0) = u_0 \in L_2(G)$ is derived in a relatively very simple way.

REFERENCES

1. Rektorys, K., On Application of Direct Variational Methods to the Solution of Parabolic Boundary Value Problems of Arbitrary Order. Czech. Math. J. 21, 1971, 318-339.
2. Rektorys, K., Variational Methods in Mathematics, Science and Engineering, 2nd Ed., D. Reidel, Dordrecht-Boston-London, 1979.
3. Rektorys, K., The Method of Discretization in Time and Partial Differential Equations, D. Reidel, Dordrecht-Boston-London, 1982.
4. Pultar, M., Solution of Abstract Hyperbolic Equations by Rothe Method. Aplikace matematiky 29, 1984, 23-39.

Part III: POLYNOMIAL AND RATIONAL APPROXIMATION METHODS

Tau Method, Collocation and Spectral Techniques

Numerical Approximation of Partial Differential Equations
E.L. Ortiz (Editor)

SOME RESULTS ABOUT THE SPECTRUM OF THE CHEBYSHEV DIFFERENCING OPERATOR

Daniele Funaro

Dipartimento di Matematica - Università di Pavia

Strada Nuova 65 - 27100 Pavia- ITALY (*)

A spectral method in space and a finite-difference scheme in time are employed to approximate the solution of the model equation: $y_t=y_x$. The operator $\partial/\partial x$ is discretized by the collocation method based on the Chebyshev nodes. The second order Runge-Kutta method is used for the operator $\partial/\partial t$. It is known that the location, in the complex plane, of the eigenvalues of the collocation matrix is crucial for the stability. A simple way of computing the coefficients of the characteristic polynomial of that matrix is shown. An explicit computation of the roots gives indications on the choice of the time step.

INTRODUCTION

In the present paper a scheme for the numerical resolution of a very simple equation will be discussed. But, as in numerical analysis often happens, the gap, between the proposal of a good computational procedure and its theoretical justification, is wide. So, even in our trivial case, a mathematically rigorous study of the dealt questions, is still not accomplished.

For a fixed T>0 and a given initial datum $y_0:]-1,1[\to\mathbb{C}$, we are concerned with finding the approximate solution $y:]-1,1[\times]0,T[\to\mathbb{C}$ of the differential equation, formally

(*) *Work performed in the research program of the Istituto di Analisi Numerica del C.N.R. - PAVIA.*
This research has been sponsored in part by the United States Army through its European Research Office. Contract N° DAJA-84-C-0035.

defined by the following system:

$$\text{(a)} \quad \begin{cases} \frac{\partial}{\partial t}y(x,t) = \frac{\partial}{\partial x}y(x,t) & \text{on }]-1,1[\times]0,T[\\ y(1,t) = 0 & t\in]0,T[\\ y(x,0) = y_0(x) & x\in]-1,1[\end{cases}$$

We shall analyse in particular, the case in which the operator $\partial/\partial x$ is discretized by the collocation method based on the Chebyshev nodes and, at the same time,for the operator $\partial/\partial t$ is used an explicit finite-difference scheme (namely the second order Runge-Kutta method). These methods are well-known and widely utilized. At present our goal is to determine what is the relation between Δt (the time discretization parameter) and N (the spatial discretization parameter) in order to attain the stability of the scheme. The analysis will lay on the knownledge of the eigenvalues of the collocation matrix. We give an explicit expression for the coefficients of the characteristic polynomial, so our analysis is reduced in determaining the position, in the complex plane, of the roots of this polynomial. Some results about the behaviour of these roots are shown theoretically, while other properties are assumed as hypotheses;nevertheless, in all the numerical tests performed, these were always satisfied.

1 - THE SEMI-DISCRETE PROBLEM

We shall begin our analysis by considering the approximation in the variable x of the problem (a) by means of the pseudo-spectral method. For a fixed integer $N \geq 2$ the collocation will be made, in the interval $[-1,1]$, on the Chebyshev nodes: $x_j = \cos(\pi j/N)$ $j=0,1,\ldots,N$. If $T_N(x) = \cos(N \arccos x)$ is the N-degree Chebyshev polynomial of the first kind, we have: $|T_N(x)|=1$ if and only if $x=x_j$ $j=0,1,\ldots,N$.

We shall denote by $\mathbb{P}_N$ the space of the polynomials $\phi:[-1,1]\to\mathbb{C}$ whose degree is not greater than N. Moreover V_N will be the subset of $\mathbb{P}_N$ of the polynomials which vanish at the point $x_0=1$. We can now define a discrete differential operator $L_N: V_N \to \mathbb{P}_{N-1}$ such that:

$$(L_N\phi)(x_j) = \phi_x(x_j) \qquad j=1,2,\ldots,N; \quad \forall\phi\in V_N.$$

The operator L_N can be represented, in a suitable basis, as a matrix. We shall choose as basis of V_N, that formed by the polynomials ℓ_i $i=1,2,\ldots,N$ defined as follows:

$$\ell_i(x_j) = \begin{cases} \delta_{ij} & i=1,2,\ldots,N-1 \\ 2\delta_{Nj} & i=N \end{cases}$$

The correspondent matrix is:

$$(1.1) \qquad M = \{m_{ij}\}_{i,j=1,2,\ldots,N} = \left\{ \frac{d}{dx}\ell_i(x) \Big|_{x=x_j} \right\}_{i,j=1,2,\ldots,N}$$

Explicitely we have (see also GOTTLIEB & TURKEL, 1983):

$$(1.2) \qquad m_{ij} = \begin{cases} \dfrac{(-1)^{i+j}}{x_i - x_j} & \text{if } i \neq j \\[2ex] \dfrac{-x_i}{2(1-x_i^2)} & \text{if } i=j\neq N \\[2ex] -\dfrac{2N^2+1}{6} & \text{if } i=j=N \end{cases}$$

We notice that it is possible to write: $M = D+A$ where D is diagonal and A is skew-symmetric.

At this point we are able to present the semi-discrete problem:

find $y_N(t)\in V_N$ $\forall t\in]0,T[$, *such that*

$$(1.3)\qquad \begin{cases} \frac{d}{dt}y_N(t) = L_N(y_N(t)) & \forall t\in]0,T[\\ y_N(0) = y_{N0} \end{cases}$$

where $y_{N0}\in V_N$ is a suitable approximation of y_0.

In order to study the approximation of the operator d/dt in (1.3), we want to analyse the behaviour of the eigenvalues of the operator L_N.

Namely we want to determine $u\in V_N$ and $\lambda\in\mathbb{C}$ such that:

$$(1.4)\qquad \begin{cases} u_x(x_j) = \lambda u(x_j) & j=1,2,\dots,N \\ u_x(x_0) = 1 \end{cases}$$

(implicitely we have $u(x_0) = 0$)

where the condition on the derivative at the extremal node has to be understood as a normalizing equation.

Setting $f(x) = T_N'(x)(x+1)/2N^2$ we have $f\in\mathbb{P}_N$ (with $f\notin\mathbb{P}_{N-1}$) and:

$$(1.5)\qquad f(x_j) = \begin{cases} 1 & \text{if } j=0 \\ 0 & \text{if } 1\leqslant j\leqslant N \end{cases}$$

Hence for u we get the relation:

$$(1.6)\qquad \begin{cases} u_x = \lambda u+f \\ u(1) = 0 \end{cases}$$

It follows immediately that $\lambda\neq 0$, otherwise it would be $f=u_x\in\mathbb{P}_{N-1}$, that is an absurd.

Hereafter we shall focus our attention on problem (1.6). By differencing the equation in (1.6) n times, we get:

$$(1.7)\qquad u^{(n+1)}(x) = \lambda u^{(n)}(x)+f^{(n)}(x) \qquad \forall n\in\mathbb{N}$$

from which:

$$(1.8)\qquad u^{(n+1)}(x) = \lambda^{n+1}u(x)+\sum_{k=0}^{n} f^{(k)}(x)\,\lambda^{n-k} .$$

Considering that $u\in\mathbb{P}_N$ if and only if $u^{(N+1)}(x)=0\ \forall x\in[-1,1]$, it follows that:

$$(1.9)\qquad \lambda^{N+1}u(x)+\sum_{k=0}^{N} f^{(k)}(x)\,\lambda^{N-k} = 0$$

which implies:

$$(1.10)\qquad u(x) = -\frac{1}{\lambda}\sum_{k=0}^{N}\frac{f^{(k)}(x)}{\lambda^{k}}$$

Finally the condition $u(1)=0$ entails:

$$(1.11)\qquad \sum_{k=0}^{N}\frac{f^{(k)}(1)}{\lambda^{k}} = 0 .$$

Now we set $c_k=f^{(k)}(1)$ $k=0,1,\ldots,N$. Then the N eigenvalues λ_i $i=1,2,\ldots,N$ which solve the problem (1.6) are the N roots of the polynomial:

$$(1.12)\qquad Q(\lambda) = \sum_{k=0}^{N} c_k\,\lambda^{N-k}$$

If we suppose for a moment that Q has N distinct roots, by (1.10) the corresponding N eigenfunctions are given by:

$$(1.13)\qquad u_i(x) = -\sum_{k=0}^{N} f^{(k)}(x)\,\lambda_i^{N-k-1} \qquad u_i\in V_N \quad i=1,2,\ldots,N .$$

Remark - If L_N admits a diagonal form then the λ_i's must be distinct. Otherwise from (1.13) the eigenfunctions would not form a basis fcr V_N.

Let's go now to determine explicitely the coefficients of Q. It is well known (see e.g. RIVLIN,1984) that T_N satisfies the differential equation:

$$(1.14)\qquad (1-x^2)T_N''(x)-xT_N'(x)+N^2T_N(x) = 0 .$$

By deriving (1.14) k times, one obtains:

$$(1.15)\qquad (1-x^2)T_N^{(k+2)}(x)-x(2k+1)T_N^{(k+1)}(x)+(n^2-k^2)T_N^{(k)}(x) = 0 \qquad k\in\mathbb{N}$$

and setting x=1 we get:

$$(1.16)\qquad \begin{cases} T_N^{(k+1)}(1) = \dfrac{N^2-k^2}{2k+1}\, T_N^{(k)}(1) & \forall k\in\mathbb{N} \\ T_N(1) = 1 . \end{cases}$$

Finally we can write:

$$(1.17)\qquad T_N^{(k+1)}(1) = \begin{cases} \prod_{j=0}^{k} \dfrac{N^2-j^2}{2j+1} & \text{if } 0\leqslant k<N \\ 0 & \text{if } k\geqslant N \end{cases}$$

On the other hand we have:

$$(1.18)\qquad f^{(k)}(x) = \frac{1}{2N^2}\left(T_N^{(k+1)}(x)(x+1)+kT_N^{(k)}(x)\right), \quad k\in\mathbb{N}$$

so that our coefficients are given by:

$$(1.19)\qquad c_k = \begin{cases} 1 & \text{if } k=0 \\ \dfrac{1}{2N^2}\,\dfrac{2N^2+k}{2k+1} \prod_{j=0}^{k-1} \dfrac{N^2-j^2}{2j+1} & \text{if } 1\leqslant k\leqslant N . \end{cases}$$

Remark - A different characterization of the eigenfunctions can be given in the following way. Observe first that the general solution of (1.6) can be written as:

$$(1.20)\qquad u(x) = -e^{\lambda x}\int_x^1 e^{-\lambda t} f(t)\, dt$$

In this case we have $u(1)=0$, but not necessarily $u\in\mathbb{P}_N$. By integrating repeatedly by parts we obtain :

$$(1.21)\qquad u(x) = \frac{e^{\lambda x}}{\lambda}\left[e^{-\lambda t}f(t)\right]_x^1 - \frac{e^{\lambda x}}{\lambda}\int_x^1 e^{-\lambda t}f'(t)\,dt =$$

$$= \frac{e^{\lambda(x-1)}}{\lambda}f(1) - \frac{f(x)}{\lambda} + \frac{1}{\lambda}\left(-e^{\lambda x}\int_x^1 e^{-\lambda t}f'(t)dt\right) =$$

$$=\ldots= -\frac{1}{\lambda}\sum_{k=0}^{N}\frac{f^{(k)}(x)}{\lambda^k} + \frac{e^{\lambda(x-1)}}{\lambda}\sum_{k=0}^{N}\frac{f^{(k)}(1)}{\lambda^k} .$$

As this point we can see that $u \in \mathbb{P}_N$ if and only if (1.11) holds. Then in place of (1.13) the following relation is also valid:

$$(1.22) \qquad u_i(x) = -e^{\lambda_i x} \int_x^1 e^{-\lambda_i t} f(t)\, dt\,, \quad i=1,2,\ldots,N\,.$$

2 - SOME PROPERTIES CONCERNING THE POLYNOMIAL Q

Some properties about Q can be proven through the knowledge of its coefficients (we recall (1.19)).
For this purpose we define B(r) as the circle in the complex plane centered in (0,0) and with the radius equal to r.
Recalling the theorem 6.4b, p.451 and the corollary 6.4k, p.457 in HENRICI,1974, we get: $\lambda_i \notin B(r_1)$ and $\lambda_i \in B(r_2)$ $i=1,2,\ldots,N$ where r_1 and r_2 are defined in the following way:

$$(2.1) \quad r_1 = \frac{1}{2} \min_{0 \leqslant k \leqslant N-1} \left| \frac{c_N}{c_k} \right|^{1/(N-k)} \quad \text{and} \quad r_2 = 2 \max_{1 \leqslant k \leqslant N} |c_k|^{1/k}\,.$$

In particular from (1.19) we obtain:

$$(2.2) \qquad \frac{c_N}{c_k} = N \frac{2k+1}{2N^2+k} \sum_{j=k}^{N-1} \frac{N^2-j^2}{2j+1} =$$

$$= \frac{N(N^2-k^2)}{2N^2+k} \prod_{j=k+1}^{N-1} \frac{N^2-j^2}{2j-1} \geqslant \frac{N(N^2-k^2)}{2N^2+k} \geqslant \frac{1}{2}$$

$$k=0,1,\ldots,N-1.$$

On the other hand we have:

$$(2.3) \qquad c_k \leqslant \prod_{j=0}^{k-1} \frac{N^2-j^2}{2j+1} \leqslant N^{2k} \qquad k=1,2,\ldots,N.$$

Hence:

$$(2.4) \qquad r_1 \geqslant \frac{1}{2} \min_{0 \leqslant k \leqslant N-1} \left(\frac{1}{2}\right)^{1/(N-k)} = \frac{1}{2}\left(\frac{1}{2}\right)^{1/N} > \frac{1}{4}$$

$$\text{and} \quad r_2 \leqslant 2N^2\,.$$

From (2.4) we can deduce that $B(r_1) \supset B(1/4)$ and that

$B(r_2) \subset B(2N^2)$. In other words we have: $1/4 < |\lambda_i| \leq 2N^2$, $i=1,2,\dots,N$.

Recalling the matrix M defined by (1.1)-(1.2) and representing the operator L_N, it follows that the characteristic polynomial of M is Q, as defined in (1.12); therefore the trace and the determinant of M are given respectively by $-c_1$ and c_N, that is:

$$(2.5)\qquad -c_1 = tr(M) = \sum_{i=0}^{N} Re\lambda_i = -\frac{1}{2N^2}\cdot\frac{2N^2+1}{3}\cdot N^2 = -\frac{2N^2+1}{6}$$

and

$$(2.6)\qquad c_N = det(M) = \left|\prod_{i=0}^{N}\lambda_i\right| = \frac{1}{2N^2}\,\frac{2N^2+N}{2N+1}\left(\prod_{j=0}^{N-1}\frac{N^2-j^2}{2j+1}\right) =$$

$$= \frac{1}{2N}\left(\prod_{j=0}^{N-1}\frac{1}{2j+1}\right)\left(\prod_{j=0}^{N-1}(N-j)\right)\left(\prod_{j=0}^{N-1}(N+j)\right) =$$

$$= \frac{1}{2N}\cdot\frac{2^{N-1}(N-1)!}{(2N-1)!}\cdot N!\cdot\frac{(2N-1)!}{(N-1)!} = 2^{N-2}(N-1)!$$

3 - TIME DISCRETIZATION FOR PROBLEM (1.3)

We go now to analyse the approximation scheme for the variable t. We consider first an equation of the kind: $y'(t)=\alpha y(t)$ with $\alpha\in\mathbb{C}$ and $Re\alpha<0$. We shall make use of the well known, second order Runge-Kutta method (also known as Modified Euler's Method). Fixed $\Delta t>0$ and given the initial datum u_0 the scheme is the following:

$$(3.1)\qquad \begin{cases} u^0 = u_0 \\ \dfrac{u^{n+\frac12}-u^n}{\Delta t/2} = \alpha u^n \\ \dfrac{u^{n+1}-u^n}{\Delta t} = \alpha u^{n+\frac12} & n>0 \end{cases}$$

or explicitely:

$$(3.2) \qquad \begin{cases} u^0 = u_0 \\ u^{n+1} = (1 + \Delta t\alpha + \dfrac{\Delta t^2\alpha^2}{2})u^n \quad , \quad n>0 \end{cases}$$

Furthermore (3.2) leads to:

$$(3.3) \qquad u^{n+1} = (1 + \Delta t\alpha + \frac{\Delta t^2\alpha^2}{2})^{n+1} u_0 \quad \forall n\in\mathbb{N} .$$

We recall briefly some classical results concerning the analysis of the stability for (3.1).

Set $\rho = \left| 1 + \Delta t\alpha + \dfrac{\Delta t^2\alpha^2}{2} \right|$; then the method is absolutely stable if and only if $\rho<1$.

Let $\alpha=\mathrm{Re}\alpha+i\mathrm{Im}\alpha=R+iI$ and suppose $R<0$; with an easy computation we can find that the values of Δt for which $\rho<1$ are those that satisfy the inequality:

$$(3.4) \qquad K(\Delta t) = \Delta t^3 M^4 + 4\Delta t^2 RM^2 + 8\Delta tR^2 + 8R < 0$$

$$\text{with } M^2 = R^2+I^2 .$$

The relation (3.4) entails an estimate of the kind; $\Delta t<\Delta t^*=\Delta t^*(\alpha)$ where $\Delta t^*\in\mathbb{R}^+$ is the only real number which satisfies $K(\Delta t^*)=0$. The algebraic expression of Δt^* is rather complicated. But, due to the following inequality;

$$(3.5) \qquad K(\Delta t) \leqslant \Delta t^3 M^4 + 4\Delta t^2 RM^2 + 8\Delta tR^2 + \frac{8R^3}{M^2} =$$

$$= (\Delta tM^2 + 2R)(\Delta t^2 M^2 + 2\Delta tR + \frac{4R^2}{M^2})$$

and by considering that $\Delta t^2 M^2+2\Delta tR+4R^2/M^2$ is a positive quantity, (3.4) holds if we choose:

$$(3.6) \qquad \Delta tM^2 + 2R < 0$$

The figure 3 shows, in the plane $\Delta t(R,I)$, the domain of stability of the method and the circle represented by the inequality (3.6).

Finally by (3.6) one has:

$$(3.7) \qquad \Delta t < \frac{-2 \operatorname{Re} \alpha}{|\alpha|^2} \leqslant \Delta t^*(\alpha)$$

which is a sufficient condition to achieve stability.
At this point we can apply the Modified Euler's Method to the differential system (1.3). By the inequality (3.7) and by classical arguments, we can state the following proposition.

Proposition - If L_N admits a diagonal form and Re $\lambda_i<0$ $i=1,2,...,N$, then by choosing:

$$(3.8) \qquad \Delta t < \min_{i=1,2,\dots,N} \frac{-2 \operatorname{Re} \lambda_i}{|\lambda_i|^2} \quad ,$$

the approximation of (1.3) by (3.1) is stable.

The explicit condition (3.8) can be substituted for (3.4) by asking it to be satisfied for every $\alpha=\lambda_i$, $i=1,2,...,N$. Inequality (3.4) is less restrictive than (3.8), but Δt is determined implicitely.

4 - REMARKS AND CONCLUSIONS

It remains to prove that the hypotheses of the proposition of the previous paragraph (i.e.: the eigenvalues of L_N are distinct and with negative real parts) are satisfied for every N. In the second paragraph some properties about the roots of Q (hence about the eigenvalues of L_N) have been proven. Nevertheless they are insufficient to our study. A polynomial whose roots have negative real parts is called *stable*. To determine if a polynomial is stable is known as *Hurwitz problem*. Various sufficient conditions have been found for that (see e.g. HENRICI,1974), but unfortunately their application to our case is hard to please. Some numerical experiments have been executed up to

N=40. The figures 1 and 2 show the position in the complex plane of the roots of Q in the cases N=11 and N=32. These tests seem to support the conjecture that Q is stable, with distinct roots, for every N.

Other experiments have been made about the choice of the time step. For a fixed N, the roots of Q are replaced in (3.4) and the correspondent Δt^* is inferred. Defined $\gamma_N = \Delta t^* N^2$, then we have stability if $\Delta t < \gamma_N / N^2$. The following table shows, for various N, the relative values of Δt^* and γ_N.

N	Δt^*	γ_N
2	1.195	8.783
4	.718	11.488
8	.195	12.480
12	.103	14.832
16	.0627	16.051
20	.0413	16.520
24	.0291	16.761
28	.0216	16.934
32	.0166	16.998
36	.0132	17.115
40	.0107	17.163

We should be tempted to say that:

(4.1) $$\Delta t^* \sim 17/N^2 .$$

A similar statement is found in GOTTLIEB & ORSZAG,1977, although the theoretical and experimental support for it follows a path which is different from the present one. However, at our knowledge, a proof for (4.1) is still to be found for every N.

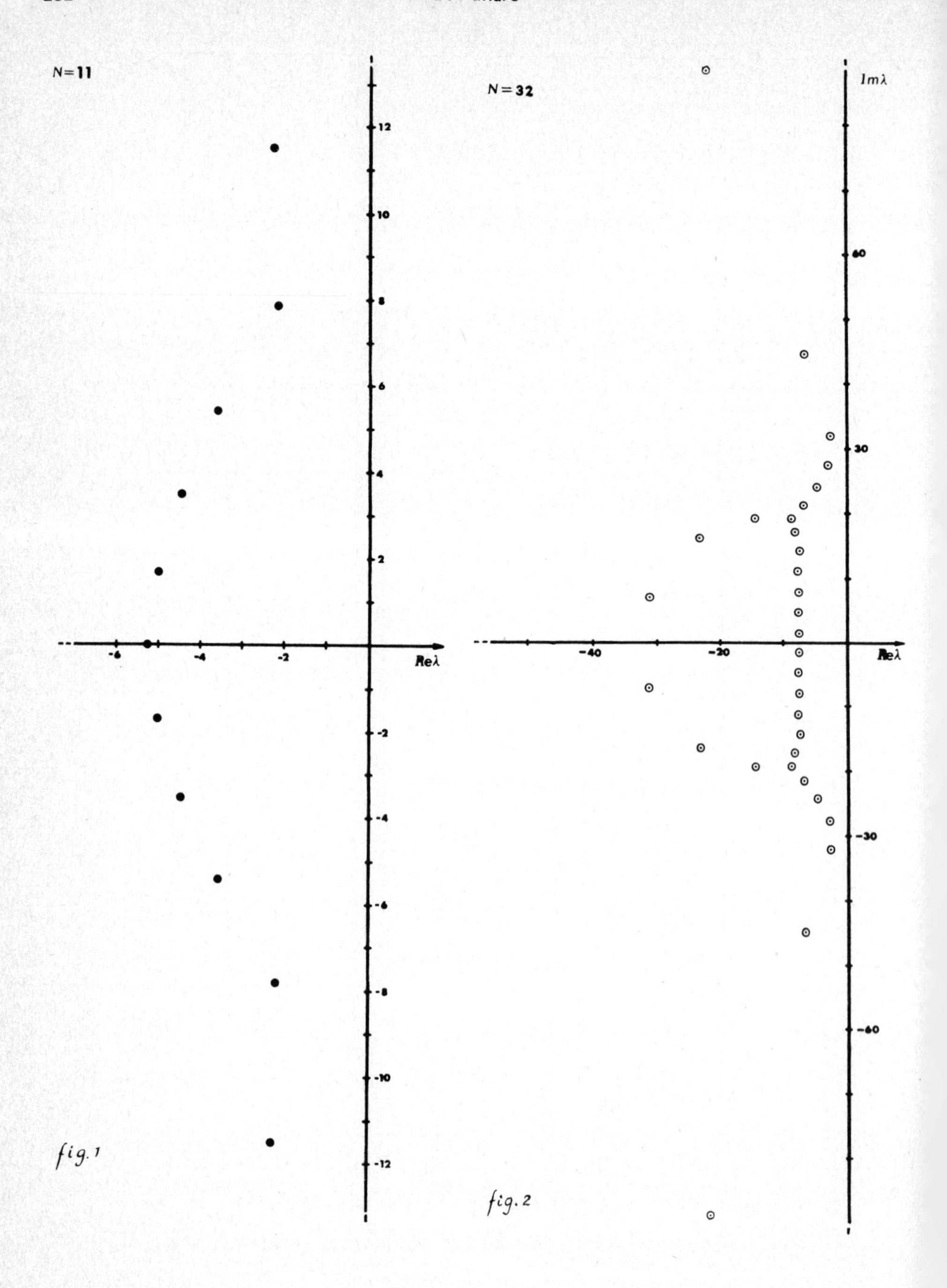
N=11
12
10
8
6
4
2
-2
-4
-6
-8
-10
-12
-6
-4
-2
Re λ
fig. 1
N=32
Im λ
60
30
-30
-60
-40
-20
Re λ
fig. 2

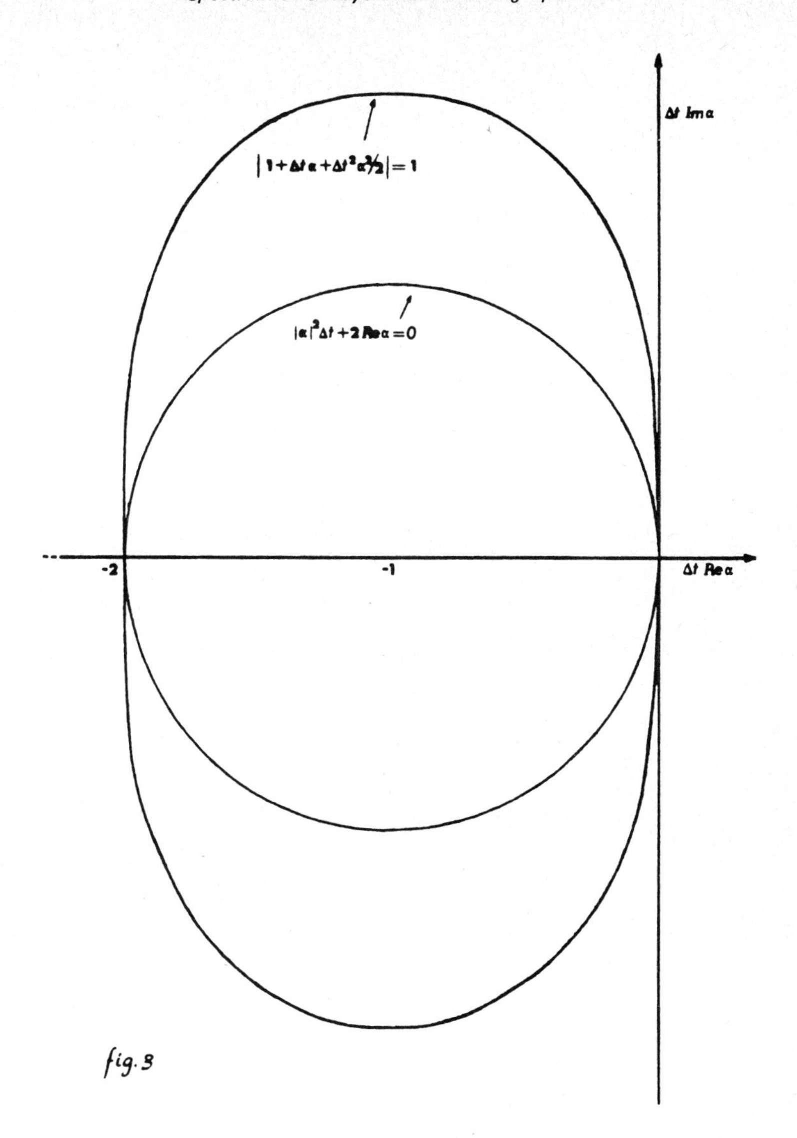

$\Delta t\ Im\,\alpha$
$|1+\Delta t\alpha+\Delta t^2\alpha^2/2|=1$
$|\alpha|^2\Delta t+2\,Re\,\alpha=0$
-2
-1
$\Delta t\ Re\,\alpha$

fig. 3

REFERENCES

1. Gottlieb D. & Orszag S.A., "*Numerical Analysis of Spectral Methods: Theory and Applications*", CBMS Regional Conference Series in Applied Mathematics 26, SIAM (Philadelphia), 1977.

2. Gottlieb D. & Turkel E., "*Spectral methods for time dependent partial differential equations*", ICASE report 83-54.

3. Henrici P., "*Applied and Computational Complex Analysis*", volumes 1,2 John Wiley & Sons (New York), 1974.

4. Rivlin T.J., "*The Chebyshev Polynomials*", John Wiley & Sons (New York), 1974.

Numerical Approximation of Partial Differential Equations
E.L. Ortiz (Editor)

ON A NEW BOUNDARY ELEMENT SPECTRAL METHOD

F.K. Hebeker
Universität-GHS, Fb. 17
D-4790 Paderborn, West Germany

An efficient numerical algorithm for partial differential equations in complicated 3-D geometries is developed in case of viscous fluid flows. The algorithm consists essentially of a combination of a boundary element method (where the resulting algebraic system is solved efficiently with a multigrid procedure) and a spectral method to treat the nonhomogeneous part in the differential equations.
The investigation reported here covers an exact mathematical foundation, a rigorous convergence analysis, and some 3-D numerical tests.

INTRODUCTION

Recently, boundary element methods have been successfully applied to a large variety of problems arising in the engineering sciences (see e.g. Brebbia [2] , Wendland [16]). Various additional fields of application may be opened if boundary element methods are combined with fast spectral methods to overcome the crucial problem of nonhomogeneous terms in the differential equations. These methods, called boundary element spectral methods, are briefly described in the present paper.

To be concrete, let us restrict ourselves to the Stokes problems of viscous hydrodynamics, but most of the proposed methods carry over easily to other problems of the applications. We consider at first the interior Stokes problem

$$- \Delta \underline{u} + \nabla p = \underline{f} \;, \quad \operatorname{div} \underline{u} = o \quad \text{in } \Omega \;, \qquad \underline{u}_{|\partial\Omega} = \underline{g} \quad \text{on the boundary } \partial\Omega \;. \tag{1}$$

Here $\underline{u}$ denotes the velocity vector and p the scalar pressure field of a viscous flow in a (smoothly bounded) 3-D cavity Ω (at constant viscosity $\nu=1$, say). Correspondingly, we have the exterior Stokes problem

$$- \Delta \underline{u} + \nabla p = \underline{f} \quad , \quad \text{div}\, \underline{u} = o \quad \text{in } \mathbb{R}^3 \setminus \overline{\Omega},$$
$$\underline{u}_{|\partial\Omega} = \underline{g} \quad , \quad \underline{u}_{\infty} = o \quad \text{at infinity.} \tag{2}$$

These Stokes problems play a central role in the development of efficient numerical algorithms for the viscous hydrodynamics, for the more general Navier Stokes equations are commonly reduced to problems similiar to (1) or (2) by means of iterative or time-stepping procedures.

The present paper continues some work previously published in (Hebeker [5],[6],[7],[8],[9]).Details and extensions of this paper will be published soon jointly with W. Borchers (Paderborn). Some financial support due to the Deutsche Forschungsgemeinschaft is gratefully acknowledged.

THE BOUNDARY INTEGRAL EQUATIONS

The Stokes differential equations form a generalized elliptic system in the sense of Agmon, Douglis, and Nirenberg. However, a simple fundamental matrix $\underline{\Gamma}$,

$$\Gamma_{ij}(x) = - \frac{1}{8\pi} \left(\frac{\delta_{ij}}{|x|} + \frac{x_i x_j}{|x|^3}\right) \quad , \quad i,j=1,2,3 \tag{3}$$

exists, and the ith column of $\underline{\Gamma}$ solves Stokes equations with the press ure function

$$\Pi_i(x) = -x_i/(4\pi|x|^3) + \text{const.} \tag{4}$$

A hydrodynamical potential theory, fundamental to any boundary element method, is available (Ladyzhenskaja [12]). In particular we are given a Green's representation formula of any regular solution of the Stokes differential system as a sum of a hydrodynamical volumn-, simple layer-, and double layer potential.

At first assume the forces $\underline{f}=\underline{o}$ in (1), (2), later on we will return to the more general case.

Then the interior problem (1) is solvable if the data satisfy

$$\int_{\partial\Omega}(\underline{g}\cdot\underline{n})\, do = 0 \tag{5}$$

(Ladyzhenskaja [12]). In this case the potential ansatz

$$\underline{u} = \underline{W\psi} \quad , \quad p \text{ analogous} \tag{6}$$

in terms of a hydrodynamical double layer potential leads (by means of the jump relations) to the boundary integral equations' system of the second kind (cf. Borchers [1])

$$(\underline{I} + 2\underline{W} + 2\underline{N})\underline{\psi} = 2\underline{g} \quad \text{on} \quad \partial\Omega \tag{7}$$

to determine the unknown surface source vector $\underline{\psi}$. Here denotes $\underline{I}$ the unit matrix, and $\underline{N}$ the boundary operator

$$\underline{N\phi}(x) = \underline{n}(x) \int_{\partial\Omega}(\underline{n}\cdot\underline{\phi})\, do \ .$$

It turns out that (7) has a unique solution $\underline{\psi}$, and consequently the corresponding potential (6) solves the problem (1).

The exterior problem (2) is unconditionally solvable (if $\underline{f}=\underline{o}$), and the potential ansatz (see Hebeker [5],[6],[7],[8],[9])

$$\underline{u} = (\underline{W} + \eta\underline{V})\underline{\psi} \quad , \quad p \text{ analogous} \tag{8}$$

in terms of a combined hydrodynamical double layer- and simple layer potential (η = const., arbitrary) produces the system

$$(\underline{I} - 2\underline{W} - 2\eta\underline{V})\underline{\psi} = - 2\underline{g} \quad \text{on} \quad \partial\Omega. \tag{9}$$

It has been shown (Hebeker [8]) that (9) has a unique solution $\underline{\psi}$ if η is chosen positive. Then the corresponding potential (8) solves the problem (2).

The surface potentials in (7) and (9) form weakly singular integrals on the boundary, hence these integral equations are well fitted for numerical purposes. For the evaluation of the potentials (6) and (8) in the space the Gaussian integral (see Ladyzhenskaja [12]) should be utilized.

THE BOUNDARY ELEMENT SPECTRAL METHOD

The Boundary Element Part

The most simple and efficient way to discretize (7) or (9) seems to be a collocation-type boundary element method. If we restrict ourselves to bodies "similiar to balls", i.e. represented in a satisfactory way by (normalized) polar coordinates:

$$x\in\partial\Omega : \quad x=F(\theta,\phi) \quad , \quad (\theta,\phi)\in [o,1]^2 , \tag{1o}$$

then we decompose $[o,1]^2$ into small quadratic elements of mesh size h. As trial functions we use globally continuous and piecewise bilinear polynomials on the parameter space $[o,1]^2$, subordinated to the given quadrangulation. Then we are looking for an approximate surface source $\underline{\psi}_h$ of this type so that the degrees of freedom are computed from the equations (7) (or (9) resp.). It has to be satisfied at the collocation nodes only, namely at all of the grid points of the mesh. Hence we are led to a linear algebraic system, containing a nonsparse but relatively small system matrix.

For details we refer to the author's previous publication : the numerical quadrature (midpoint rule, or 2×2 Gaussian rule) as well as the multigrid iteration procedure for the linear system has been investigated by a rigorous convergence analysis in (Hebeker [8]).

The Spectral Part

The usual treatment of the general case $\underline{f} \neq \underline{o}$ by means of volumn potentials is extremely expensive. Hence we propose here a spectral method as an alternative. Consider the interior problem (1). We split this linear problem:

$$\underline{u} = \underline{u}_1 + \underline{u}_2 \quad , \quad p = p_1 + p_2 \quad , \tag{11}$$

where $(\underline{u}_1,p_1)$ is <u>any</u> solution of Stokes differential equations, but $(\underline{u}_2,p_2)$ solves (1) with $\underline{f} = \underline{o}$ and the boundary conditions

$$\underline{u}_{2|\partial\Omega} = \underline{g} - \underline{u}_{1|\partial\Omega} \ . \tag{12}$$

Hence we merely have to determine any $\underline{u}_1$, p_1 .

Extend $\underline{f}$ smoothly to a cube C containing Ω so that $\underline{f}_{|\partial C} = \underline{o}$. A complete system of eigensolutions of Stokes equations in a cube with periodic boundary conditions has been constructed in (Borchers [1]). Hence we can project explicitly any Sobolev space (subjected to the periodicity condition) onto its divergence-free subspace.

Therefore, assume $\underline{f}$ divergence-free and periodic in C . Let

$$\underline{f}_o = \frac{1}{|C|} \int_C \underline{f} \, dx \ .$$

Then split:

$$\underline{u}_1 = \underline{u}_3 + \underline{u}_4 \ , \quad p_1 = p_3 + p_4 \ . \tag{13}$$

Here $(\underline{u}_3, p_3)$ is obtained from the Stokes problem

$$- \Delta \underline{u}_3 + \nabla p_3 = \underline{f} - \underline{f}_o \ , \quad \operatorname{div} \underline{u}_3 = o \quad \text{in } C \tag{14}$$

with periodic boundary conditions, and $(\underline{u}_4, p_4)$ solves

$$- \Delta \underline{u}_4 + \nabla p_4 = \underline{f}_o \ , \quad \operatorname{div} \underline{u}_4 = o \quad \text{in } C \ . \tag{15}$$

In (Borchers [1]) it is proved that (14) is solvable (since $\int_C (\underline{f} - \underline{f}_o) \, dx = \underline{o}$) , and any solution is given as a spectral series (which allows to use heavily the FFT). On the other hand, it is easy to construct a particular solution of (15).

ON COMPUTATIONAL RESULTS

Many numerical results in 3-D have been obtained by the author to test the proposed boundary element methods (Hebeker [5],[6],[7],[8], [9]) . They include the viscous drag of a sphere (test against Stokes' exact formula) , the flow around an irregular body, Oseen's flow, or the nonlinear and nonstationary Navier Stokes flow. It turns out that - using a mesh size of $h=1/16$ which leads to invert a nonsparse 867×867 matrix - the mean relative error is commonly of about few percents. In particular, the viscous drag of a sphere is estimated up to an error of o.7% .

Some encouraging numerical examples corresponding to the problem (14) of the spectral part are contained in (Borchers [1]). At present the combination of both software packages to obtain the full boundary element spectral method is under implementation.

REFERENCES

1. Borchers W., Eine Fourier-Spektralmethode für das Stokes-Resolventenproblem, Submitted for publication, Univ. of Paderborn 1985, 13pp.
2. Brebbia C.A., et al. (ed.), *Boundary Element Methods*, Springer, Berlin 1983.
3. Fischer T.M. and Rosenberger R., A Boundary Integral Method for the Numerical Computation of the Forces Exerted on a Sphere in Viscous Incompressible Flows Near a Plane Wall, Preprint, Univ. of Darmstadt 1985, 30pp.
4. Hackbusch W., Die schnelle Auflösung der Fredholm'schen Integralgleichung zweiter Art, *Beitr. Numer. Math.* 9 (1982), 47-62.
5. Hebeker F.K., A Theorem of Faxén and the Boundary Integral Method for 3-D Viscous Incompressible Flows, Techn. Rep., Univ. of Paderborn 1982, 14pp.
6. Hebeker F.K., A Boundary Integral Approach to Compute the 3-D Oseen's Flow Past a Moving Body, Conf. on *Numerical Mathods in Fluid Dynamics* (Roma 1983), M. Pandolfi and R. Piva (eds.), Vieweg, Braunschweig 1984, 124-130.
7. Hebeker F.K., A Boundary Element Method for Stokes Equations in 3-D Exterior Domains, Conf. on *The Mathematics of Finite Elements*

and Applications (Uxbridge U.K. 1984), J.R. Whiteman (ed.), Academic Press, London 1985, 257-263.

8. Hebeker F.K., Efficient Boundary Element Methods for 3-D Exterior Viscous Flows, Submitted for publication, Univ. of Paderborn 1985, 35pp.

9. Hebeker F.K., Efficient Boundary Element Methods for 3-D Viscous Flows, Conf. on *Boundary Element Methods in Engineering* (Como Italy 1985), C.A. Brebbia (ed.), to appear, 8pp.

1o. Hsiao G.C. and Kopp P. and Wendland W.L., A Galerkin Collocation Method for Some Integral Equations of the First Kind, *Computing* 25 (198o), 89-13o.

11. Hsiao G.C. and Kopp P. and Wendland W.L., Some Applications of a Galerkin Collocation Method for Boundary Integral Equations of the First Kind, *Math. Meth. Appl. Sci.* 6 (1984), 28o-325.

12. Ladyzhenskaja O.A., *The Mathematical Theory of Viscous Incompressible Flow*, Gordon and Breach, New York 1969.

13. Novak Z., Use of the Multigrid Method for the Laplacean Problems in 3-D, Conf. on *Multigrid Methods* (Köln 1981), W. Hackbusch and U. Trottenberg (eds.), Springer, Berlin 1982, 576-598.

14. Quarteroni A. and Maday Y., Spectral and Pseudospectral Approximations of the Navier Stokes Equations, *SIAM J. Numer. Anal.* 19 (1982), 761-78o.

15. Wendland W.L., On the Asymptotic Convergence of Some Boundary Element Methods, Conf. on *The Mathematical Theory of Finite Elements and Applications*(Uxbridge U.K. 1981), J.R. Whiteman (ed.), Academic Press, London 1982, 281-312.

16. Wendland W.L., Boundary Element Methods and Their Asymptotic Convergence, CISM Summer School on *Theoretical Acoustics and Numerical Techniques* (Udine Italy 1982), P. Filippi (ed.), Springer, Wien

1983, 135-216.

17. Zhu J., A Boundary Integral Equation Method for the Stationary Stokes Problem in 3-D, Conf. on *Boundary Element Methods* (Hiroshima 1983), C.A. Brebbia (ed.), Springer, Berlin 1983, 283-292.

Numerical Approximation of Partial Differential Equations
E.L. Ortiz (Editor)

ON RATIONAL APPROXIMATION TO SEMIGROUPS OF LINEAR OPERATORS

Arturo Ribagorda
Carlos Vega
Polytechnic University of Madrid
Madrid, Spain

The evolution problem, whose operator is the infinitesimal generator of a operator's semigroup, is considered.
A new kind of rational approximation to operator semigroups is presented.
These approximations lead to stable schemes.

1. Let $(B; \|\bullet\|)$ a Banach space, A a time-independent linear operator with $D(A) \subset B$ and $u_o \in D(A)$.

We shall consider the Cauchy's problem

$$\frac{du}{dt} = Au, \qquad u(0) = u_o$$

and an approximate scheme for it given by

$$v(kh) = [r(ha)]^k u_o; \quad k = 1, 2, \ldots; \quad t = nh, \tag{1}$$

where $r(z)$ is a rational approximation to the exponential function e^z and v_o is an approximation to the element u_o.

The scheme (1) is called A-*acceptable* if $|r(z)| < 1$ for $\operatorname{Re} z < 0$ and it is called *stable* if there exists $C = C(t)$ independent of k such that $\|r^k(hA)\| \leqslant C$.

It is known that the schemes (1) is not stable in the general case (Brenner and Thomee, 1979; Hersh and Kato, 1979).

In this paper we propose a class of operators (*P-operators*) and a class of rational approximation (*M-approximations*) which lead to stable schemes.

2. We shall consider operators A which are infinitesimal generators of strongly continuous semigroups of operators $\{T(t, A); t \geqslant 0\}$ of type ω. This means that

a) $T(t, A) \in L(B)$ for all $t \geqslant 0$;

b) $T(t+s, A) = T(t, A) \cdot T(s, A)$ for all $t, s \geqslant 0$;

c) $T(0, A) = I$;

d) $T(t, A)x$ is continuous abstract function on $t \geq 0$ for all $x \in B$;

e) there exist real constants $M > 0$ and w such that $\|T(t, A)\| \leq \leq M e^{wt}$ for all $t \geq 0$;

f) $Ax = \lim_{t \to 0^+} \frac{T(t, A)x - x}{t} = T'(0, A)x$.

In this case we shall write $A \in C_o(w)$.

A operator $A \in C_o(w)$ is called *P-operator* if

$$A = \lambda_0 P + A_1, \tag{2}$$

where $P = P^2$ is bounded projection operator, λ_0 is a real number, $A_1 \in C_o(w_1)$ with $w_1 < \lambda_0$ and $PA_1x = A_1Px = 0$ for all $x \in D(A)$.

A rational approximation $r(z)$ of exponential function e^z is called *M-approximation* if it is A-acceptable and if

$$\text{Max}\ \{|r(z)|: \ \text{Re}\ z = \xi\} = |r(\xi)|.$$

3. We shall explain now our main result.

Theorem. Let A a P-operator and let $r(z)$ a M-approximation. Then the scheme (1) is stable.

Proof. If ρ and σ design the spectral radius and the spectrum respectivelly, we have according the theorem of spectral transformation

$$\rho[r(hA)] = \sup\ \{|\lambda|: \lambda \in \sigma[r(hA)]\} = \sup\ \{|r(hz)|: z \in \sigma(A)\}.$$

It is known that the spectrum of a operator $A \in C_o(w)$ belongs to the half-plane $\text{Re}\ z < w$. Furthermore, if A is a P-operator, it is possible to verify that λ_0 is an eigenvalue of A and that if μ is another eigenvalue of A, then $\text{Re}\ \mu < \lambda_0$. Therefore

$$\rho[r(hA)] = \sup\{|r(hz)|: \text{Re}\ z \leq \lambda_0\}.$$

From (2) we deduce

$$T(t, A) = P e^{\lambda_0 t} + S(t, A_1),$$

where $PS = SP = 0$. Applying the Laplace transformation, we obtain

$$R(\lambda, A) = P \frac{1}{\lambda - \lambda_0} + (I - P)\, R(\lambda, A_1).$$

Therefore

$$r(hA) = \frac{1}{2\pi i} \oint_C r(h\lambda)\, R(\lambda, A)\, d\lambda = P\, r(h\lambda_0) + (I - P)\, r(hA_1) \tag{3}$$

and it is easy to proof now that all requirements of P-operator is verified by $r(hA)$. Finally, it is possible proof that if A is P-operator, then there exists a norm $|||\cdot|||$ which is equivalent to the initial norm $\|\cdot\|$ and such that $\rho(A) = |||A|||$. Therefore, we have

$$|||r(hA)||| = \sup \{|r(hz)|: \text{Re } z \leqslant \lambda_0\}.$$

Let us suppose that $w \leqslant 0$. Then $\lambda_0 \leqslant 0$ also. As $r(z)$ is A-acceptable, we have, applying the principle of the maximus modulus,

$$|||r(hA)||| = \max \{|r(hz)|: \text{Re } z = \lambda_0\} = |r(h\lambda_0)| < 1,$$

where we use that $r(z)$ is a M-approximation. Due the equivalence of the norms $|||\cdot|||$ and $\|\cdot\|$, we obtain $\|r(hA)\| < C$ and

$$\|r^k(hA)\| < C,$$

that is, the scheme (1) is stable.

If $w > 0$, we cannot use the formula (3), because now the spectrum of A belongs to half-plane $\text{Re } z < w$ with $w > 0$ and the formula (3) takes place only if the spectrum of A belongs to the half-plane $\text{Re } z < 0$ in which is analytical the function $r(z)$. By this reason, in the case $w > 0$ we put

$$r(hA) = r(hB) + hw f_h(hB),$$

where $B = A - Iw$ and $f_h = \frac{1}{hw}[r(hA) - r(hB)]$. If $a(t)$ and $b(t)$ are bounded measures such that

$$r(z) = \int_0^\infty e^{zt}\, da(t) \quad \text{and} \quad f(z) = \int_0^\infty e^{zt}\, db(t),$$

we can write, using the operational calculus of Hille and Phillips (Hille and Phillips, 1974),

$$r(hA) = \int_0^\infty T(t, A)\, da(t); \qquad r(hB) = \int_0^\infty T(t, B)\, da(t);$$

$$f_h(hB) = \int_0^\infty T(t, B)\, db(t).$$

We have now

$$|||r(hA)||| \leqslant |||r(hB)||| + hw\, |||f(hB)|||. \tag{4}$$

As $B \in C_0(w_2)$ with $w_2 < 0$, we can proof that

$$|||r(hB)||| < 1. \tag{5}$$

On the other hand, as $r(z)$ is A-acceptable, we have

$$f_h(z) = \frac{1}{hw}[r(z + hw) - r(z)] = \frac{P(z)}{Q(z)} \quad \text{with degree } P(z) < \text{degree } Q(z).$$

By this reason, for $z = \xi i + \eta$, we have $f_h(\xi i) \in L^2(R)$ and $f_h'(\xi i) \in L^2(R)$, so that $f_h(\xi i) \in \hat{M}$ and $f_h(z) \in \tilde{M}$. Here M is the Banach space of the bounded measure $a(t)$ with the norm $\|a(t)\| = \int_{-\infty}^{+\infty} d\,|a|\,(t)$;

$\hat{M}$ is the algebra of the Fourier transformation $\hat{a}(z) = \int_{-\infty}^{+\infty} e^{itz}\, d\,a(t)$ with the norm $\|\hat{a}\| = \|a\|$ and $\tilde{M}$ is the algebra of the Laplace transformation $\tilde{a}(z) = \int_0^{\infty} e^{zt}\, d\,a(t)$ of $a(t)$. As $f_h(z) \in \tilde{M}$, we have

$$|||f_h(hB)||| \leqslant C. \tag{6}$$

Introducing (5) and (6) in (4), we obtain $|||r(hA)||| \leqslant 1 + h\,w\,C$, so that $|||r^k(hA)||| \leqslant (1 + h\,w\,C)^k < e^{wCt}$ and consequently $\|r^k(hA)\| < C$, that is, the scheme (1) is stable again.

4. Before terminate we observe that the operator corresponding to a differential equations system of the stiff kind, the Laplace operator and the Boltzman operator are P-operators.

Besides, the Pade approximations situated below the principal diagonal of the Pade table are M-approximations.

REFERENCES

1. Brenner, P. and Thomee, V. On rational approximations of semigroups, SIAM J. Numer. Anal., vol. 16, n° 4 (1979), pg. 683-694.
2. Hersh, R. and Kato, T. High-accuracy stable difference schemes for well-posed initial value problems, SIAM J. Numer. Anal., vol. 16, n° 4 (1979), pg. 670-682.
3. Hille, E. and Phillips, R. Functional Analysis and semigroups, Amer. Math. Soc., Providence, Rhode Island, 1974.

Numerical Approximation of Partial Differential Equations
E.L. Ortiz (Editor)

A QUICK SURVEY OF RECENT DEVELOPMENTS AND APPLICATIONS OF THE τ-METHOD

Manuel R. de J. da Silva

Faculdade de Ciências da Universidade do Porto

Porto, Portugal

The τ-method was first conceived by C. Lanczos in 1938 to construct polynomial approximations to the solution y of a given problem involving an equation of the form $Dy = P$, where P is an algebraic polynomial and D a linear algebraic or ordinary differential operator with polynomial coefficients, together with some suplementary (initial, boundary or mixed) conditions.

The basic idea is as follows : With the given equation we associate a neighbouring one, $Dy_n = P + Z_m$, where Z_m is a conveniently chosen polynomial perturbation. Z_m is usually chosen to be a linear combination of Tchebysheff or Legendre polynomials with free coefficients, called the τ-parameters, which are to be determined in such a way that the τ-approximant y_n of y is the unique polynomial solution to the perturbed equation and satisfies exactly the given suplementary conditions.

The τ-method has recently been made amenable to error analysis and to computer programming by E. Ortiz and his co-workers at Imperial College, University of London. Since then, it has been successfully applied to so many problems in Numerical and Functional Analysis that we should no longer call it a method but a philosophy instead. In support of this we give a brief survey of the theoretical approaches to and applications of the τ-method with emphasis on those authors' recent work on the numerical treatment of differential equations.

1. Given an equation of the form

$$Dy(x) = 0 \tag{1}$$

(we shall see in §2.2 that there is no loss in generality in taking this equation to be homogeneous), where D is a linear (differential, integral or algebraic) operator mapping the linear space $\mathbb{P}$ of algebraic polynomials into itself, such as, for instance, a linear differential operator with polynomial coefficients, its integrated forms as in (Coleman, 1976), (Fox, 1962), and (Ortiz, 1974), or a polynomial multiplier, and given $\nu \geq 0$ suplementary (initial, boundary or mixed) conditions through linear combinations of function and derivative values of $y = y(x)$, say

$$f_j(y) = \sigma_j \quad , \quad j = 1(1)\,\nu \quad , \tag{2}$$

where f_j are linear functionals, Lanczos' τ-method may be used to construct polynomial or rational approximations $y_n = y_n(x)$ to the solution $y(x)$ of (1)-(2) in a form suitable for numerical evaluation. The basic philosophy is to perturb the given problem through the addition to the r.h.s. of (1) of an mth degree polynomial $Z_m(x)$ chosen so that $y_n(x)$ is the unique polynomial solution of the following τ-problem:

$$Dy_n(x) = Z_m(x) \equiv \sum_{j=0}^{m} z_j^{(m)} x^j \tag{3}$$

$$f_j(y_n) = \sigma_j \quad , \quad j = 1(1)\,\nu \quad . \tag{4}$$

The choice of $Z_m(x)$ in (3) depends on the given suplementary conditions, the very nature of D, and the approximation properties that $y_n(x)$ is to have. Therefore, $Z_m(x)$ is to contain ν parameters to be determined to ensure that $y_n(x)$ satisfies conditions (4). By the very nature of D we mean intrinsic properties such as its height h, its range R_D and its polynomial kernel K_D. The height of D is given by $h = \sup_{n \in \mathbb{N}_0} \{\partial(Dx^n) - n\}$, where $\mathbb{N}_0 = \{0,1,2,\ldots\}$ and ∂ stands for degree of, and thus $Z_m(x)$ is to have degree $m = n+h$. The range of D is not, in general, the whole of $\mathbb{P}$. In this case, there is to exist an s-dimensional residual subspace R_S complementary to R_D:

$$\mathbb{P} = R_D \oplus R_S \quad , \quad R_D \cap R_S = \{0\} \quad , \tag{5}$$

$$R_S = \text{span}\,\{x^k : k \in S\} \quad , \quad S = \{k \in \mathbb{N}_0 : x^k \notin R_D\} \; .$$

So, $Z_m(x)$ is also to contain s parameters to be determined so that no component of $Z_m(x)$ lies in R_S, i.e.,

$$z_j^{(m)} = 0 \quad , \quad j \in S. \tag{6}$$

K_D is the k_o-dimensional linear subspace generated by the exact polynomial solutions of the given equation (1).

With no loss in generality, we take the approximation range to be $0 \leq x \leq 1$. The error function, $\eta_n(x) = y(x) - y_n(x)$, is such that

$$D\eta_n(x) = -Z_m(x) \quad , \quad f_j(\eta_n) = 0 \quad , \quad j = 1(1)\nu \quad . \tag{7}$$

In the hope that the smallness of $D\eta_n(x)$ will imply that of $\eta_n(x)$, we take

$$Z_m(x) = \sum_{i=0}^{r} \tau_i^{(n)} \rho_{m-i}(x) \quad (8) \quad \text{or} \quad Z_m(x) = \sum_{i=0}^{r} \tau_i^{(n)} x^i \rho_{m-r}(x) \ , \quad (9)$$

where $r = \nu + s - k_o - 1$, the $\tau_i^{(n)}$, $i = 0(1)\,r$, are the τ-parameters to be determined and $\rho_k(x)$ is the kth degree Tchebysheff or Legendre polynomial according as $y_n(x)$ is required to be a good approximation to $y(x)$ over the whole range $0 \leq x \leq 1$ or at $x = 1$. Details and applications of this can be found in (Lanczos, 1957, 1973), (Kizner, 1966), and (Ortiz, 1975 a) b)).

The existence and uniqueness questions for the τ-approximation problem are reduced to the existence and uniqueness of the τ-parameters. We have $\nu + s$ linear algebraic equations, namely ν suplementary conditions in (4) and s compatibility conditions in (6), to determine $\nu + s$ unknowns, namely $\nu + s - k_o$ τ-parameters in (8) or (9) and k_o parameters associated with K_D, the identification of which may be made as in (Llorente et Ortiz, 1968) or (Ortiz, 1969). It is shown in (Crisci and Ortiz, 1981) that the $\tau_i^{(n)}$, $i = 0(1)\,r$, are uniquely determined and tend to zero exponentially in terms of a certain function of n.

If $Z_m(x)$ is chosen as in (8), then $y_n(x)$ is, in general, very close to the best uniform approximation of $y(x)$ by algebraic polynomials of degree $\leq n$, as shown by (Rivlin and Weiss, 1968), (Rivlin, 1974), (Freilich and Ortiz, 1975, 1982), (Namasivayam and Ortiz, 1985 a) b)). As we shall see in §3, there are classes of problems for which the τ-method is optimal in the sense that the rate of convergence of the τ- approximants to the exact solution of the given problem is the same as that of the best uniform approximation of that solution by algebraic polynomials.

2. There are essentially two approaches to the numerical solution of the τ- problem in §1 (3) - (4) : The operational approach as in (Ortiz and Samara, 1980, 1981) and (da Silva, 1982) and that in terms of canonical polynomials as in (Lanczos, 1952, 1957) and (Ortiz, 1969, 1974).

2.1. Ortiz - Samara's approach is essentially based on the use of the matrix operators μ and η, which are such that $x \cdot \underline{x} = \mu \underline{x}$ and $d\underline{x} / dx = \eta \underline{x}$, where $\underline{x} = (1,x,x^2,\ldots)'$ and the dash stands for transposition, in terms of which any linear differential operator D as in §1(1) is easily represented. It is called operational in the sense that it reduces the given problem in §1 (1) - (2) to a purely algebraic (approximate) one. There is in (da Silva, 1982) an elementary recursive implementation of the operational approach relying chiefly on common knowledge on systems of linear algebraic equations.

2.2. For a linear operator D as in §1 (1), Lanczos' canonical polynomials $Q_k(x)$ are formally defined by

$$DQ_k(x) = x^k \quad , \quad k \in \mathbb{N}_o \ . \tag{1}$$

If R_D is not the whole of $\mathbb{P}$, then Lanczos' definition is inconsistent for the indices k in the set S defined in §1 (5). Since the number s of those indices is finite, we may easily get rid of the ghostly polynomials $Q_k(x)$, $k \in S$, either by using the technique given in §1 (6), or by using Ortiz' redefining equation

$$DQ_k(x) = x^k + R_k(x) \quad , \quad k \in \mathbb{N}_o \ , \tag{2}$$

where $R_k \in R_S$, based on the fact in §1 (5) that every element of $\mathbb{P}$

is uniquely decomposed into the sum of an element in R_D with another in R_S . For $k \notin S$, equation (1) is consistent and so $R_k(x) \equiv 0$. For $k \in S$, the r.h.s. of (2) lies in R_S , which is absurd, unless $R_k(x) = -x^k$.

Canonical and residual polynomials may be computed recursively as in (Ortiz, 1969): writing

$$Dx^n = \sum_{j=0}^{m} \Pi_{nj} x^j = \Pi_{nm} x^m + \sum_{j=0}^{m-1} \Pi_{nj}(DQ_j(x) - R_j(x)) ,$$

i.e.,

$$D(x^n - \sum_{j=0}^{m-1} \Pi_{nj} Q_j(x)) = \Pi_{nm} x^m - \sum_{j=0}^{m-1} \Pi_{nj} R_j(x) ,$$

and assuming that $\Pi_{nm} \neq 0$ and that $Q_j(x)$ and $R_j(x)$, $j = 0(1)\, m-1$, have already been computed, then

$$Q_m(x) = (x^n - \sum_{j=0}^{m-1} \Pi_{nj} Q_j(x)) / \Pi_{nm} ,$$

$$R_m(x) = -(\sum_{j=0}^{m-1} \Pi_{nj} R_j(x)) / \Pi_{nm} , \quad n \in \mathbb{N}_0 .$$

Canonical polynomials are, in fact, equivalence classes of polynomials, module K_D , but this is a technical point, the details of which are to be found in (Ortiz, 1969).

Having computed the index set S , the first m Lanczos - Ortiz' canonical polynomials, and the k_o generators $P_j(x)$ of K_D , the solution $y_n(x)$ of the τ - problem in §1 (3) - (4) is immediately at hand. Indeed, writing

$$Dy_n(x) = \sum_{j=0}^{m} z_j^{(m)} x^j = \sum_{j=0}^{m} z_j^{(m)} (DQ_j(x) - R_j(x)) ,$$

i.e.,

$$D(y_n(x) - \sum_{j=0}^{m} z_j^{(m)} Q_j(x)) = - \sum_{j=0}^{m} z_j^{(m)} R_j(x) ,$$

we see that the r.h.s. is in R_S , which is absurd, unless $\sum_{j=0}^{m} z_j^{(m)} R_j(x) \equiv 0$, i.e., $z_j^{(m)} = 0$, $j \in S$, showing that the use of residuals is numerically equivalent to the formal use of undefined canonical polynomials and the subsequent cancelation of their coefficients.

These are the compatibility conditions of the τ - problem whose solution is

$$y_n(x) = \sum_{j=0}^{m} z_j^{(m)} Q_j(x) + \sum_{j=1}^{k_o} c_j P_j(x) \quad . \tag{3}$$

In addition to being an efficient basis for the representation of τ - solutions, canonical polynomials have a number of useful properties : They satisfy a self-starting recurrence relation involving at most $\nu + h$ consecutive terms. They are independent of the order n of the actual τ - approximant and so, if we need $y_{n+1}(x)$ after $y_n(x)$ has been constructed, only $Q_{m+1}(x)$ has to be computed. They are independent of the given suplementary conditions, hence initial and boundary problems are treated alike. They are also independent of the approximation interval and so piecewise polynomial and rational τ - approximations to differentiable functions are easily constructed as in (Luke, 1969, 1975) and (Ortiz, 1975 a) b)). As for the existence of undefined canonical polynomials, that is not a weakness of their representation capability, but a consequence of structural properties of D .

The τ - solution of a given nonhomogeneous equation, $DY(x) = P_N(x)$, $P_N(x) = \sum_{j=0}^{N} p_j x^j$, which is the exact polynomial solution of the perturbed equation $DY_n(x) = P_N(x) + Z_m(x)$, with $Z_m(x)$ chosen as before, is obtained at once by adding to the τ - solution (3) of the corresponding homogeneous equation the r.h.s. of the given one with $Q_j(x)$ replacing x^j , $j = 0(1)\ N$, i.e.,

$$Y_n(x) = y_n(x) + \sum_{j=0}^{N} p_j Q_j(x) \quad .$$

3. Our original intention was to give here as complete an account as possible of old and new applications of the τ - method, but, due to lack of time and space, we shall not go far beyond the mere mentioning of a few.

3.1. The τ-method was firstly applied to approximation of functions. It was extensively used by (Lanczos, 1938, 1952, 1957) to construct near minimax polynomial approximations via economization of power series and asymptotic expansions for high precision tabulation of special functions of mathematical physics. It was also extensively used by (Lanczos, 1957, 1973) and (Luke, 1969, 1975) to construct rational approximations to differentiable functions, by (Ortiz, 1972, 1974) to generate approximate expansions in arbitrary polynomial series, by (Ortiz, 1975 a) b)) to produce piecewise polynomial and rational approximations, and by (Freilich and Ortiz, 1975, 1982) to give simultaneous approximations of a function and its derivatives.

3.2. The τ-method has been extensively used to produce numerical solutions to linear FDEs, ODEs, and SODEs. Polynomial τ-solutions to some linear FDEs are given in (Ortiz, 1978), (da Silva, 1982), and (Ortiz and Samara, 1983 a)). A number of applications to IVPs, two-point and multipoint BVPs in linear ODEs, even of high order, may be found in (Ortiz, Purser, and Canizares, 1972), (Onumanyi, Ortiz, and Samara, 1981), and (Onumanyi and Ortiz, 1982). There are in (Wragg, 1966) polynomial τ-solutions to linear ODEs with two-point boundary conditions resulting from finite difference approximations to Stefan problems and in (Ortiz, 1980) polynomial and rational τ-solutions to boundary layer problems.

Canonical polynomials have been extended to linear SODEs by (Freilich and Ortiz, 1975, 1982) and (Namasivayam and Ortiz, 1985 a)), who have shown that, for linear SODEs with constant coefficients, the components of the τ-error vector and the corresponding best uniform polynomial approximation constants are of the same order. Ortiz' step-by-step formulation of the τ-method has also been extended by (Crisci and

Russo, 1982, 1983) to produce one-step integration methods which are A- stable for every order.

3.3. Both approaches to the τ-method have been extended to EVPs , $D_\lambda y(x) = 0$, $a \leq x \leq b$, $f_j(y) = 0$, $j = 1(1)\ \nu$, with $D_\lambda = \sum_{r=0}^{t} \lambda^r D_r$, where D_r is a linear differential operator of order ν with polynomial coefficients. Both approaches lead to the same system of $\nu + s$ linear homogeneous algebraic equations whose determinant must vanish to produce a nontrivial solution. Details and applications of the approach in terms of canonical polynomials can be found in (Chaves and Ortiz , 1968) and (Ortiz, 1974). Extensive numerical experiments with the operational approach are reported by (Ortiz an Samara, 1983 b)) and by (Liu and Ortiz, 1982 a) b)), who have found, for Mathieu's and Schrodinger's EVPs, numerical solutions of remarkable accuracy.

Ortiz- Samara's method has also been used with success by (Liu, Ortiz, and Pun, 1985) to give numerical solutions to EVPs in linear PDEs, to Steklov's EVPs in particular.

3.4. Following (Ortiz, 1978), nonlinear problems can be reduced to a sequence of linear ones and each of these can be treated with the τ-method. We let $y^{(0)}(x)$ be a polynomial satisfying the given suplementary conditions and insert it in the nonlinear terms of the given operator to get a linear equation with polynomial coefficients, which we perturb to get a τ-solution $y^{(1)}(x)$, and start again with $y^{(1)}(x)$ instead of $y^{(0)}(x)$. If this iteration process, which we may represent by $y^{(k+1)}(x) = T(y^{(k)}(x))$, $k = 0,1...$, is contractive, then its fixed point will be the solution to the given nonlinear problem. Extensive numerical examples on the application of the τ-method to nonlinear ODEs can be found in (Onumanyi, Ortiz, and Samara, 1981), (Ortiz, 1978),

(Ortiz and Samara, 1983 a)) and (Ortiz and Dinh, 1985), where it is shown that, for equations of Riccati's type, the rate of convergence is quadratic.

3.5. The operational approach to the τ - method has been extended by (Ortiz and Samara, 1980, 1984) to linear PDEs with bivariate polynomial coefficients. The coefficients of a bivariate τ - approximant are determined by means of a reduced set of matrix operations as those referred to in §2.1 and involve no discretization of the variables, no approximate quadratures, and no special trial functions. They have solved biharmonic and parabolic PDEs with a variety of suplementary conditions and some of the corresponding normalized τ - error surfaces exhibit a remarkable oscillatory behaviour.

Ortiz - Samara's method has also been used with success by (Ortiz and Pun, 1985) to produce approximate solutions to nonlinear PDEs as an extension to the bidimensional case of the Ortiz' linearization technique referred to in §3.4. Results of high accuracy were obtained for elliptic, parabolic, and hyperbolic problems, even when polynomial τ - approximants of a relatively low degree were used.

The τ - method in terms of canonical polynomials has been combined with the well-known method of lines to give an efficient hybrid τ - lines method for the numerical treatment of linear PDEs. There are in (El Misiery and Ortiz, 1985) numerical solutions of the biharmonic equation which compete most favourably with those obtained by a specially designed finite element method. Canonical polynomials have also been extended to linear PDEs with constant coefficients by (Namasivayam and Ortiz, 1985 b)), who have shown that, for such PDEs with constant initial conditions, the τ - method is optimal in the sense referred to in §1.

REFERENCES

1. CHAVES T. and ORTIZ E.L. , On the Numerical Solution of Two Point Boundary Value Problems for Linear Differential Equations , Z. Angew. Math. Mech., 48 , 1968, 415 - 418.
2. COLEMAN J.P. , The Lanczos Tau - Method , J. Inst. Maths. Applics, 17 , 1976, 85 - 97.
3. CRISCI M.R. and ORTIZ E.L. , Existence and Convergence Results for the Numerical Solution of Differential Equations with the Tau - Method , NAS Res. Rep. 1 - 16 , Imperial College, Univ. of London , 1981.
4. CRISCI M.R. and RUSSO E. , A - stability of a Class of Methods for the Numerical Integration of Certain Linear Systems of Ordinary Differential Equations , Math. Comp. 38 , 1982, 431 - 435.
5. CRISCI M.R. and RUSSO E. , An Extension of Ortiz' Recursive Formulation of the Tau-Method to Certain Linear Systems of Ordinary Differential Equations , Math. Comp. 41 , 1983, 27 - 42.
6. FOX L. , Chebyshev Methods for Ordinary Differential Equations , Comput. J. 4 , 1962, 318 - 331.
7. FREILICH J.H. and ORTIZ E.L. , Simultaneous Approximation of a Function and its Derivative with the Tau - Method , NAS Res. Rep. 1 - 45 , Imperial College, Univ. of London, 1975.
8. FREILICH J.H. and ORTIZ E.L. , Numerical Solution of Systems of Ordinary Differential Equations with the Tau - Method : An Error Analysis , Math. Comp. 39 , 1982, 467 - 479.
9. KIZNER W. , Error Curves for Lanczos' "Selected Points" Method , Comput. J. 8 , 1966, 372 - 382.
10. LANCZOS C. , Trigonometric Interpolation of Empirical and Analytical Functions , J. Math. Phys. 17 , 1938, 123 - 199.
11. LANCZOS C. , Tables of Chebyshev Polynomials $S_n(x)$ and $C_n(x)$ Introduction , Nat. Bur. Standards Applied Math. Ser. 9 , U.S. Govt. Printing Office, Washington , 1952.
12. LANCZOS C. , Applied Analysis , Pitman, London , 1957.
13. LANCZOS C. , Legendre Versus Chebyshev Polynomials , in Topics in Numerical Analysis, ed. J.J.H. Miller, AP, London , 1973, 191 - 201.
14. LIU K.M. and ORTIZ E.L. , Approximation of Eigenvalues Defined by Ordinary Differential Equations with the Tau-Method, in Matrix Pencils, ed. B. Kagstrom and A. Ruhe , LNM nº 973, Springer - Verlag, Berlin , 1982 a), 90 - 102.
15. LIU K.M. and ORTIZ E.L. , Eigenvalue Problems for Singularly Perturbed Differential Equations , in Computational and Asymptotic Methods for Boundary and Interior Layers, ed. J.J.H. Miller, Boole Press, Dublin , 1982 b), 324 - 329.
16. LIU K.M., ORTIZ E.L., and PUN K.-S. , Numerical Solution of Steklov's Partial Differential Equation Eigenvalue Problem with the Tau-Method, NAS Res. Rep., Imperial College, Univ. of London, 1985.
17. LLORENTE P. et ORTIZ E.L. , Sur Quelques Aspects Algébriques D'une Méthode D'Approximation De M. Lanczos , Math. Notae , XXI , 1968, 17 - 23.

18. LUKE Y. , The Special Functions and their Approximations, AP., N.Y. , 1969.

19. LUKE Y. , Mathematical Functions and their Approximations, AP, London , 1975.

20. EL MISIERY A.E.M. and ORTIZ E.L. , Numerical Solution of Regular and Singular Biharmonic Problems with the Tau Lines Method , NAS Res. Rep. , Imperial College, Univ. of London, 1985.

21. NAMASIVAYAM S. and ORTIZ E.L. , On Figures Generated by Normalized Tau Approximation Error Curves , NAS Res. Rep. , Imperial College, Univ. of London, 1985 a).

22. NAMASIVAYAM S. and ORTIZ E.L. , Best Approximation and the Numerical Solution of Partial Differential Equations with the Tau-Method , NAS Res. Rep. , Imperial College, Univ. of London, 1985 b) , to appear in Port. Math.

23. ONUMANYI P. and ORTIZ E.L. , Numerical Solution of High Order Boundary Value Problems for Ordinary Differential Equations with an Estimation of the Error , Internat. J. Numer. Meth. Eng. 18 , 1982, 775 - 781.

24. ONUMANYI P., ORTIZ E.L., and SAMARA H. , Software for a Method of Finite Approximations for the Numerical Solutions of Differential Equations , Appl. Math. Modelling 5 , 1981, 282 - 286.

25. ORTIZ E.L. , The Tau-Method , SIAM J. Numer. Anal. 6 , 1969, 480 - 492.

26. ORTIZ E.L. , A Recursive Method for the Approximate Expansion of Functions in a Series of Polynomials , Comp. Phys. Comm. 4 , 1972, 151 - 156.

27. ORTIZ E.L. , Canonical Polynomials in the Lanczos Tau-Method , in Studies in Numerical Analysis, ed. B.K.P. Scaife, AP, London, 1974, 73 - 93.

28. ORTIZ E.L. , Step by Step Tau-Method - Part I. Piecewise Polynomial Approximations , Comp. & Maths. with Appls. 1, 1975 a), 381 - 392.

29. ORTIZ E.L. , Sur Quelques Nouvelles Applications de la Method Tau , in Analyse et Contrôle de Systèmes, ed. IRIA, Paris, 1975 b), 247 - 257.

30. ORTIZ E.L. , On the Numerical Solution of Nonlinear and Functional Differential Equations with the Tau-Method, in Numerical Treatment of Differential Equations in Applications, ed. R. Ansorge and W. Tornig , LNM nº 679, Springer - Verlag, Berlin, 1978, 127 - 139.

31. ORTIZ E.L. , Polynomial and Rational Approximations of Boundary Layer Problems with the Tau-Method , in Boundary and Interior Layers - Computational and Asymptotic Methods, ed. J.J.H. Miller, Boole Press, Dublin, 1980, 387 - 391.

32. ORTIZ E.L. and DINH A.P.-N. , On the Convergence of the Tau - Method for Nonlinear Differential Equations of Riccati's Type , J. Nonlinear Analysis, Methods & Applications 9 , 1985, 53 - 60.

33. ORTIZ E.L. and PUN K.-S. , Numerical Solution of Nonlinear Partial Differential Equations with the Tau - Method , NAS Res. Rep. , Imperial College, Univ. of London, 1985.

34. ORTIZ E.L., PURSER W.F.C., and CANIZARES F.J.R. , Automation of the Tau-Method , Numerical Analysis Res. Rep. 1 , Imperial College, Univ. of London, 1972.

35. ORTIZ E.L. and SAMARA H. , A new operational Approach to the Numerical Solution of Differential Equations in Terms of Polynomials , in Innovative Numerical Analysis in Applied Engineering Sciences, ed. R. Shaw et al, Univ. Press of Virginia, Charlotteville, 1980, 643-652.

36. ORTIZ E.L. and SAMARA H. , An operational Approach to the Tau-Method for the Numerical Solution of Nonlinear Differential Equations , Computing 27 , 1981, 15-25.

37. ORTIZ E.L. and SAMARA H. , Matrix Displacement Mappings in the Numerical Solution of Functional and Nonlinear Differential Equations with the Tau-Method , Num. Funct. Anal. and Optimiz. 6 , 1983 a), 379-398.

38. ORTIZ E.L. and SAMARA H. , Numerical Solution of Differential Eigenvalue Problems with an Operational Approach to the Tau-Method , Computing 31 , 1983 a), 95-103.

39. ORTIZ E.L. and SAMARA H. , Numerical Solution of Partial Differential Equations with Variable Coefficients with an Operational Approach to the Tau-Method , Comp. & Maths. with Appls. 10 , 1984, 5-13.

40. da SILVA M.R. , LACALGEBRA Versions of Lanczos' Tau-Method for the Numerical Solution of Differential Equations , Port. Math. 41 , 1982, 295-316.

41. RIVLIN T.J. and WEISS B. , Lanczos' τ-method and Polynomial Approximation in the Plane , J. Math. Anal. Applics. 22 , 1968, 402-417.

42. RIVLIN T.J. , The Chebyshev Polynomials , Wiley, N.Y., 1984.

43. WRAGG A. , The Use of Lanczos τ-methods in the Numerical Solution of a Stefan Problem , Comput. J. 9 , 1966, 106-109.

Numerical Approximation of Partial Differential Equations
E.L. Ortiz (Editor)

CUBIC AND QUINTIC SPLINE EXTRAPOLLATED COLLOCATION METHODS FOR TWO-POINT BOUNDARY VALUE PROBLEMS

Maria Joana Soares

Área de Matemática - Universidade do Minho
Braga, Portugal

A modified collocation method for the solution of second order two-point boundary value problems due to (Daniel and Swartz, 1975) is considered, and a connection between this method and certain results on the a posteriori correction of odd degree interpolatory splines is established. This leads to the computation of improved derivative approximations, and to the generalization of the method of Daniel and Swartz.

1. Introduction

This paper is concerned with the extrapollated collocation method (ECM) of (Daniel and Swartz, 1975) for solving linear second order two point boundary value problems of the form,

$$Ly \equiv y^{(2)}(x)+e_1(x)y^{(1)}(x)+e_0(x)y(x)=f(x); \quad a \le x \le b, \tag{1.1a}$$

$$\sum_{j=0}^{1} [\alpha_{ij}y^{(j)}(a)+\beta_{ij}y^{(j)}(b)] = 0 \ ; \quad i=1,2, \tag{1.1b}$$

where α_{ij}, β_{ij} are constants. We assume throughont that the following conditions hold :

C1. The boundary value problem has a unique solution $y \in C^6[a,b]$.

C2. The functions $e_i; i=0,1,$ and f are in $C[a,b]$.

C3. The equation $y^{(2)}=0$ subject to the boundary conditions (1.1b) has only the trivial solution.

Also, for notational convenience, we denote the boundary conditions (1.1b) by

$$By = 0. \tag{1.1c}$$

Let s be the cubic spline interpolating the solution y of (1.1) at the knots

$$x_i = a+ih; \quad i=0(1)n \quad , \ h=(b-a)/n, \tag{1.2 }$$

and satisfying the end-conditions

$$s_i^{(2)}=y_i^{(2)}-\frac{h^2}{12}y_i^{(4)}\ ;\quad i=0,n. \tag{1.3}$$

Then, the form of the end-conditions of s and the assumption C1, concerning the continuity of y, imply that:

$$y_i^{(1)}=s_i^{(1)}+0(h^4)\ ;\ i=0(1)n, \tag{1.4}$$

$$y_o^{(2)}=\frac{1}{12}(14s_o^{(2)}-5s_1^{(2)}+4s_2^{(2)}-s_3^{(2)})+0(h^4), \tag{1.5a}$$

$$y_i^{(2)}=\frac{1}{12}(s_{i-1}^{(2)}+10s_i^{(2)}+s_{i+1}^{(2)})+0(h^4)\ ;\quad i=1(1)n-1, \tag{1.5b}$$

and

$$y_n^{(2)}=\frac{1}{12}(14s_n^{(2)}-5s_{n-1}^{(2)}+4s_{n-2}^{(2)}-s_{n-3}^{(2)})+0(h^4); \tag{1.5c}$$

see, e.g. (Lucas,1983) and (Daniel + Swartz, 1975) .

These spline approximations are used in the ECM as follows.

The differential equation (1.1a) implies that

$$Ly_i=f_i\ ;\quad i=0(1)n, \tag{1.6}$$

and therefore, the substitution of the approximations (1.4) - (1.5) into the boundary conditions (1.1b) and into the Eqs.(1.6) gives, respectively

$$BS=0(h^4), \tag{1.7a}$$

and

$$Q_n\underline{s}^{(2)}+\Delta_{n,1}\underline{s}^{(1)}+\Delta_{n,o}\underline{s}=\underline{f}+0(h^4)\ , \tag{1.7b}$$

where:

(i) $\underline{s}^{(j)}$; j=0,1,2 and $\underline{f}$ are the (n+1)-dimensional column vectors

$$\underline{s}^{(j)}=(s_i^{(j)})_{i=o}^{n+2}\ ;\ j=o,1,2\quad\text{and}\quad \underline{f}=(f_i)_{i=o}^{n+2}\ , \tag{1.8}$$

(ii) $\Delta_{n,j}$;j=o,1 are the (n+1)×(n+1) diagonal matrices

$$\Delta_{n,j}=\text{diag}\ (e_j(x_o),\ldots,\ e_j(x_n))\ ;\ j=o,1, \tag{1.9}$$

and

(iii)

$$Q_n = 1/12 \begin{bmatrix} 14 & -5 \;\; 4 \;\; -1 \;\; 0 \;\; \cdots & 0 \\ 1 & & 0 \\ \cdot & & \cdot \\ \cdot & & \cdot \\ \cdot & T_n & \cdot \\ \cdot & & \cdot \\ 0 & & 1 \\ 0 & \cdots 0 \;\; -1 \;\; 4 \;\; -5 & 14 \end{bmatrix} \qquad (1.10a)$$

where $T_n=(t_{ij})$ is the $(n-1)\times(n-1)$ tridiagonal matrix with

$$t_{ij} = \begin{cases} 10, & \text{if } i=j, \\ 1, & \text{if } |i-j|=1. \end{cases} \qquad (1.10b)$$

The extrapollated collocation method of (Daniel and Swartz, 1975) follows directly from the above. That is, the ECM spline $\tilde{s}$, approximating the solution y of (1.1), is obtained by simply dropping the $O(h^4)$ terms from (1.7). In other words, $\tilde{s}$ satisfies

$$Q_n \underline{\tilde{s}}^{(2)} + \Delta_{n,1} \underline{\tilde{s}}^{(1)} + \Delta_{n,} \underline{\tilde{s}} = \underline{f} \qquad (1.11a)$$

and

$$B\tilde{S}=0, \qquad (1.11b)$$

and these constitute an $(n+3)\times(n+3)$ linear system for determining the $(n+3)$ unknown parameters of the spline $\tilde{s}$. Regarding convergence, it is shown in (Daniel and Swartz, (1975)) that, under the assumptions C1-C3, the ECM is well-defined and that

$$\|\tilde{s}^{(j)} - y^{(j)}\| \;\; O(h^{4-j}) \;\; ; \; j=0(1)3. \qquad (1.12)$$

That is, the ECM is an $O(h^4)$ method for the solution of boundary value problems of the form (1.1). (In (1.12) and throughout this paper, $\|.\|$ denotes the $L_\infty[a,b]$ function norm).

The objectives of this note are as follows.

(i) To show that improved derivative approximations $\tilde{s}_c^{(j)}$ to $y^{(j)}$; $j=1(1)4$, satisfying

$$|\tilde{s}_c^{(j)}(x) - y^{(j)}(x)| = O(h^r); \qquad x \in [a,b]$$

with (1.13)

$$r=\min\{6-j,4\}; \quad j=1(1)4,$$

can be obtained easily from the parameters of the ECM spline $\tilde{s}$, by using certain recent results concerning the a posteriori correction of odd degree interpolatory splines.

(ii) To indicate briefly how the a posteriori correction results can be used to generalize the ECM to various other high order methods for the solution of fourth order boundary value problems, by means of quintic splines.

2. Corrected approximations

As in Section 1, we assume that the boundary value problem (1.1) has a unique solution $y \in C^6[a,b]$, and we let s be the cubic spline interpolating y at the equally-spaced knots (1.2) and satisfying the end-conditions (1.3). Also, for $x \in [x_i, x_{i+1}]$; $i=0(1)n-1$, we define the "corrected" approximations Y_M ; $M=1,2$, by

$$Y_M(x) = s(x) + \sum_{m=0}^{M-1} \left\{ \frac{h^{4+m}}{(4+m)!} d_i^{(4+m)} \cdot P_m\left(\frac{x-x_i}{h}\right) \right\}, \tag{2.1}$$

where

$$P_0(x) = x^4 - 2x^3 + x^2 \quad , \quad P_1(x) = x^5 - \frac{5}{3}x^3 + \frac{2}{3}x, \tag{2.2}$$

and $d_i^{(4+m)}$; $m=0,1$, are approximations to the derivatives $y_i^{(4+m)}$;$m=0,1$, respectively. These derivative approximations are linear combinations of values of $s^{(2)}$, and are defined as follows:

(i) $$d_i^{(4)} = \frac{1}{h^2} \delta^2 s_i^{(2)} ; \; i=1(1)n-1, \tag{2.3a}$$

and

$$d_0^{(4)} = \begin{cases} d_1^{(4)}, & \text{it } M=1, \\ 2d_1^{(4)} - d_2^{(4)}, & \text{if } M=2, \end{cases} \tag{2.3b}$$

(ii) $$d_i^{(5)} = \frac{1}{2h^3} \{\delta^2 s_{i+1}^{(2)} - \delta^2 s_{i-1}^{(2)}\} ; \; i=2(1)n-2, \tag{2.4a}$$

$$d_0^{(5)} = d_1^{(5)} = \frac{1}{h} \{d_2^{(4)} - d_1^{(4)}\}, \tag{2.4b}$$

and

$$d_{n-1}^{(5)} = \frac{1}{h} \{d_{n-1}^{(4)} - d_{n-2}^{(4)}\} . \tag{2.4c}$$

Then, it can be shown that, for $0 \le j \le 4$ and $M=0,1,2$,

$$y^{(j)}(x)=Y_M^{(j)}(x)+O(h^{4-j+M}) \quad ; \quad x\in[x_i,x_{i+1}], \quad i=1(1)n-1, \tag{2.5a}$$

where we used Y_0 to denote the spline s, i.e.

$$Y_0 := s. \tag{2.5b}$$

For the non-periodic cubic spline s considered here, the above result is established in (Papamichael and Soares, 1985 (a)). However, (2.5) is essentially a special case of a much more general result of (Lucas, 1983), concerning the a posteriori correction of odd degree periodic splines on equally-spaced knots.

The following direct consequences of (2.5) are of special interest here.

(i) The zeros of the polynomials $P_0^{(j)}$; $j=1(1)3$, give the points in $[x_i,x_{i+1}]$ where the derivatives $s^{(j)}$; $j=1(1)3$, of s display superconvergence. That is,

$$y^{(j)}(x_i+\mu_j h)=Y_1^{(j)}(x_i+\mu_j h)+O(h^{5-j})$$

$$= s^{(j)}(x_i+\mu_j h)+O(h^{5-j}) \quad ; \quad i=0(1)n-1 , \tag{2.6}$$

where μ_j ; $j=1(1)3$, denote respectively the zeros of the polynomials $P_0^{(j)}$; $j=1(1)3$ in $[0,1]$. These zeros are as follows :

$$\mu_1=0,1/2,\ 1 , \quad \mu_2=(3\pm\sqrt{3})/6 \quad \text{and} \quad \mu_3=1/2. \tag{2.7}$$

In particular,

$$y^{(1)}(x_i)= Y_1^{(1)}(x_i)+ O(h^4)$$

$$= s^{(1)}(x_i)+ O(h^4) \quad ; \quad i=0(1)n. \tag{2.8}$$

(ii) Formula (2.5) with $x=x_i$ gives improved approximations to the derivatives $y_i^{(j)}$; $j=1(1)4$, in terms of the values of the derivatives of s at the knots. For example, we have the following :

$$y_0^{(2)}=Y_2^{(2)}(x_0)+O(h^4)$$

$$=\frac{1}{12}(14s_0^{(2)}-5s_1^{(2)}+4s_2^{(2)}-s_3^{(2)})+O(h^4), \tag{2.9a}$$

$$y_i^{(2)} = Y_2^{(2)}(x_i) + O(h^4)$$

$$= \frac{1}{12}(s_{i-1}^{(2)} + 10s_i^{(2)} + s_{i+1}^{(2)}) + O(h^4) \ ; \quad i=1(1)n-1, \tag{2.9b}$$

and

$$y_n^{(2)} = Y_2^{(2)}(x_n) + O(h^4)$$

$$= \frac{1}{12}(14s_n^{(2)} - 5s_{n-1}^{(2)} + 4s_{n-2}^{(2)} - s_{n-3}^{(2)}) + O(h^4). \tag{2.9c}$$

Since the approximations (2.8) and (2.9) are precisely those given by (1.4) and (1.5), it follows that the ECM may be regarded as a method where in Eqs. (1.6), the derivatives $y_i^{(2)}$ and $y_i^{(1)}$ are replaced respectively, by the corrected approximations $Y_2^{(2)}(x_i)$ and $Y_1^{(1)}(x_i)$.

Regarding convergence, Daniel and Swartz show that, under the assumptions C1 - C3,

$$||\tilde{s}^{(j)} - s^{(j)}|| = O(h^4) \ ; \ j=0(1)2, \tag{2.10}$$

where, as before, $\tilde{s}$ is the ECM spline defined by the Eqs.(1.11). Hence, by using the triangle inequality, they show that

$$||\tilde{s}^{(j)} - y^{(j)}|| = O(h^{4-j}) \ ; \ j=0(1)3, \tag{2.11}$$

and also that the spline $\tilde{s}$ displays superconvergence of the form (2.6) at the special points (2.7) , i.e. that

$$y^{(j)}(x_i + \mu_j h) = \tilde{s}^{(j)}(x_i + \mu_j h) + O(h^{5-j}); \ i=0(1)n-1, \tag{2.12a}$$

where

$$\mu_1 = 0, 1/2, 1 \ , \ \mu_2 = (3 \pm \sqrt{3})/6, \quad \mu_3 = 1/2 \ ; \tag{2.12b}$$

see (Daniel and Swartz, 1975, §4). However, by constructing corrected approximations of the form (2.3), it is possible to obtain improved approximations to $y^{(j)}$ at any point $x \in [a,b]$, as follows.

Let $\tilde{Y}_M$; M=1,2, denote the corrected approximation corresponding to the ECM spline $\tilde{s}$. That is, $\tilde{Y}_M$ is obtained from (2.3) by replacing s and $d_i^{(4+m)}$; m=0,1, respectively by $\tilde{s}$ and $\tilde{d}_i^{(4+m)}$; m=0,1 , where $\tilde{d}_i^{(4+m)}$

denote the approximations (2.3)-(2.4) with s replaced by $\tilde{s}$. Then, for $0\le j\le 4$ and $M=1,2$, (2.5) gives

$$y^{(j)}(x) - \tilde{Y}_M^{(j)}(x) = y^{(j)}(x) - Y_M^{(j)}(x) + Y_M^{(j)}(x) - \tilde{Y}_M^{(j)}(x)$$

$$= Y_M^{(j)}(x) - \tilde{Y}_M^{(j)}(x) + 0(h^{4-j+M});$$

$$x\in[x_i, x_{i+1}] \ ; \ i=0(1)n-1, \qquad (2.13a)$$

where

$$Y_M^{(j)}(x) - \tilde{Y}_M^{(j)}(x) = s^{(j)}(x) - \tilde{s}^{(j)}(x) + \sum_{m=0}^{M-1} \{\frac{h^{4-j+m}}{(4+m)!}(d_i^{(4+m)} - \tilde{d}_i^{(4+m)}) \times P_m^{(j)}(\frac{x-x_i}{h})\}. \qquad (2.13b)$$

Also, from (2.10) and the definitions of the approximations $d_i^{(4+m)}$ and $\tilde{d}_i^{(4+m)}$, we have that

$$||s^{(3)} - \tilde{s}^{(3)}|| = 0(h^3), \qquad (2.14)$$

$$\left.\begin{array}{l} |d_i^{(4)} - \tilde{d}_i^{(4)}| = 0(h^2), \\ \text{and} \\ |d_i^{(5)} - \tilde{d}_i^{(5)}| = 0(h) \ ; \quad i=1(1)n-1. \end{array}\right\} \qquad (2.15)$$

Therefore, by combining the results (2.10) and (2.13)-(2.15), we obtain, for $0\le j\le 4$ and $M=1,2$,

$$y^{(j)}(x) = \tilde{Y}_M^{(j)}(x) + 0(h^r) \quad ; \quad x\in[x_i, x_{i+1}], \ i=0(1)n-1, \qquad (2.16a)$$

where

$$r=\min\{4-j+M, 4\}. \qquad (2.16b)$$

The result (2.16) implies that the best order of approximation to y is that given by (2.11), i.e.

$$y(x) = \tilde{s}(x) + 0(h^4), \quad x\in[a,b]. \qquad (2.17)$$

However, for the derivatives $y^{(j)}$; $j=1(1)4$, improved orders of approximation can be obtained by using one or two correction terms, as follows.

$$y^{(1)}(x)=\tilde{Y}_1^{(1)}(x)+O(h^4),$$

$$y^{(j)}(x)=\tilde{Y}_2^{(j)}(x)+O(h^{6-j}) \;;\; j=2(1)4, \quad x\in[a,b]. \tag{2.18}$$

To illustrate the improvements in accuracy that can be achieved by using the corrected approximations $\tilde{Y}_M^{(j)}$, we consider the following problem, which is taken from (Daniel and Swartz, 1975, Problem 1,p.172), and which has the analytical solution

$$y(x)=\frac{1}{1+4x^2}.$$

Problem 1

$$y^{(2)}(x)+\frac{16x}{1+4x^2}y^{(1)}(x)+\frac{8}{1+4x^2}y(x)=0, \quad 0\le x\le 1,$$

$$y(0)=1, \qquad y(1)=0.2.$$

The numerical results presented are estimates of the maxinum errors

$$\max_{0\le i\le n-1}\left\{\max_{x\in[x_{i-1},x_i]}\left|y^{(j)}(x)-\tilde{Y}_M^{(j)}(x)\right|\right\},$$

obtained by sampling $y^{(j)}-\tilde{Y}_M^{(j)}$ at 800 equally spaced points in [0,1]. We denote these estimates by $E_M^{\{j\}}(n)$ and in Table 1 we list the values of $E_M^{\{j\}}(80)$. We also list the computed values

$$R_M^{\{j\}}(40)=\log_2\{E_M^{\{j\}}(40)/E_M^{\{j\}}(80)\},$$

giving the observed rates of convergence of the corrected approximations. (As before, $\tilde{Y}_0:=\tilde{s}$, i.e., $E_0^{\{j\}}(n)$; j=1(1)4 are estimates of

$$\max_{0\le i\le n-1}\left\{\max_{x\in[x_i,x_{i+1}]}\left|y^{(j)}(x)-\tilde{s}^{(j)}(x)\right|\right\})$$

Table 1

	M=0	M=1	M=2
J=0	3.44 E-8 4.1	2.82 E-8 4.1	2.73 E-8 4.1
J=1	6.03 E-6 3.0	6.31 E-7 4.3	3.11 E-7 4.8
J=2	5.08 E-3 2.1	2.90 E-4 3.4	3.92 E-5 3.9
J=3	2.41 E 0 1.0	1.61 E-1 2.2	1.93 E-2 2.9
J=4		3.61 E+1 1.1	5.82 E 0 1.9

Top entries : Values of $E_M^{\{J\}}(80)$

Bottom entries: Values of $R_M^{\{J\}}(40)$

Theoretical Rates: $R_M^{\{J\}} = \min\{4-J+M,4\}$.

3. Generalizations

The results of (Lucas, 1983) concerning the a posteriori correction of odd degree periodic splines can also be extended easily to non periodic quintic splines. That is, given an interpolatory quintic spline s with equally spaced knots (1.2), it is possible to construct corrected approximations Y_M;M=1,2,3, involving one, two or there correction terms. These corrected approximations are of the form:

$$Y_M(x)=s(x)+\sum_{m=0}^{M-1}\left\{\frac{h^{6+m}}{(6+m)!}d_i^{(6+m)}P_m\left(\frac{x-x_i}{h}\right)\right\};\ x\in[x_i,x_{i+1}],\quad i=0(1)n-1, \tag{3.1}$$

where now the polynomials P_m;m=0,1,2 are

$$P_0(x)=x^6-3x^5+\frac{5}{2}x^4-\frac{1}{2}x^2,\quad P_1(x)=x^7-\frac{7}{2}x^5+\frac{7}{2}x^3-x,$$
$$P_2(x)=x^8-7x^4+6x^2, \tag{3.2}$$

and the derivative approximations $d_i^{(6+m)}$; m=0,1,2, are in terms of values of $s^{(4)}$ as follows :

(i) $d_i^{(6)}=\frac{1}{h^2}\delta^2 s_i^{(4)}$; i=1(1)n-1, (3.3a) and

$$d_0^{(6)} = \begin{cases} d_1^{(6)} , & \text{if } M=1, \\ 2d_1^{(6)} - d_2^{(6)} , & \text{if } M=2, \\ 3d_1^{(6)} - 3d_2^{(6)} + d_3^{(6)} , & \text{if } M=3 . \end{cases} \tag{3.3b}$$

(ii) $$d_i^{(7)} = \frac{1}{2h} \{d_{i+1}^{(6)} - d_{i-1}^{(6)}\} \; ; \; i=1(1)n-2, \tag{3.4a}$$

$$d_0^{(7)} = \begin{cases} d_1^{(7)} , & \text{if } M=2, \\ 2d_1^{(7)} - d_2^{(7)} , & \text{if } M=3, \end{cases} \tag{3.4b}$$

and

$$d_{n-1}^{(7)} = \begin{cases} \frac{1}{h}(d_{n-1}^{(6)} - d_{n-2}^{(6)}) , & \text{if } M=2, \\ \frac{1}{2h}(3d_{n-1}^{(6)} - 4d_{n-2}^{(6)} + d_{n-3}^{(6)}) , & \text{if } M=3. \end{cases} \tag{3.4c}$$

(iii) $$d_i^{(8)} = \frac{1}{h^2} \{d_{i+1}^{(6)} - 2d_i^{(6)} + d_{i-1}^{(6)}\} \; ; \; i=1(1)n-2, \tag{3.5a}$$

$$d_0^{(8)} = d_1^{(8)} , \tag{3.5b}$$

and

$$d_{n-1}^{(8)} = d_{n-2}^{(8)} . \tag{3.5c}$$

In the quintic case, corresponding to (2.5) we have, for $0 \le j \le 6$ and $M=0,1,2,3$,

$$y^{(j)}(x) = Y_M^{(j)}(x) + O(h^{6-j-+M}) \; ; \; x \in [x_i, x_{i+1}], \quad i=0(1)n-1 , \tag{3.6}$$

provided that y is sufficiently smooth and the end-conditions of s are chosen appropriately ; see (Papamichael and Soares, 1985)

Consider the fourth order boundary value problem

$$Ly(x) \equiv y^{(4)}(x) + \sum_{j=0}^{3} e_j(x) y^{(j)}(x) = f(x) \; ; \quad a \le j \le b, \tag{3.7a}$$

$$\sum_{j=0}^{3} [\alpha_{ij} \, y^{(j)}(a) + \beta_{ij} \, y^{(j)}(b)] = 0 \; ; \quad i=1(1)4. \tag{3.7b}$$

Then, by analogy with the ECM , the replacement of the derivatives $y_i^{(j)}$, in the equations

$$Ly_i = f_i \; ; \quad i=0(1)n, \tag{3.8}$$

and in the boundary conditions (3.7b) , by approximations of the form

$Y_M^{(j)}(x_i)$ leads to modified collocation methods for the solution of fourth order boundary value problems of the form (3.7). Full details regarding the derivation of such methods can be found in (Papamichael and Soares, 1985(b)), where also a unified convergence analysis for modified collocation methods is presented. Finally, we note that the extrapollated collocation method of (Daniel and Swartz, 1975) also applies to second order non-linear boundary value problems. Similarly, the generalizations of the ECM also extend to non-linear boundary value problems.

This is a report of joint work with N. Papamichael. Dept. of Maths. and Stats., Brunel University.

REFERENCES

1. Daniel, J.W. and Swartz, B.K., Extrapollated Collocation for Two-Point Boundary Value Problems using Cubic Splines, *J.Inst.Maths. Applics 16 (1975), 161-174.*

2. Lucas, T.R., A Posteriori Improvements for Interpolating Periodic Splines, *Maths Comput. 40(1983), 243-251.*

3. Papamichael, N. and Soares, M.J.(a), A Posteriori Corrections for Cubic and Quintic Interpolating Splines at Equally-Spaced knots, TR/13/85, Brunel University (1985).

4. Papamichael, N. and Soares, M.J.(b), Modified Collocation Methods for Two-Point Boundary Value Problems Using Cubic and Quintic Splines, TR Brunel University (1985?).

Numerical Approximation of Partial Differential Equations
E.L. Ortiz (Editor)

Piecewise Polynomial Approximations for Cauchy Singular Integrodifferential Equations

Peter Linz
University of California
Davis, California, USA

Cauchy singular integral equations arise in a variety of applications; consequently their numerical solution is of considerable interest. In addition to various classical approaches, such as orthogonal polynomial approximations and Gaussian quadrature methods, approximations by piecewise polynomials have been studied. This paper shows how piecewise polynomial approximations can be used to solve singular integrodifferential equations.

1. Introduction

Because of their importance in practical applications there has been a considerable recent interest in the numerical solution of singular integral equations of the form

$$a(x)y(x) + b(x) \fint_{-1}^{1} \frac{k(x,t)y(t)}{x-t}\, dt = g(x), \qquad -1 < x < 1. \tag{1}$$

The most widely studied numerical methods for this equation are global, based either on approximations by orthogonal polynomials or on Gaussian quadrature. Such methods work well if all functions, such as the solution and the kernel in equation (1), are well-behaved. In many practical situations this may not be the case - an observation which prompted the development of a another class of methods based on piecewise polynomial approximations. The underlying idea was introduced in (Gerasoulis and Srivastav 1981) using piecewise linear functions and later generalized (Gerasoulis 1982, Jen and Srivastav 1981) to higher degree polynomials. In this paper we propose a modification of the

Gerasoulis -Srivastav method that is suitable for the solution of integrodifferential equations.

2. A Class of Piecewise Polynomial Approximation Methods

We consider the singular integral equation of the first kind

$$⨍_{-1}^{1} \frac{y(t)}{x-t} dt + \int_{-1}^{1} k(x,t)y(t)dt = g(x). \tag{2}$$

This equation has index 1, hence y(x) behaves like $(1-x^2)^{-1/2}$ near $|x| = 1$. Furthermore, to assure the unique solvability of (2) we need a subsidiary condition which we will choose as

$$\int_{-1}^{1} y(t)dt = C. \tag{3}$$

To account for the singularity in the solution, we introduce the new function

$$u(x) = \sqrt{1-x^2}\, y(x), \tag{4}$$

and rewrite (2) and (3) as

$$⨍_{-1}^{1} \frac{u(t)}{\sqrt{1-t^2}\,(x-t)} dt + \int_{-1}^{1} \frac{k(x,t)u(t)}{\sqrt{1-t^2}} dt = g(x), \tag{5}$$

$$\int_{-1}^{1} \frac{u(t)}{\sqrt{1-t^2}} dt = C . \tag{6}$$

We then choose a set of expansion functions $\phi_i(x)$

and approximate u(x) by

$$u_n(x) = \sum_{i=1}^{n} c_i\phi_i(x). \tag{7}$$

The coefficients c_i will be determined by imposing n conditions; one of these can come from (6), the others are obtained by satisfying (5) in some approximate sense. To be more specific, let us write

$$\xi_i(x) = \fint_{-1}^{1} \frac{\phi_i(t)}{\sqrt{1-t^2}\ (x-t)}\, dt, \tag{8}$$

$$\psi_i(x) = \int_{-1}^{1} \frac{k(x,t)\phi_i(t)}{\sqrt{1-t^2}}\, dt, \tag{9}$$

$$\beta_i = \int_{-1}^{1} \frac{\phi_i(t)}{\sqrt{1-t^2}}\, dt\ . \tag{10}$$

We next choose n-1 linear functionals L_i and require that

$$L_i(u_n(x)) = L_i(g(x)), \qquad i = 1,2,\ldots,n-1, \tag{11}$$

which leads us to a linear system for the coefficients c_i.

This framework is completely general. By choosing various ϕ_i and L_i we can generate a great variety of different methods. Whether a proposed method is useful depends, in part, on how easy it is to evaluate the functions ξ_i, ψ_i, β_i, and the linear functionals on them.

One class of methods is obtained by choosing piecewise polynomials for the ϕ_i. For this case, the functions ξ_i and β_i can be given in closed form. For example, to obtain piecewise linear approximations on a mesh $-1 = t_1 < t_2 < \dots < t_n = 1$, $h_i = t_{i+1} - t_i$, we use as ϕ_i the familiar hat functions

$$
\begin{aligned}
\phi_i(x) &= (x-t_{i-1})/h_{i-1}, && t_{i-1} \leq x \leq t_i, \\
&= (t_{i+1}-x)/h_i, && t_i \leq x \leq t_{i+1}, \qquad (12)\\
&= 0, && \text{otherwise} .
\end{aligned}
$$

In this case, ξ_i and β_i can be evaluated explicitly as

$$
\xi_i(x) = \frac{1-\delta_{i1}}{h_{i-1}} \{m_{i-1} + (x-t_{i-1})\ell_{i-1}(x)\} + \frac{1-\delta_{in}}{h_i} \{(t_{i+1}-x)\ell_i(x) - m_i\}, \tag{13}
$$

$$
\beta_i = \frac{1-\delta_{i1}}{h_{i-1}}\{\sqrt{1-t_{i-1}^2} - \sqrt{1-t_i^2} - t_{i-1}m_{i-1}\} + \frac{1-\delta_{in}}{h_i}\{\sqrt{1-t_{i+1}^2} - \sqrt{1-t_i^2} - t_{i+1}m_i\}. \tag{14}
$$

Here δ_{ij} is the Kronecker delta and

$$
m_i = \sin^{-1}t_i - \sin^{-1}t_{i+1}, \tag{15}
$$

$$
\ell_i(x) = \frac{1}{\sqrt{1-x^2}} \ln \frac{(x \tan \theta_{i+1}/2 - 1 - \sqrt{1-x^2})(x \tan \theta_i/2 - 1 + \sqrt{1-x^2}\ )}{(x \tan \theta_{i+1}/2 - 1 + \sqrt{1-x^2})(x \tan \theta_i/2 - 1 - \sqrt{1-x^2}\ )}, \tag{16}
$$

$$
\theta_i = \sin^{-1}t_i. \tag{17}
$$

In the Gerasoulis-Srivastav method, the ψi are evaluated approximately by replacing k(x,t) with a simple form, say a constant or a low order polynomial. This works well if k(x,t) is smooth; if it is not, a different approach is needed. For integrodifferential equations, k(x,t) is discontinuous, but has a rather special form, making the evaluation of the ψ_i a simple matter.

To complete the definition of a specific method we must pick the functionals L_i. One choice is collocation, where we take a set of distinct points $t_1^*, t_2^*, \ldots, t_{n-1}^*$ and let $L_i(f(x)) = f(t_i^*)$. Collocation methods are simple to use, but our theoretical understanding of their convergence is still incomplete. Alternatives are Galerkin's method or least-squares methods for which stability and convergence can be guaranteed under certain conditions (see Linz 1984).

3. The Solution of Integrodifferential Equations

Although our method is quite general, we consider in detail only the well-known Prandtl's integrodifferential equation

$$a(x)y(x) + \fint_{-1}^{1} \frac{y'(t)}{x-t}\,dt = g(x), \tag{18}$$

with subsidiary conditions

$$y(-1) = y(1) = 0. \tag{19}$$

Piecewise polynomial approximations are constructed readily if we first transform (18) into an integral equation. Using (19) we get the equation

$$a(x)\int_{-1}^{x} y'(t)dt + \fint_{-1}^{1} \frac{y'(t)}{x-t}\,dt = g(x), \tag{20}$$

as well as the subsidiary condition

$$\int_{-1}^{1} y'(t)dt = 0. \tag{21}$$

It is known from the theory of singular integral equations that $y(x)$ behaves like $(1-x^2)^{1/2}$ near the ends of the interval; consequently $y'(x)$ has a singularity of the form $(1-x^2)^{-1/2}$ at $|x| = 1$. Thus, if we write

$$u(x) = \sqrt{1-x^2}\, y'(x), \tag{22}$$

then $u(x)$ satisfies

$$⨍_{-1}^{1} \frac{u(t)}{\sqrt{1-t^2}\,(x-t)}\, dt + a(x) \int_{-1}^{x} \frac{u(t)}{\sqrt{1-t^2}}\, dt = g(x), \tag{23}$$

$$\int_{-1}^{1} \frac{u(t)}{\sqrt{1-t^2}}\, dt = 0. \tag{24}$$

For piecewise polynomial approximations the functions ψ_i can be evaluated explicitly. In the simple case of piecewise linear approximations we have

$$\psi_i(x) = a(x)\, \gamma_i(x), \tag{25}$$

where

$$\gamma_i(x) = \int_{-1}^{x} \frac{\phi_i(t)}{\sqrt{1-t^2}}\, dt$$

$$= 0, \qquad x < t_{i-1},$$

$$= \frac{1-\delta_{i1}}{h_{i-1}} \{\sqrt{1-t_{i-1}^2} - \sqrt{1-z_i^2} - t_{i-1}(\sin^{-1} z_1 - \sin^{-1} t_{i-1})\}$$

$$(26)$$

$$+ \frac{\nu(x)}{h_i} \{t_{i+1}(\sin^{-1} z_2 - \sin^{-1} t_i) - \sqrt{1-t_i^2} + \sqrt{1-z_2^2}\}, \quad x > t_{i-1}$$

with $z_1 = \min(x,t_i)$, $z_2 = \min(x,t_{i+1})$,

$$\nu(x) = 0, \quad x < t_i,$$
$$= 1, \quad x > t_i.$$

From an approximation $u_n(x)$ to $u(t)$ we can get an approximation to $y(x)$ by

$$y_n(x) = \int_{-1}^{x} \frac{u_n(t)}{\sqrt{1-t^2}}\, dt$$

$$= \sum_{i=1}^{n} c_i \int_{-1}^{x} \frac{\phi_i(t)}{\sqrt{1-t^2}}\, dt$$

$$= \sum_{i=1}^{n} c_i \gamma_i(x). \tag{27}$$

Example. As a numerical test case, we consider the equation

$$y(x) - \frac{1}{2\pi} ⨍_{-1}^{1} \frac{y'(t)}{t-x}\, dt = 1, \tag{28}$$

The results were computed using piecewise linear approximations with constant step size $h = 2/(n-1)$ and collocation at $t_i^* = t_i + h/2$. Approximate values of y, for several values of h, are shown below. The convergence of the numerical results is evident

t \ h	.25	.125	.0625
0	.69705	.69703	.69705
.2	.68933	.68935	.68937
.4	.66393	.66399	.66400
.6	.61186	.61160	.61164
.8	.50449	.50240	.50233

REFERENCES

Gerasoulis, A., and R. P. Srivastav, "A method for the numerical solution of singular integral equations with a principal value integral," Int. J. Engrg. Sci. 19 (1981), 1293-1298.

Gerasoulis, A., 1982, "The use of piecewise quadratic polynomials for the solution of singular integral equations of Cauchy type," Comp. & Maths. with Appls. 8 (1982), 15-22.

Jen, E., and R. P. Srivastav, "Cubic splines and approximate solution of singular integral equations," Math. Comp. 37 (1981), 417-423.

Linz, P., "Stability analysis for the numerical solution of singular integral equations," in "Numerical Solution of Singular Integral Equations," A. Gerasoulis and R. Vichnevetsky (eds), IMACS, 1984.

Part IV: VARIATIONAL METHODS AND SPECIAL TECHNIQUES

Variational Methods

Conformal Transformation Methods and Asymptotic Techniques

Numerical Approximation of Partial Differential Equations
E.L. Ortiz (Editor)

METHODS OF COMPUTATION OF CRITICAL POINTS OF NONLINEAR FUNCTIONALS

Alexander Eydeland
University of Massachusetts at Amherst
Amherst, U.S.A.

In this paper the method of transformation of the objective functional is presented as a general approach to a problem of finding critical points of nonlinear functionals. Particular attention is given to non-convex problems, and constrained problems. Specific applications are discussed including numerical solution of the minimal surface obstacle problem, computation of periodic solutions of nonlinear Hamiltonian systems and solving semilinear boundary value problems. Different aspects of numerical implementation of the method of transformation of the objective functional, such as global convergence, mesh refinement for nonlinear problems, etc., are discussed.

1. In this paper we develop numerical procedures for solving the following general problem: *Find a critical or an extremum point of the functional*

$$\Phi(u) = \Sigma_{i=1}^{m} \int_{D_i} f_i(x,u,\nabla u)dx \ , \quad u \in B \ , \tag{1}$$

where B is a subset of a Hilbert space H. The choice of B and H depends on the nature of the problem and of the constraints. The sets D_i are convex subsets of $\mathbb{R}^n$.

Numerical procedures that we propose for finding critical or extremum points of the functional (1) are based on one method, which we call a method of nonlinear transformation of the objective functional (Eydeland, 1984). The idea of the method is as follows.

Assume that there exist continuously differentiable functions $g_i(x,z,\overline{p},\alpha)$, $i = 1,\ldots,m$, $z \in \mathbb{R}^1$, $\overline{p} \in \mathbb{R}^n$, $\alpha \in \mathbb{R}^1$, which are convex in $z,\overline{p},\alpha$ for every fixed x and such that the composition functions $g_i(x,u,\nabla u,f_i(x,u,\nabla u))$ are simpler than the functions $f_i(x,u,\nabla u)$ from (1). The term "simpler" means here that these compositions are linear functions or quadratic functions, or any functions such that a minimum point of the functional

$$\Sigma_{i=1}^{m} \int_{D_i} \{\beta_i(x) g_i(x,u,\nabla u, f_i(x,u,\nabla u)) + (\text{quadratic and linear terms in } u,\nabla u)\}dx \tag{2}$$

over a set $B_1 \subset B$ can be found faster and by more efficient numerical methods than critical or extremum points of the functional (1) for any choice of coefficients $\beta_i(x)$ from a certain set of functions (usually it is a set of non-negative continuous functions). The subset B_1 is dense in B and the functional (2) is differentiable in B_1 for every choice of the coefficients β_i .

According to the method of nonlinear transformation of the objective functional we consider the following iterative procedure:

$u^0 \in B_1 \subset B$;

for $k = 0,1,\ldots$ u^{k+1} *is a point of the global minimum over* B_1 *of the functional*

$$\Gamma^k(u) = \Sigma_i \int_{D_i} \{[g_i(x,u,\nabla u, f_i(x,u,\nabla u)) - u \frac{\partial g_i}{\partial z}(x,u^k,\nabla u^k,\alpha_i^k(x)) - \nabla u \frac{\partial g_i}{\partial \bar{p}}(x,u^k,\nabla u^k,\alpha_i^k(x))] / \frac{\partial g_i}{\partial \alpha}(x,u^k,\nabla u^k,\alpha_i^k(x))\}dx \ . \tag{3}$$

The functions $\alpha_i^k(x)$ *are chosen in the following way. If there exists a contant* $\delta > 0$ *such that* $\frac{\partial g_i}{\partial \alpha}(x,w,\nabla w, f_i(x,w,\nabla w)) \geq \delta$ *for every* $w \in B$, $x \in D_i$, $i = 1,\ldots,m$, *then* $\alpha_i^k(x) = f_i(x,u^k,\nabla u^k)$. *Otherwise*

$\alpha_i^k(x) = \max \{\delta^k, \frac{\partial g_i}{\partial \alpha}(x,u^k,\nabla u^k, f_i(x,u^k,\nabla u^k))\}$. The sequence $\{\delta^k\}$, $k = 0,1,\ldots,$ monotonically converges to zero and satisfies certain conditions on the rate of decay, e.g. see (Eydeland, 1985).

In this paper the transforming functions $g_i(x,z,\bar{p},\alpha)$ always satisfy the following conditions.

$$g_i(x,z,\bar{p},\alpha) \text{ are convex in the variables } (z,\bar{p},\alpha) \in \mathbb{R}^{n+2}$$

$$\text{for any fixed } x \in D_i \text{ and } \frac{\partial g_i}{\partial \alpha}(x,z,\bar{p},\alpha) \geq 0 \tag{4}$$

$$\text{and may vanish only when } \alpha = 0 \ .$$

For the sequence $\{u^k\}$ generated by the procedure (3) we can immediately prove the following convergence result.

THEOREM. If the functional $\Phi(u)$ is bounded from below in B and if the sets D_i are bounded in $\mathbb{R}^n$ then the sequence $\Phi(u^k)$, $k = 0,1,\ldots,$ converges.

Proof. We introduce a functional $\phi(u,u^k)$, where $u \in B_1$, $u^k \in B_1$:

$$\begin{aligned}\phi(u,u^k) = \Sigma_i \int_{D_i} \{\alpha_i^k(x) + [g_i(x,u,\nabla u,f_i(x,u,\nabla u)) - g_i(x,u^k,\nabla u^k,\alpha_i^k(x))\\ - (u-u^k)\frac{\partial g_i}{\partial z}(x,u^k,\nabla u^k,\alpha_i^k(x)) - (\nabla u-\nabla u^k)\frac{\partial g_i}{\partial \bar{p}}(x,u^k,\nabla u^k,\alpha_i^k(x))]/\\ \frac{\partial g_i}{\partial \alpha}(x,u^k,\nabla u^k,\alpha_i^k(x))\}dx \ .\end{aligned}$$

By (3)

$$\phi(u^{k+1},u^k) \leq \phi(u^k,u^k) \ . \tag{5}$$

By convexity of the functions g_i

$$\phi(u^{k+1},u^{k+1}) \geq \Sigma_i \int_{D_i} f_i(x,u^{k+1},\nabla u^{k+1})dx = \Phi(u^{k+1}) \tag{6}$$

Therefore

$$\begin{aligned}\phi(u^{k+1},u^k) \leq \phi(u^{k+1},u^{k+1}) + \Sigma_i \int_{D_i} \{[g_i(x,u^{k+1},\nabla u^{k+1},f_i(x,u^{k+1},\nabla u^{k+1}))\\ - g_i(x,u^k,\nabla u^k,\alpha_i^k(x)) - (u^{k+1}-u^k)\frac{\partial g_i}{\partial z}(x,u^k,\nabla u^k,\alpha_i^k(x))\\ - (\nabla u^{k+1}-\nabla u^k)\frac{\partial g_i}{\partial \bar{p}}(x,u^k,\nabla u^k,\alpha_i^k(x)) - (f_i(x,u^{k+1},\nabla u^{k+1}) - \alpha_i^k(x)) \cdot\\ \frac{\partial g_i}{\partial \alpha}(x,u^k,\nabla u^k,\alpha_i^k(x))]/\frac{\partial g_i}{\partial \alpha}(x,u^k,\nabla u^k,\alpha_i^k(x)\}dx \leq \phi(u^{k+1},u^{k+1}) \ .\end{aligned} \tag{7}$$

The last inequality above follows from the conditions (4). Combining (5) and (7) we obtain the sequence of inequalities

$$(8) \qquad \ldots \leq \phi(u^{k+1},u^{k+1}) \leq \phi(u^{k+1},u^{k}) \leq \phi(u^{k},u^{k}) \leq \ldots .$$

By (6) and by the boundedness of $\Phi(u)$ we obtain that the sequence $\phi(u^k,u^k)$ converges. For bounded D_i $\phi(u^k,u^k) = \Phi(u^k) + O(\delta^k)$. Since $\delta^k \downarrow 0$ we have that the sequence $\Phi(u^k)$ converges. The theorem is proved.

The techniques developed in (Eydeland, 1984; Eydeland, 1984; Eydeland, 1985) and other works can be used to show that the sequence $\{u^k\}$ itself converges in some norm to a solution of the problem (1).

Note that the functional (3) is of the form (2) and therefore, by the assumption above, its minimum can be easily found.

2. A natural question arises, "How large is the class of functionals (1) for which we can find simplifying convex transformations g_i ?" In this section we show that this class is large. We consider the following cases.

i) The functional $\Phi(u)$ from (1) is convex and the matrices of second derivations $D^2_{z,\bar{p}} f_i$ with respect to the variables $z,\bar{p}$ are bounded for every fixed $x \in D_i$, $i = 1,\ldots,m$, i.e. $\|D^2_{z,\bar{p}} f_i\| \leq \gamma_i$, where γ_i are positive constants. Consider now the functions $g_i(x,z,\bar{p},\alpha) = \alpha - f_i(x,u,\nabla u) + \langle H_i(x)\xi,\xi\rangle$, where by ξ we denote a pair $(z,\bar{p})$. These functions g_i transform the functional $\Phi(u)$ into a quadratic functional since $g_i(x,u,\nabla u,f_i(x,u,\nabla u)) = Q(u,\nabla u)$, where Q is a quadratic form in $\mathbb{R}^{n+1}$. Positive definite matrices $H_i(x)$ are chosen in such a way that the condition (4) is satisfied, i.e. the functions g_i are convex. This means that $\|H_i(x)\| \geq \gamma_i$, $i = 1,\ldots,m$.

Example 1. Consider the problem (1) with the functional

$$\Phi(u) = \int_D \{\tfrac{1}{2}|\nabla u|^2 + \Psi(u)\}dx \, , \tag{9}$$

where $\Psi(u) = \int_0^u \psi(\zeta)d\zeta$ is a convex non-negative functional. We want to minimize the functional (9) over a hyperspace $B = \{u \in W_1^2(D) \, , \, u = \phi$ on $\partial D\}$. This problem corresponds to the Dirichlet problem for the equation

$$-\Delta u + \psi(u) = 0 \quad \text{in} \quad D \, , \quad u = \phi \quad \text{on} \quad \partial D \, . \tag{10}$$

The functional $\Phi(u)$ in (9) is of the form (1) with $f_1(\nabla u) = \frac{1}{2}|\nabla u|^2$, $f_2(u) = \Psi(u)$. If we take $g_1(\alpha) = \alpha$, $g_2(x,u,\alpha) = \alpha - \Psi(u) + \frac{1}{2} s^k \Psi''(u^k(x))u^2$, where s^k is a positive number, then as can be easily shown the procedure (3) is an iteration of Newton's method with step adjustment for the equation (10). The value $1/s^k$ is a step size. We note that it follows from the above that in order to guarantee the convergence s^k should be chosen in such a way that the function $g_2(x,u,\alpha)$ is convex in u for all $u \in B$ and for any fixed $x \in D$.

Another example of this type of transforming functions can be found in (Eydeland, 1984).

ii) Very frequently encountered in applications is the functional $\Phi(u)$ from (1) with the functions $f_i(x,u,\nabla u) = G_i(Q_i(u,\nabla u) + L_i(u,\nabla u))$, $i = 1,\ldots,m$, where $Q_i(u,\nabla u)$ are quadratic forms in $u,\nabla u$, $L_i(u,\nabla u)$ are linear functions in $u,\nabla u$. The functions $G_i(\alpha)$ are such that the functions $g_i(\alpha) = G_i^{-1}(\alpha)$ are convex. Then the compositions $g_i(f_i)$ are quadratic functions and the procedure (3) can be successfully applied.

Example 2. We consider the minimal surface problem with

$$\Phi(u) = \int_D \sqrt{1+|\nabla u|^2}\, dx \ , \quad u \in B \tag{11}$$

$$B = \{u \in W_1^1(D) \ , \quad u = \phi \ \text{ on } \ \partial D\} \ .$$

The functional $\Phi(u)$ is of the form (1) with $f_1(\nabla u) = (1+|\nabla u|^2)^{\frac{1}{2}}$. We take the transforming function $g_1(\alpha) = \alpha^2$. Then $g_1(f_1(\nabla u)) = 1+|\nabla u|^2$. We choose $B_1 = W_1^2(D)$. The procedure (3) in this case can be written as

u^0 *is an arbitrary function from* $W_1^2(D)$; *for* $k = 0,1,\ldots$.
u^{k+1} *is a point of minimum of a quadratic functional*

$$\Gamma^k(u) = \int_D \{|\nabla u|^2/\sqrt{1+|\nabla u^k|^2}\}dx \ , \ u \in W_1^2(D) \ , \quad u = \phi \ \text{ on } \ \partial D \ .$$

As was proved in (Eydeland, 1984), the sequence $\{u^k\}$ converges to the minimum point u^* of the functional (11) in the W_1^1-norm, provided that $u^* \in C^1(\bar{D})$.

Another example can be found in the Thomas-Fermi theory, where $G_i(\alpha) = \alpha^{5/3}$, see (Lieb and Simon, 1977).

iii) The functional $\Phi(u)$ in (1) is such that the functions $f_i(x,u,\nabla u) = -F_i(x,u,\nabla u)$, where $F_i(x,u,\nabla u)$ are convex functions in $u,\nabla u$ for every fixed x, $i = 1,\ldots,m$. The set $B = \{u \in H : \Psi(u) = R\}$, where Ψ is a quadratic functional and R is a positive constant. This problem corresponds to a nonlinear eigenvalue problem: $\Phi'(u) + \lambda\Psi'(u) = 0$, $u \in H$. As transforming functions in this case we take the convex functions $g_i(x,u,\nabla u,\alpha) = \alpha + F_i(x,u,\nabla u)$. Then the procedure (3) is reduced to solving simple linear problems.

Example 3. We consider a second order dynamical system

$$(12) \qquad \ddot{u} + \nabla V(u) = 0 \ , \quad u \in \mathbb{R}^n \ .$$

Let $V(u)$ be convex and even. Our goal is to find a periodic solution of (12). Following (Berger, 1970) we replace (12) with a problem of finding a minimum point $u^*(t)$ of the functional

$$(13) \qquad \Phi(u) = \int_0^\pi - V(u)dt$$

subject to constraints

$$(14) \qquad \int_0^\pi |\dot{u}|^2 dt = R \ , \quad u(0) = u(\pi) = 0 \ ,$$

where R is a positive constant. It is clear that the vector function $\bar{u}(t) = u^*(t/\sqrt{\lambda})$ for $t \in [0,\pi]$, $\bar{u}(t) = -u^*(-t/\sqrt{\lambda})$ for $t \in [-\pi,0]$, where λ is the Lagrange multiplier for the problem (13-14) is a periodic solution of (12) with the period $2\pi/\sqrt{\lambda}$. For the transformation $g_1(u,\alpha) = \alpha + V(u)$ the procedure (3) reduces to solving simple problems

$$\ddot{u}^{k+1} + \lambda^{k+1} V(u^k) = 0 \ , \quad u(0) = u(\pi) = 0 \ , \quad \int |\dot{u}^{k+1}|^2 dt = R \ .$$

This is a generalization of the inverse power method. If the function $V(u)$ is not convex we still can transform $\Phi(u)$ into quadratic functional, see (Eydeland, 1984). Since the problem (13-14) may have the whole set Ω_R^* of solutions, convergence results for the sequence $\{u^k\}$, which is generated by (3), are of the form $\mathrm{dist}(u^k,\Omega_R^*) \to 0$ where the distance is calculated either in $(C^0[0,\pi]^n$ - norm or in $\overset{\circ}{W}{}_1^2[0,\pi])^n$ - norm .

Another classical example arises in the differential geometry. We want to maximize

$$\Phi(u) = \int_D - \mathrm{Cosh}\ u\ dx \ , \quad u = \phi \ \text{ on } \ \partial D \ ,$$

subject to

$$\int_D |\nabla u|^2 = R .$$

J. Spruck in (Spruck, 1985) discusses this problem in detail and proposes a procedure which solves this problem. This procedure is precisely the procedure (3) with $g_1(u,\alpha) = \alpha + \text{Cosh } u$.

iii) The functional $\Phi(u) = \Phi_0(u) + \int_D A_\pm (L(x,u,\nabla u))dx$.

Here $\Phi_0(u)$ is a functional for which we already know the simplifying functions. The function $A_-(\xi) = \xi^p$ for $\xi \leq 0$, $1 \leq p \leq 2$, $A_-(\xi) = 0$ for $\xi > 0$; $A_+(\xi) = A_-(-\xi)$. The function L is linear in $u, \nabla u$ for every fixed x . The functionals of this type usually arise in the constrained problems which are solved by penalty-type methods. The case $p = 2$ corresponds to a differentiable penalty function, $p = 1$ to a non-differentiable. Another source of the problems with this functional is the free-boundary problems, see (Friedman, 1982).

Simplifying transformations for the second part of $\Phi(u)$ can be taken in the form $g_1(x,u,\nabla u,\alpha) = [\alpha + A_\mp(L(x,u,\nabla u))]^{2/p}$ or $g_1(x,u,\nabla u,\alpha) = \alpha^{2/p} + [A_\mp(L(x,u,\nabla u))]^{2/p}$. In both cases the composition $g_1(x,u,\nabla u,A_\pm(L(x,u,\nabla u)))$ is a quadratic function.

Example 4. Consider a minimal surface obstacle problem. We have to minimize the functional

$$\Phi(u) = \int_D \sqrt{1+|\nabla u|^2}\, dx , \quad u \geq \Psi(x) \geq 0 , \quad u \in W_1^2(D) , \quad u = \phi \text{ on } \partial D$$

We solve this problem by the penalty method with a non-differentiable penalty functional. The problem now is to minimize the functional

$$\Phi_\rho(u) = \int_D \sqrt{1+|\nabla u|^2}\, dx + \int_D \rho A_-(u-\Psi(x))dx ,$$

where $A_-(\xi) = \xi$ for $\xi \leq 0$, $A_-(\xi) \equiv 0$ for $\xi > 0$, ρ is a fixed constant such that the minimum point for Φ_ρ is also a solution of the original obstacle problem.

It is clear that if we take $g_0(\alpha) = \alpha^2$ and $g_1(x,u,\alpha) = (\alpha+\rho A_+(u-\Psi(x))^2$ then the procedure (3) for the functional Φ_ρ reduces to a sequence of simple linear problems. The resulting sequence $\{u_\rho^k\}$ converges to the solution of the obstacle problem from any initial guess, see (Eydeland, 1985).

3. In this section we discuss some aspects of numerical implementation of the procedure (3). First we mention that the important feature of this procedure is that it converges globally, i.e. it converges from arbitrary initial guess. Second, we note that we can quite naturally incorporate into the procedure (3) changes of the grid size. Suppose the function u_{2h}^k is defined on a coarse grid with a grid size $2h$. What will correspond to this function on the grid with a step h ? It is clear that if we apply (3) with $u_h^k \equiv u^{k+1}$ and $u^k \equiv u_{2h}^k$ (u_{2h}^k here should be considered as a function on the fine grid) then u_h^k will not destroy the monotone convergence of $\phi(u^k,u^k)$ from the theorem above. Thus all convergence results proved on one grid will hold (possibly with different constants) for multigrid procedures. The third advantage of the method is that we discretize the procedure (3) which is in most cases of linear problems. Therefore the discretization does not constitute a serious problem which it could have been had we started with discretization of (1) directly.

REFERENCES

1. Berger M., On periodic solutions of second order Hamiltonian systems, J. Math. Analysis, 29, 1970, pp. 512-522.
2. Eydeland A., A method of solving nonlinear variational problems by nonlinear transformation of the objective functional, Part I, Numerische Math., 43, 1984, pp. 59-82.
3. Eydeland A., A method of calculating periodic solutions of second order dynamical systems, Math. Research Center Report #2768, University of Wisconsin, Madison, 1984.

4. Eydeland A., A method of solving nonlinear variational problems by nonlinear transformation of the objective functional, Part II, Numerische Math., to appear.
5. Eydeland A., Computational procedures for non-convex variational problems, 1985, to appear.
6. Friedman A., Variational principles and free-boundary problems, Wiley, New York, 1982.
7. Lieb E. H. and Simon B., The Thomas-Fermi theory of atoms, molecules and solids, Adv. in Math., 23, 1977, pp. 22-117.
8. Spruck J., Article to appear.

Numerical Approximation of Partial Differential Equations
E.L. Ortiz (Editor)

NUMERICAL SOLUTIONS OF DEGENERATE AND PSEUDOPARABOLIC VARIATIONAL INEQUALITIES

Francesco Scarpini
Department of Mathematics
University "La Sapienza"
Rome, Italy

We study a model of variational inequality of degenerate elliptic-parabolic type.We prove an existence and uniqueness theorem,apply the finite element method and verify the convergence of the approximate solutions.Then we consider a problem with an internal obstacle related to the pseudoparabolic operator governing the heat propagation in non-simple elastic materials (Ting T.W. 1974),in this case we obtain a more regular solution and an error estimate.At the end of this paper we study an application to the theory of seepage of homogeneous liquid in fissured rocks(Zheltov I.V.,Kochina I.N., Barenblatt G.I. 1960).

INTRODUCTION.In recent years degenerate and pseudoparabolic variational inequalities were studied from theoretical point of view (Carrol R.W., Showalter R.E. 1976;Kuttler K.L. 1984).I am considering here the numerical aspects of these problems with the aim of costructing approximate convergent solutions and of obtaining in some more regular cases,the error estimates.

FIRST PART

1.1 NOTATIONS. We adopt the usual notations in the area of evolution differential problems (Lions J.L. 1969) and the summation convention on repeated indices.We call:

-Ω an open bounded set in R^n with boundary $\Gamma=\partial\Omega$ regular enough,

-$]0,T[$ the bounded interval of time t,

-$Q=]0,T[\times\Omega$, $\Sigma=]0,T[\times\Gamma$,

-$b_{r,s}(x)\xi_r\xi_s$ the quadratic form where $b_{r,s}\geq 0, b_{r,s}\in L^\infty(\Omega)$ $(r,s=1,\ldots,n)$,

-$b(u,v)=\int_\Omega[b_{r,s}\partial_r u\partial_s v + b_o uv]dx$ where $b_o\geq 0$, $b_o\in L^\infty(\Omega)$,

-$Bu=-\partial_s(b_{r,s}\partial_r u)+b_o u$ the operator related to $b(u,v)$,

-$a_{r,s}(x)\xi_r\xi_s$ the quadratic form where $a_{r,s}=a_{s,r}\in L^\infty(\Omega)$ such that $a_{r,s}(x)\xi_r\cdot\xi_s\geq\alpha|\xi|^2$, $\forall\xi\in R^n$, a.e. $x\in\Omega,\alpha>0$

- $a(u,v)=\int_\Omega [a_{rs}\partial_r u \partial_s v + a_0 uv]\,dx$, where $a_0 \geq 0$, $a_0 \in L^\infty(\Omega)$,
- $Au = -\partial_s(a_{r,s}\partial_r u) + a_0 u$ the operator related to $a(u,v)$,
- $H=L^2(\Omega)$, $(u,v)=(u,v)_H=\int_\Omega uv\,dx$, $|u|=(u,u)^{1/2}$
- $V=H_0^1(\Omega)=\{v:\partial^\alpha v \in L^2(\Omega), |\alpha| \leq 1 \ / \ v|_\Gamma = 0$ a.e.$\}$, $((u,v))=(u,v)_V=\int_\Omega \mathrm{grad}u.\mathrm{grad}v\,dx$
- $||u||=((u,u))^{1/2}$
- $W=H^2(\Omega)\cap V=\{v:\partial^\alpha v \in L^2(\Omega), |\alpha| \leq 2 \ / v|_\Gamma = 0$ a.e.$\}$, $(((u,v)))=(u,v)_W=\int_\Omega \Delta u.\Delta v\,dx$
  $|||u|||=(((u,u)))^{1/2}$ where Δ is the Laplace operator,
- $||v||_b=(b(v,v))^{1/2}$ the seminorm corresponding to $b(.,.)$,
- $K(t)=\{v(t):v(t)\in V \ / \ v(t)\geq\psi(t)$ a.e. in Ω, $\forall t\in]0,T[\}$,
- $K=\{v:v\in L^2(0,T;V) \ / \ v\geq\psi$ a.e. in $Q\}$,
- $1/\varepsilon\beta(v)=-1/\varepsilon(v-\psi)^-$ the penalization operator from V to V bounded, monotone, Lipschitz continuous, where $\varepsilon\in]0,1[$, $v^-=0 \vee (-v)$.

1.2. A DEGENERATE EVOLUTION PROBLEM

We study the following problem:

$$(P)\begin{cases} \text{to find } u(t)\in K(t): \\ b(u'(t),v(t)-u(t))+a(u(t),v(t)-u(t))\geq (f(t),v(t)-u(t)), \forall v(t)\in K(t) \\ Bu(0)=0 \text{ a.e. in } \Omega \end{cases}$$

f, ψ are the data of (P), v' is the time derivative in generalized sense. From (P) by integrating we obtain the following problem:

$$(\mathcal{P})\begin{cases} \text{to find } u\in K \\ \int_0^T [b(u'(t),v(t)-u(t))+a(u(t),v(t)-u(t))]dt\geq\int_0^T (f(t),v(t)-u(t))dt, \forall v\in K \\ Bu(0)=0 \text{ a.e. in } \Omega. \end{cases}$$

THE PENALIZED PROBLEM

We consider the following penalized problem:

$$(P_\varepsilon)\begin{cases} \text{to find } u\in V: \\ \varepsilon(u'_\varepsilon(t),v(t))+b(u'_\varepsilon(t),v(t))+a(u_\varepsilon(t),v(t))+1/\varepsilon(\beta u_\varepsilon(t),v(t))=(f(t),v(t)) \\ u_\varepsilon(0)=0 \text{ a.e. in } \Omega \qquad \forall t\in]0,T[, v(t)\in V \end{cases}$$

where: $f(t)\in V'$ (the dual space of V); ψ is a regular function.

In correspondence to the perturbed bilinear form $b_\varepsilon(u,v)=b(u,v)+\varepsilon(u,v)$ in (P_ε) we use the following norms:

$$||v||_{b_\varepsilon}=(b_\varepsilon(v,v))^{1/2} \quad ; \quad ||v||_{L^\infty(0,T;b_\varepsilon)}=\operatorname*{ess.sup}_{t\in[0,T]}||v(t)||_{b_\varepsilon}.$$

By considering the sequences of eigenvalues $\{\lambda_i\}_{i\in N}$ and of ortonormal eigenvectors $\{v_i\}_{i\in N}$ of problem:

$$(P_\lambda)\begin{cases} Av=\lambda v \text{ in } \Omega \\ v|_\Gamma=0 \end{cases}$$

We construct the ordinary non linear differential system:

(1.2.1) $b_\varepsilon(u'_{\varepsilon,m}(t),v_j)+a(u_{\varepsilon,m}(t),v_j)+1/\varepsilon(\beta u_{\varepsilon,m}(t),v_j) = (f(t),v_j), j=0,\ldots m$ $t\varepsilon[0,T]$

(1.2.2) $u_{\varepsilon,m}(0)=0$

(1.2.3) $u_{\varepsilon,m}(t,x) = U^m_i(t)v_i(x)$

and obtain some "a priori estimates" for its solution. We prove in this way the following

THEOREM I. Let

(1.2.4) $f,f'\varepsilon L^2(Q)$

(1.2.5) $\psi(0)\leq 0, \psi|_\Sigma \leq 0; \psi,\psi'\varepsilon L^\infty(0,T;H^1(\Omega)); \psi''\varepsilon L^2(0,T;H^1(\Omega))$

then there exists a unique solution u_ε of (P_ε) such that:

(1.2.6) $u_\varepsilon, u'_\varepsilon \varepsilon\ L^2(0,T;V)\cap L^\infty(0,T;b_\varepsilon)$.

Furthermore by passing to limit for $\varepsilon\to 0$ we prove:

THEOREM II. Under the hypotheses (1.2.4),(1.2.5) there exists a unique solution of (P) such that:

(1.2.7) $u\varepsilon L^\infty(0,T;V)$; $u'\varepsilon L^2(0,T;V)$.

1.3. A PSEUDO-HEAT CONDUCTION PROBLEM WITH INTERNAL OBSTACLE.

In this section we consider a pseudoparabolic problem of Sobolev type, with an internal obstacle depending on time and obtain a more regular solution. The problem is originated from a modified heat operator governing the states of conductive temperature u and thermodinamic temperature $u-\Delta u$ in an elastic body (Ting T.W. 1974) by inserting a constraint. Here

(1.3.1) $B=I+A$, $A=-\Delta$

and we call $(P_A),(\mathcal{P}_A)$ the problems corresponding to $(P),(\mathcal{P})$ of section 1.2.

By pursuing in the method of section 1.2 and by taking in account of the more regular case (1.3.1) we can costruct some stronger a priori estimates and prove the:

THEOREM III. We suppose

(1.3.2) $f,f'\varepsilon L^2(Q)$

(1.3.3) $\psi|_\Sigma = 0, \psi\varepsilon L^\infty(0,T;W), \psi'\varepsilon L^\infty(0,T;V), \psi''\varepsilon L^2(0,T;V)$

then there exists a unique solution u of $(\mathcal{P}_A)$ such that:

(1.3.4) $u\varepsilon L^\infty(0,T;W)$, $u'\varepsilon L^\infty(0,T;V)$.

1.4. THE SEEPAGE INEQUALITY.

The problem of homogeneous liquids seepage in fissured rocks, was studied by Zheltov I.V., Kochina I.N., Barenblatt G.I. in 1960 for the first time, by classifying it as a Sobolev type equation.

We consider now the corresponding variational inequality with an obstacle on the boundary, derived from this problem. We adopt the following notations that complete and modify those of section 1.1.:

$-\Gamma = \Gamma_1 \cup \Gamma_2$; Γ_1 is the impermeable rocky wall ($\text{mis}\,\Gamma_1 > 0$); Γ_2 is the porous part of wall; $\Sigma_1 =]0,T[\times\Gamma_1$, $\Sigma_2 =]0,T[\times\Gamma_2$;

- Q is a fissure whose dimension is considered prevalent in comparison with dimension of rock pores;

$-\gamma_0$ is the trace operator of zero order: $\gamma_0 v = v|_\Gamma$;

$-\gamma_1$ " " " " " " " first " " : $\gamma_1 v = \partial_n v|_\Gamma$;

-n is the unit normal vector;

$-A = -\Delta$

$-V = \{v : v \in H^1(\Omega) \,/\, \gamma_0 v|_{\Gamma_1} = 0 \text{ a.e.}\}$;

$-K(t) = \{\tilde{v}(t) : \tilde{v}(t) = v(t) + \eta v'(t); v(t), v'(t) \in V \,/\, \gamma_0 \tilde{v}|_{\Gamma_2} \geq 0 \text{ a.e.}\}$;

$-K = \{\tilde{v} : \tilde{v} = v + \eta v'; v, v' \in L^2(0,T;V) / \gamma_0 \tilde{v}|_{\Sigma_2} \geq 0 \text{ a.e.}\}$;

$-\eta$ is a positive coefficient.

We suppose $f \in C(0,T;H)$ such that $f(0)=0$ a.e. in Ω and consider the following strong problem:

$$(P)\begin{cases} u' + A(\eta u' + u) = f \text{ a.e. in } Q \\ \gamma_0 u|_{\Sigma_1} = 0;\ \gamma_0(\eta u' + u)|_{\Sigma_2} \geq 0;\ \gamma_1(\eta u' + u)|_{\Sigma_2} \geq 0; \\ \gamma_0(\eta u' + u)\gamma_1(\eta u' + u)|_{\Sigma_2} = 0 \\ u(0) = 0 \text{ a.e. in } \Omega \end{cases}$$

The phisical meaning of (P) is the following: η represents a characteristic of fissured rocks obtained by an averaging process. η tending to zero implies the increase of fissuring and the decrease of rock block. When $\eta = 0$ we have the classical evolution operator for the seepage under elastic condition and a simplified Signorini problem. $u + \eta u'$ is the pression of liquid in Q. $(u + \eta u')|_\Gamma$ is the flow of liquid through Γ.

We put $\tilde{u} = \eta u' + u$ and we can easily verify that (P) is equivalent to the following evolution variational problem:

$$(E.V.P.)\begin{cases} \text{to find } \tilde{u} \in K \text{ such that:} \\ \int_0^T [a(\tilde{u}(t), \tilde{v}(t) - \tilde{u}(t)) + (u'(t), \tilde{v}(t) - \tilde{u}(t)) - (f(t), \tilde{v}(t) - \tilde{u}(t))]dt \geq 0 \quad \forall\, \tilde{v} \in K \\ u(0) = 0 \text{ a.e. in } \Omega. \end{cases}$$

By considering a base $\{v_i\}_{i \in N}$ of the separable space V, dense in V we can solve in the usual way the penalized problem:

$$(\mathcal{P}_\varepsilon)\begin{cases} \text{to find } \tilde{u}_\varepsilon \in L^2(0,T;V): \\ \int_0^T [\, a(\tilde{u}_\varepsilon(t),\tilde{v}(t))+(u'(t),\tilde{v}(t))+1/\varepsilon(\gamma_0\beta\tilde{u}_\varepsilon(t),\gamma_0\tilde{v}(t))_{\Gamma_2}-(f(t),\tilde{v}(t))\,]dt=0 \\ u_\varepsilon(0)=0 \text{ a.e. in } \Omega \qquad\qquad \forall\ \tilde{v}\in L^2(0,T;V) \end{cases}$$

where $\beta v=-v^-$ and prove the following:

THEOREM IV. Let $f\in C(0,T;H)$, $f(0)=0$ a.e. in Ω then $(\mathcal{P}_\varepsilon)$ has a unique solution such that

$$(1.4.1) \qquad u_\varepsilon \in L^\infty(0,T;V) \quad , \quad u'_\varepsilon \in L^2(0,T;V).$$

Then passing to limit for $\varepsilon\to 0$ we have:

THEOREM V. Under the hypotheses $f\in C(0,T;H)$, $f(0)=0$ a.e. in Ω the problem (E.V.P.) has a unique solution such that

$$(1.4.2) \qquad u \in L^\infty(0,T;V) \quad , \quad u'\in L^2(0,T;V).$$

SECOND PART

2.1. AN APPROXIMATION OF DEGENERATE PROBLEM OF SECTION 1.2.

For semplicity we suppose $\Omega\subset\mathbb{R}^2$ be a convex polygon. By following (Scarpini F., Vivaldi M.A. 1978) we construct the finite dimensional space $\overset{\circ}{S}_{k,h}(Q)\subset BV(0,T;V)$ of function of bounded variation with values in V. k and h are the respective discretization parameters of time and space variable. We decompose $[0,T]$ in subintervals:

$$(2.1.1) \qquad [t_p,t_{p+1}[\,,\ p=0,\dots m-1 \text{ of length } K=T/m$$

and we consider the associated characteristic functions:

$$(2.1.2) \qquad \theta_p^k(t)\begin{cases} =1 \ ,t\in[t_p,t_{p+1}[\\ =0 \ ,t\notin[t_p,t_{p+1}[\quad p=0,\dots m-1 \end{cases} ; \quad \theta_m^k(t)\begin{cases} =1, t=T \\ =0, t\in[0,T[. \end{cases}$$

Furthermore we consider a "regular triangulation" $\mathcal{T}_h$ (Ciarlet P.G. 1978) of Ω related to parameter h and call x_i, with $i\in I_0=\{1,\dots N_0\}$, the internal nodes of Ω. Let $\phi_i^h(x)$ be the finite affine element corresponding to $\mathcal{T}_h$:

$$(2.1.3) \qquad \phi_i^h(x) \quad \text{is affine on triangles } \overset{\circ}{T}\in\mathcal{T}_h,$$

$$(2.1.4) \qquad \phi_i^h(x_j)=\delta_{i,j} \quad (\delta_{i,j} \text{ is the Kronecker delta})$$

and we define the space

$$(2.1.5) \qquad \overset{\circ}{S}_{k,h}=\{\theta_p^k(t)\phi_i^h(x)\}_{i\in I_0}^{p=0,\dots m} \quad ;$$

the interpolate function

$$(2.1.6) \qquad v_I(t,x)=\theta_p^k(t)\phi_i^h(x)v(t_p,x_i) \ ,\forall\ v\in \overset{\circ}{C}(\bar{Q});$$

the partial interpolates

(2.1.7) $v_{I_1}(t,x) = \theta_p^k(t)v(t_p,x)$ in time variable

(2.1.8) $v_{I_2}(t,x) = \phi_i^h(x)v(t,x_i)$ in space variable.

Obviously we have

(2.1.9) $v_I(t,x) = (v_{I_1}(t,x))_{I_2}$.

$K_{k,h}$ is the convex set

(2.1.10) $K_{k,h}=\{ v_{k,h}: v_{k,h}\epsilon S^o_{k,h} \ / \ v_{k,h}\geq\psi_I \text{ in } Q\}$.

With reference to problem ($\mathcal{P}$) of section 1.2 we consider the following approximate problem:

$$(\mathcal{P}_{k,h})\quad \begin{cases} \text{to find } u_{k,h}\epsilon K_{k,h} \\ \int_0^T [b(u'_{k,h}(t),v_{k,h}(t)-u_{k,h}(t)) + a(u_{k,h}(t),v_{k,h}(t)-u_{k,h}(t))]dt\geq \\ \int_0^T(f(t),v_{k,h}(t)-u_{k,h}(t))dt \quad ,\forall\ v_{k,h}\epsilon K_{k,h} \ ;Bu_{k,h}(0)=0 \end{cases}$$

where $v'_{k,h}$ is the time derivative of $v_{k,h}$ in the sense of function of bouded variation and integral is meant in Stieltjes sense.It is well-known that:

(2.1.11) $\int_0^T z_{k,h}(t)v'_{k,h}(t)dt = \sum_{p=1}^{m} z_{k,h}(t_p)(v_{k,h}(t_p)-v_{k,h}(t_{p-1}))$.

We set the following notations:

$$u_{k,h}(t,x) = U_i^p\theta_p^k(t)\phi_i^h(x) \ , \ U=\{U^p\}^{p=0,\dots m} =\{U_i^p\}_{i\epsilon I_o}^{p=0,\dots m}$$

$$v_{k,h}(t,x) = V_i^p\theta_p^k(t)\phi_i^h(x) \ , \ V=\{V^p\}^{p=0\dots m} =\{V_i^p\}_{i\epsilon I_0}^{p=0,\dots m}$$

$$B=\{b_{i,j}\}_{i,j\epsilon I_o} =\{b(\phi_i^h,\phi_j^h)\}_{i,j\epsilon I_o} \quad , \quad A=\{a_{i,j}\}_{i,j\epsilon I_o} =\{a(\phi_i^h,\phi_j^h)\}_{i,j\epsilon I_o}$$

$$C^p=\{C_i^p\}_{i\epsilon I_o} \ ,C_i^p =\int_{t_p}^{t_{p+1}} \theta_p^k(t)(f(t),\phi_i^h)dt \ ,p=0,\dots m-1.$$

By using these notations we transform ($\mathcal{P}_{k,h}$) into the following algebraic problem:

$$(2.1.12)\quad \begin{cases} U^p\geq\Psi^p \ , \ U^m\geq\Psi^m \\ (B(U^p-U^{p-1})+(kAU^p-C^p)(V^p-U^p)+B(U^m-U^{m-1})(V^m-U^m)\geq 0 \ ,\forall\ V^p\geq\Psi^p \\ V^m\geq\Psi^m \qquad\qquad\qquad p=1,\dots m-1 \end{cases}$$

and then into the complementarity system:

$$(S.C.)_p \begin{cases} U^p\geq\Psi^p \\ M^p=B(U^p-U^{p-1})+kAU^p-C^p\geq 0 \\ M^p(U^p-\Psi^p) = 0 \quad ,p=1,\dots m-1 \end{cases} \qquad (S.C.)_m \begin{cases} U^m\geq \Psi^m \\ M^m=B(U^m-U^{m-1})\geq 0 \\ M^m(U^m-\Psi^m)=0 \end{cases}$$

The matrix B+kA in $(S.C)_p$ is positive definite.(S.C) has a unique solution which can be computed by well-known algorithms (Glowinski R,Lions J.L.,Tremolieres,1976;Mosco U,Scarpini F.1975;Scarpini F.1975).

2.2 THE CONVERGENCE OF DISCRETE SOLUTION.

By using the above mentioned discretization we can prove the:

THEOREM VI. Let u be the solution of $(\mathcal{P})$ such that:

(2.2.1) $u \in L^{\infty}(0,T;V)$, $u' \in L^2(0,T;V)$.

Let $u_{k,h}$ be the solution of $(\mathcal{P}_{k,h})$, then under additional condition:

(2.2.2) $\psi(t',x) \leq \psi(t,x)$ if $t \leq t'$

we have

(2.2.3) $\lim u_{k,h} = u$ weakly in $L^2(0,T;V)$.

EXAMPLE. Let $b_{r,s}=0$ in Ω, b_o be the characteristic function:

(2.2.4) $$b_o(x)\begin{cases} =1 & \text{if } x \in \bar{\Omega}_1 \\ =0 & \text{if } x \in \bar{\Omega}_2 = \Omega - \bar{\Omega}_1. \end{cases}$$

The problem $(\mathcal{P})$ of section 1.2 in this case, is an evolution problem in $Q_1 =]0,T[\times \Omega_1$, an elliptic problem in $Q_2 =]0,T[\times \Omega_2$. If $\bar{\Omega}$ and $\bar{\Omega}_1$ are convex polygons and $\mathcal{T}^h = \mathcal{T}^h_1 \cup \mathcal{T}^h_2$ an adapted triangulation, we have by means of Ths.II,VI:

$\lim u_{k,h} = u$ weakly in $L^2(0,T;V)$.

2.3. THE PSEUDOPARABOLIC PROBLEM OF SECTION 1.3 .ERROR ESTIMATE.

We can prove the strong convergence $\lim u_{k,h} = u$ strongly in $L^2(0,T;V)$ is linked to the possibility of constructing a sequence of functions $v_{k,h}$ belonging to $K_{k,h}$ such that $\lim v_{k,h} = u$ strongly in $L^{\infty}(0,T;V)$ and thus, after all, to the proof of a regularity theorem better than Th.II in section 1.2, for example Th.III in section 1.3.

THEOREM VII. Let u be a solution of ( ) such that:

$$u \in L^{\infty}(0,T;W) \quad , \ u' \in L^2(0,T;V).$$

Let $u_{k,h}$ be the solution of $(\mathcal{P}_{k,h})$, then under the condition (2.2.2) we have

$$\lim u_{k,h} = u \text{ strongly in } L^2(0,T;V).$$

Owing to Th.VII in the present case the strong convergence holds. We can prove also the following:

LEMMA I. Let $u \in L^{\infty}(0,T;W)$, $u' \in L^2(0,T;V)$ then the following interpolation error estimates hold (Scarpini F., Vivaldi M.A.1978):

(2.3.1) $\|u-u_I\|^2_{L^2(Q)} \leq C(h^4+k^2)(\|u'\|^2_{L^2(Q)} + \|u\|^2_{L^{\infty}(0,T;W)})$

(2.3.2) $\|u-u_I\|^2_{L^2(0,T;V)} \leq C(h^2+k^2)(\|u'\|^2_{L^2(0,T;V)} + \|u\|^2_{L^{\infty}(0,T;W)})$

(2.3.3) $\|(u-u_I)(t_p)\|^2 \leq Ch^2\|u\|^2_{L^{\infty}(0,T;W)}$.

By using the interpolation estimates in lemma I we can prove the:

THEOREM VIII. Let u be the solution of $(\mathcal{P}_A)$ such that:

(2.3.4) $u \in L^{\infty}(0,T;W)$, $u' \in L^2(0,T;V)$

and $u_{k,h}$ the solution of $((\mathcal{P}_{k,h})_A)$, then we have the following approximation error estimate:

(2.3.5) $\|u-u_{k,h}(T)\| + \left(\sum_{p=1}^{m} \|u_{k,h}(t_p)-u_{k,h}(t_{p-1})\|^2\right)^{1/2} + \|u-u_{k,h}\|_{L^2(0,T;V)} \leq C(h^{1/2}+k^{1/2})$ where C is independent on solutions.

2.4. THE SEEPAGE INEQUALITY OF SECTION 1.4. COMPLEMENTARITY SYSTEMS.

At first we write the complementarity system which is equivalent to (E.V.P.):

$$
(C.S)\quad \begin{cases} u|_{\Sigma_1}=0,\ u|_{\Sigma_2}\geq 0; u(0)=0 \text{ a.e. in } \Omega \\ \int_0^T [a(\tilde{u}(t),\tilde{v}(t))+(u'(t),\tilde{v}(t))-(f(t),\tilde{v}(t))]dt \geq 0,\ \forall \tilde{v}\in K \\ \int_0^T [\|\tilde{u}(t)\|^2+(u'(t),\tilde{u}(t))-(f(t),\tilde{u}(t))]dt = 0. \end{cases}
$$

We set the following additional notations:

$-I=I_0 \cup I_\partial=\{1,\dots,N_0,\dots,N\}$ the set of indexes of nodes in Ω and on $\Gamma=\partial\Omega$;

$-I_\partial=I_1 \cup I_2$ where $I_1=\{N_0+1,\dots N_1\}$ is the set of indexes of nodes on Γ_1, $I_2=\{N_1+1,\dots,N\}$ is the set of indexes on Γ_2.

$-\ A=\{a_{i,j}\}_{I\times I}=\{a(\phi_i^h,\phi_j^h)\}_{I\times I}$, $B=\{b_{i,j}\}_{I\times I}=\{1/h^2(\phi_i^h,\phi_j^h)\}_{I\times I}$

$-\ \mathcal{A}=(\eta+k)A+h^2B$, $\mathring{\mathcal{A}}=\eta A+h^2B$, $\tilde{U}^p=(\eta+k)U^p-\eta U^{p-1}$.

By applying finite affine element method to (C.S) and using the previous notations, we obtain the following systems:

$$
(C.S)_p \begin{cases} U^p_{I_1}=0\ ,\tilde{U}^p_{I_2}\geq 0 \\ M^p_{I_0}=\mathcal{A}_{I_0I_0}U^p_{I_0}+\mathcal{A}_{I_0I_2}U^p_{I_2}-(C^p_{I_0}+\mathring{\mathcal{A}}_{I_0I_0}U^{p-1}_{I_0}+\mathring{\mathcal{A}}_{I_0I_2}U^{p-1}_{I_2})=0 \\ M^p_{I_2}=\mathcal{A}_{I_2I_0}U^p_{I_0}+\mathcal{A}_{I_2I_2}U^p_{I_2}-(C^p_{I_2}+\mathring{\mathcal{A}}_{I_2I_0}U^{p-1}_{I_0}+\mathring{\mathcal{A}}_{I_2I_2}U^{p-1}_{I_2})\geq 0 \\ M^p_{I_2}U^p_{I_2}=0 \end{cases}
$$

$$
(C.S.)_m \begin{cases} U^m_{I_1}=0,\ U^m_{I_2}=U^{m-1}_{I_2}\geq 0 \\ M^m_{I_0}=\mathring{\mathcal{A}}_{I_0I_0}(U^m_{I_0}-U^{m-1}_{I_0})+\mathring{\mathcal{A}}_{I_0I_2}(U^m_{I_2}-U^{m-1}_{I_2})=0 \\ M^m_{I_2}=\mathring{\mathcal{A}}_{I_2I_0}(U^m_{I_0}-U^{m-1}_{I_0})+\mathring{\mathcal{A}}_{I_2I_2}(U^m_{I_2}-U^{m-1}_{I_2})\geq 0 \\ M^m_{I_2}(U^m_{I_2}-U^{m-1}_{I_2})=0 \end{cases}
$$

endowed with a unique solution U_I.

REFERENCES

1. Brezis H. "On some degenerate non linear parabolic equation" Proc. Symp.Pure Math.18,Amer.Math.Soc.(1970),28-38
2. Brezis U.,Rosenkranz W.,Singer B. "On degenerate elliptic-parabolic equation occurring in the theory of probability".Comm.Pure Appl.Math. 24 (1971).
3. Carrol R.W.,Showalter R.E. "Singular and degenerate Cauchy Problems" Academic Press N.Y.(1976)
4. Ciarlet P.G. " The Finite Element Method for Elliptic Problems"North-Holland,Amsterdam (1978).
5. Glowinski R.,Lions G.L.,Tremolieres R. "Analyse Numérique des Inéquations Variationnelles" Dunod,Paris (1976) Vol.I,II.
6. Kuttler K.L. "Degenerate Variational Inequalities of Evolution" Nonlinear Analysis 8.8 (1984) 837-850
7. Lions J.L. "Quelques méthodes de résolution des problémes aux limites non linéaires" Dunod,Paris (1969)
8. Lions J.L. "Approximation Numérique des Inéquations d'évolution" in "Constructive Aspects of Functional Analysis" Corso C.I.M.E.,Erice 27 Giugno-7 Luglio 1971.Ed.Cremonese,Roma 1973.
9. Mosco U.,Scarpini F. "Complementarity Systems and Approximation of Variational Inequalities" R.A.I.R.O.,R.I.(1975) 83-104
10. Scarpini F."Some Algorithms Solving the Unilateral Dirichlet Problem with Two Constraints"Calcolo 12 (1975),113-149.
11. Scarpini F.,Vivaldi M.A. "Evaluation de l'erreur d'approximation pour une inéquation parabolique relative aux convexes dépendant du temps" Applied Math. and Optim.(1978),121-138.
12. Ting T.W. " A Cooling Process According to Two-Temperature Theory of Heat Conduction" J.of Math.Anal. and Appl.45 (1974),23-31

13. Zheltov IV.,Kochina I.N.,Barenblatt G.I."Basic concepts in the theory of seepage of homogeneous liquids in fissured rocks (strata)" J.Appl.Math.Mech. 24,5 (1960) 1286-1303.

Numerical Approximation of Partial Differential Equations
E.L. Ortiz (Editor)

NUMERICAL CONFORMAL MAPPING TECHNIQUES FOR THE SOLUTION OF TWO-DIMENSIONAL LAPLACIAN BOUNDARY VALUE PROBLEMS

N. Papamichael, M.K. Warby***

*Department of Mathematics and Statistics

**Institute of Computational Mathematics

Brunel University

Uxbridge, Middlesex, UB8 3PH, England.

We consider the use of a numerical conformal mapping technique for the solution of Laplacian boundary value problems. In particular, we report recent developments concerning the application of the Bergman kernel method for numerical conformal mapping.

1. Introduction

Let Ω be a finite simply-connected domain with boundary $\partial\Omega$ in the complex z-plane $(z = x + iy)$, where $\partial\Omega$ is a closed piecewise analytic Jordan curve. Also, let A_j; $j = 1(1)4$, be four points in counterclockwise order on $\partial\Omega$, and let R_H denote the rectangle

$$R_H = \{(\xi,\eta) : 0 < \xi < 1 \ , \ 0 < \eta < H\} \ , \tag{1.1}$$

in the w-plane $(w = \xi + i\eta)$. Then, it follows from the Riemann mapping theorem that, for a certain H, there exists a unique conformal map $F: \Omega \rightarrow R_H$, which takes the four boundary points A_j; $j = 1(1)4$, respectively onto the four vertices (0,0),(1,0),(1,H) and (0,H) of R_H. It also follows that, for a given Ω, the height H of the rectangle is determined completely by the positions of the points A_j on $\partial\Omega$, i.e.

$$H = H\{A_1,A_2,A_3,A_4\} \ . \tag{1.2}$$

In fact, H is an important domain functional known as the conformal module of the quadrilateral defined by the points A_j; $j = 1(1)4$; see (Gaier, 1974;1979).

Let now f be the function which maps conformally Ω onto the unit disk

$$D = \{\zeta : |\zeta| < 1\} , \tag{1.3}$$

so that

$$f(\hat{z}) = 0 \quad \text{and} \quad f'(\hat{z}) > 0 , \tag{1.4}$$

where $\hat{z}$ is some fixed point in Ω. Then, the mapping $F: \Omega \rightarrow R_H$ can be constructed by the composition of $f: \Omega \rightarrow D$ and a simple Schwarz-Christoffel transformation $S: D \rightarrow R_H$. That is,

$$F = S \circ f , \tag{1.5}$$

where S can be expressed as

$$S = S_3 \circ S_2 \circ S_1 , \tag{1.6a}$$

with

$$S_1(t) = i(1+t)/(1-t) , \tag{1.6b}$$

$$S_2(t) = \{(\alpha_2-\alpha_4)/(\alpha_2-\alpha_1)\}.\{(t-\alpha_1)/(t-\alpha_4)\} , \tag{1.6c}$$

and

$$S_3(t) = \{sn^{-1}(t^{\frac{1}{2}},k)\}/K(k) , \quad k = \{S_2(\alpha_3)\}^{-\frac{1}{2}} . \tag{1.6d}$$

In (1.6c) and (1.6d), α_j; j=1(1)4, are the images of the points A_j; j=1(1)4, under the transformation $S_1 \circ f$, and sn(.,k), K(k) denote respectively the Jacobian elliptic sine and the complete elliptic integral of the first kind, each with modulus k. Also, the conformal module (1.2), i.e. the height H of the rectangle R_H, is given quite simply by

$$H = K\{(1-k^2)^{\frac{1}{2}}\}/K(k) ; \tag{1.7}$$

see e.g. (Bowman, 1953). (Observe that since the bilinear transformation S_1 maps the unit circle onto the real axis, the α_j in (1.6c)-(1.6d) are all real.)

It follows easily from the above that the main computational requirement for the construction of S, by means of (1.6), is the calculation of two incomplete elliptic integrals of the first kind, for each transformed point; see (Papamichael and Symm, 1975;p.180). For this reason, the problem of determining the conformal map $F: \Omega \rightarrow R_H$ may be regarded as solved once the function $f: \Omega \rightarrow D$ is found. In particular, if f is known exactly, then (1.5)-(1.6) give essentially the exact conformal map F.

In this paper, we consider the problem of determining approximations to F, by means of (1.5)-(1.6), in cases where the mapping function f is not known exactly. Our specific objectives are as follows:

(i) To illustrate the practical significance of the conformal map F, by considering a number of simple applications involving the solution of Laplacian boundary value problems.

(ii) To report recent developments concerning the application of the Bergman kernel method (BKM), for determining approximations to the mapping function $f: \Omega \to D$.

(iii) To present numerical examples illustrating the use of the BKM, for computing approximations to the mapping $F: \Omega \to R_H$.

2. Applications of the conformal map $F: \Omega \to R_H$

With the notation of Section 1, let Ω represent a "thin" plate of homogeneous electrically conducting material of resistivity κ, and suppose that constant voltages V_1 and V_2 are applied respectively to the boundary segments $\overline{A_1A_2}$ and $\overline{A_3A_4}$, whilst the remainder of $\partial\Omega$ is insulated. Let I be the total current between $\overline{A_1A_2}$ and $\overline{A_3A_4}$, and consider the problem of determining the geometric resistance

$$r(\Omega) = (V_2 - V_1)/\kappa I , \tag{2.1}$$

of Ω. This problem may of course be solved by determining the solution of Laplace's equation

$$\nabla^2 V = 0 , \quad \text{in } \Omega , \tag{2.2a}$$

subject to the Dirichlet boundary conditions

$$V = V_1 , \text{ on } \overline{A_1A_2} ; \quad V = V_2 , \text{ on } \overline{A_3A_4} , \tag{2.2b}$$

and the Neumann conditions

$$\partial V/\partial n = 0 , \text{ on } (\overline{A_2A_3}) \cup (\overline{A_4A_1}) . \tag{2.2c}$$

Once V is found, $r(\Omega)$ may be determined from (2.1), after first computing I as a line integral of ∇V along any curve running from $\overline{A_2A_3}$ to $\overline{A_4A_1}$.

Although the mixed boundary value problem (2.2) appears to be rather simple, its solution by standard numerical techniques may present serious difficulties, due to the geometry of Ω and/or the presence of boundary singularities. (For example, if $\partial\Omega$ is smooth then (2.2) has serious singularities at the four points A_j; $j = 1(1)4$, where the boundary conditions change from Dirichlet to Neumann.) By contrast, if the conformal map $F: \Omega \rightarrow R_H$ is available then the geometric resistance of Ω is given, quite simply, by the conformal module $H\{A_1,A_2,A_3,A_4\}$, i.e.

$$r(\Omega) = H , \tag{2.3}$$

where H is the height of the rectangle R_H; see (Bowman, 1953; Gaier, 1974; Papamichael and Kokkinos, 1981; Trefethen, 1984). Another application of the conformal map F, in which the value of H is of special significance, concerns the determination of the capacitance of symmetric ring condensers; see (Campbell, 1975; Gaier, 1979; Papamichael and Kokkinos, 1981).

Suppose now that the solution V of the boundary value problem (2.2) is required at some point $P \equiv (x,y) \in \overline{\Omega}$. Let $P' \equiv (\xi,\eta) \in \overline{R}_H$ be the image of P under the mapping F. Then, because the Laplace equation and the boundary conditions (2.2b)-(2.2c) are conformally invariant,

$$V(P) = \{(V_2 - V_1)\eta\}/H + V_1 . \tag{2.4}$$

That is, the solution of (2.2) at any point $P \in \overline{\Omega}$ is known immediately, once H and the imaginary co-ordinate of the point P' are found. Of course, if the conformal map F is applied to Laplacian problems with more general boundary conditions than those of (2.2), then it is unlikely that the transformed problem in R_H will have a simple analytical solution. Moreover, in such applications, the situation is complicated further by the fact that the map F itself may introduce singularities at the images of boundary points, where the original problem did not have singular behaviour. However, there are many cases where the conformal map $\Omega \rightarrow R_H$ can be used to overcome the difficulties associated with the numerical solution of problems involving curved boundaries and boundary singularities, and to produce solutions of high accuracy. Several such applications are considered in (Papamichael and Sideridis, 1979a), where the conformal transformation procedure is applied not only to Laplacian problems, but also to problems involving elliptic equations of the form

$$\nabla^2 V(x,y) = a(x,y)V(x,y) + b(x,y) \ . \tag{2.5}$$

3. The Bergman kernel method (BKM), for determining approximations to $f: \Omega \to D$

With the notation of Section 1, let $L_2(\Omega)$ be the Hilbert space of all square integrable analytic functions in Ω, denote by $\langle .,. \rangle$ the inner product of $L_2(\Omega)$, i.e.

$$\langle u,v \rangle = \iint_\Omega u(z)\overline{v(z)}dxdy \ , \tag{3.1}$$

and let $K(.,\hat{z})$, $\hat{z} \in \Omega$, be the Bergman kernel function of Ω. This kernel function is characterized uniquely by the reproducing property

$$\langle \eta, K(.,\hat{z}) \rangle = \eta(\hat{z}) \ , \quad \forall \ \eta \in L_2(\Omega) \ , \tag{3.2}$$

and is related to the mapping function $f: \Omega \to D$, normalized by (1.4), by means of

$$f'(z) = \sqrt{\pi}\ K(z,\hat{z})/\{K(\hat{z},\hat{z})\}^{\frac{1}{2}} \ ; \tag{3.3}$$

see e.g. (Bergman, 1970; Gaier, 1964; Gaier, 1980).

In the BKM the approximation to f is obtained from (3.3), after first approximating the kernel $K(.,\hat{z})$ by a finite Fourier series sum. If η_j; $j = 1,2,...,$ is a complete set of $L_2(\Omega)$ then the main steps of the method may be described as follows.

The set $\{\eta_j\}_{j=1}^n$ is orthonormalized, by means of the Gram-Schmidt process, to give the orthonormal set $\{\eta_j^*\}_{j=1}^n$. Then, because of (3.2),

$$\begin{aligned} K_n(z,\hat{z}) &= \sum_{j=1}^n \langle K(.,\hat{z}),\eta_j^* \rangle \eta_j^*(z) \\ &= \sum_{j=1}^n \overline{\eta_j^*(\hat{z})}\eta_j^*(z) \ , \end{aligned} \tag{3.4}$$

is the nth partial Fourier sum of the kernel function, and hence, from (3.3),

$$f_n(z) = \{\pi/K_n(\hat{z},\hat{z})\}^{\frac{1}{2}} \int_0^z K_n(t,\hat{z})dt \tag{3.5}$$

is the nth BKM approximation to the mapping function f. That is, the BKM approximates f by an explicit formula of the type

$$f_n(z) = \sum_{j=1}^n c_j \mu_j(z) \ , \tag{3.6}$$

where the μ_j are integrals of the basis functions η_j. This is, of course, an important advantage of the method.

The application of the BKM has received considerable attention in the last few years, and some of the recent developments concerning the implementation of the method are summarized in the following remarks.

(i) The Gram-Schmidt process is numerically unstable and, for this reason, the basis set $\{\eta_j\}$ must be chosen so that the resulting approximating series (3.6) converges rapidly. This can be achieved as proposed in (Levin et al, 1978) and (Papamichael and Kokkinos, 1981), by using an "augmented" basis set formed by introducing into the "monomial" set,

$$(z-\hat{z})^{j-1} \quad ; \quad j=1,2,\ldots , \tag{3.7}$$

functions that reflect the dominant singularities of $K(.,\hat{z})$, or equivalently of f', on $\partial\Omega$ and in $\text{compl}(\overline{\Omega})$. The problem of determining these singularities of f' and of constructing appropriate augmented basis sets, is considered in the two references cited above and, much more fully, in the two recent papers by (Papamichael, Warby and Hough, 1983;1984).

(ii) The main computation involved in the construction of the BKM approximation (3.6) occurs in the orthonormalization of the basis set $\{\eta_j\}_{j=1}^{n}$ and, in particular, in the evaluation of the inner products

$$\langle\eta_r,\eta_s\rangle = \iint_\Omega \eta_r(z)\overline{\eta_s(z)}\,dxdy \,. \tag{3.8}$$

By using a Green's formula, these inner products can be expressed as

$$\langle\eta_r,\eta_s\rangle = \frac{1}{2i}\int_{\partial\Omega} \eta_r(z)\overline{\mu_s(z)}\,dz \,, \tag{3.9}$$

where $\mu_s' = \eta_s$. The contour integrals (3.9) can then be computed accurately, by choosing a suitable parametrization of $\partial\Omega$, and using Gauss-Legendre quadrature. Full details of this quadrature procedure can be found in (Levin et al, 1978; Papamichael and Kokkinos, 1981).

(iii) The principle of maximum modulus suggests a reliable method for determining an estimate E_n of the maximum error in the modulus of the approximation f_n. This estimate is given by

$$E_n = \max_j \Big| 1 - |f_n(z_j)| \Big| \,, \tag{3.10}$$

where $\{z_j\}$ is a set of boundary test points. Furthermore, the comparison of successive error estimates E_n provides a simple termination criterion for determining the "optimum" number n = Nopt of basis functions, which gives maximum accuracy in the sense described in (Levin et al, 1978; p.177).

(iv) Regarding convergence, we certainly have that

$$\lim_{n\to\infty} \| K_n(.,\hat{z}) - K(.,\hat{z}) \| = 0 , \tag{3.11}$$

and, in the space $L_2(\Omega)$, this norm convergence implies that $K_n(z,\hat{z}) \to K(z,\hat{z})$, uniformly in every compact subset of Ω. Hence, we have that $f_n(z) \to f(z)$ uniformly in every compact subset of Ω. Furthermore, the two books of (Gaier, 1964; 1980) and the recent papers by (Simonenko, 1978) and (Kulikov, 1981) contain a number of important results, which establish the uniform convergences in $\bar{\Omega} = \Omega \cup \partial\Omega$ of the BKM approximations corresponding to the use of the monomial basis set (3.7). Finally, the convergence properties of the BKM are investigated in (Papamichael and Warby, 1984) where, in particular, the results contained in the above references by Gaier, Simonenko and Kulikov are used to provide some theoretical explanation of the improvement in accuracy which is achieved, when the set (3.7) is augmented by the introduction of appropriate singular functions.

(v) The stability properties of the BKM are studied in (Papamichael and Warby, 1984) where, in particular, a geometrical characterization of the degree of instability in the orthonormalization of the monomial set (3.7) is established.

4. Numerical examples

In this section we consider four examples where, in each case, Ω is a domain bounded by the three straight lines $x = 0$, $y = 0$, $x = 1$, and an analytic curve with cartesian equation $y = \phi(x)$, where ϕ is positive in $[0,1]$. In each case, the points A_j; $j = 1(1)4$, are chosen to be the four corners of Ω, i.e.

$$\Omega = \{(x,y) : 0 < x < 1 \ , \ 0 < y < \phi(x)\} \tag{4.1a}$$

and

$$A_1 \equiv (0,0) \ , \ A_2 = (1,0) \ , \ A_3 = (1,\phi(1)) \ , \ A_4 \equiv (0,\phi(0)) \ . \tag{4.1b}$$

All our examples are taken from a recent paper by (Challis and Burley, 1982), who consider the mapping of domains of the special form (4.1), in connection with the solution of problems of viscous flow in wavy channels. More specifically, Challis and Burley consider the mapping $F^{[-1]} : R_H \to \Omega$, and use a Fourier series method to approximate the function $G = F^{[-1]}$ by an expansion of the form

$$G_N(w) = w + \sum_{j=1}^{N} a_j \sin(j\pi w) \ . \tag{4.2}$$

As is indicated in (Gutknecht, 1984,§9(c)), their Fourier series technique is related closely to the method of Garrick, which is described in detail in (Gaier, 1964; pp.194-207).

Here, in each example, we use the composite transformation (1.5), and approximate $F: \Omega \to R_H$ by

$$F_n = S \circ f_n \ , \tag{4.3}$$

where:

(a) f_n is a BKM approximation to the function $f: \Omega \to D$.

(b) The Schwarz-Christoffel transformation is performed, in the form (1.6), by using a subroutine due to (Papamichael and Sideridis, 1979b).

For the purposes of comparison, we list for each Ω two approximations to the conformal module $H\{A_1,A_2,A_3,A_4\}$, obtained by using respectively (4.3) and the Fourier series method of (Challis and Burley, 1982). Also, for each Ω, we present error estimates illustrating the considerable improvement in accuracy which is achieved, when the BKM monomial basis (3.7) is augmented by the introduction of appropriate singular functions; see Section 3.<u>(i)</u>.

In presenting the numerical results we use the following abbreviations:

(i) <u>BKM/MB</u>: Results obtained by using the BKM with monomial basis (3.7).

(ii) BKM/AB: Results obtained by using the BKM with augmented basis, constructed as described in (Papamichael et al, 1984); see Section 3(i). (The augmented basis sets used in our examples are listed in (Warby, 1984, §6.3)).

(iii) Nopt: Optimum number of BKM basis functions; see Section 3(iii).

(iv) E_{Nopt}: Estimate of the maximum error in the modulus of the BKM approximation f_{Nopt}; see Section 3(iii).

(v) H_{Nopt}: Approximation to H obtained, from (4.3), by using the BKM/AB approximation f_{Nopt}.

(vi) $\tilde{H}$: Approximation to H obtained by using the Fourier series method of (Challis and Burley, 1982). (It should be observed that the approximate conformal modules computed by Challis and Burley are not sufficiently accurate for the purposes of comparison. For this reason, the values of $\tilde{H}$ listed below were computed by using fast Fourier transforms, and increasing the number of terms in (4.2) until convergence occurred to the number of figures quoted.)

In each example, we identify the geometry of Ω by listing the function ϕ.

Example 1. $\phi(x) = 0.25 + 0.2\text{sech}^2(2.5x)$.

BKM/MB: Nopt = 30, $E_{30} = 1.0 \times 10^{-2}$.

BKM/AB: Nopt = 20, $E_{20} = 4.8 \times 10^{-6}$, $H_{20} = 0.312\ 443$.

$\tilde{H} = 0.312\ 444$.

Example 2. $\phi(x) = 1.5 + x$.

BKM/MB: Nopt = 16, $E_{16} = 4.0 \times 10^{-3}$

BKM/AB: Nopt = 19, $E_{19} = 3.4 \times 10^{-6}$, $H_{19} = 1.779\ 361$.

$\tilde{H} = 1.779\ 360$.

Exact value of H is: 1.779 359 959; see (Bowman, 1935).

Example 3. $\phi(x) = 2.0 - 0.5\cos\pi x$.

BKM/MB: Nopt = 25, $E_{25} = 4.8 \times 10^{-3}$

BKM/AB: Nopt = 23, $E_{23} = 3.2 \times 10^{-5}$, $H_{23} = 1.728\ 200$.

$\widetilde{H} = 1.728\ 173$.

Example 4. $\phi(x) = 1.25 - (x - 0.25)^2$.

BKM/MA: Nopt = 30, $E_{30} = 4.5 \times 10^{-3}$.

BKM/AB: Nopt = 23, $E_{23} = 7.1 \times 10^{-6}$, $H_{23} = 0.970\ 511$.

$\widetilde{H} = 0.970\ 513$.

REFERENCES

1. Bergman, S., The kernel function and conformal mapping, Math. Surveys 5, A.M.S., Providence, R.I., 1970.
2. Bowman, F., Notes on two dimensional electric field problems. Proc. London Math. Soc. 39 (1935) 205-215.
3. Bowman, F., Introduction to elliptic functions, English University Press, London, 1953.
4. Campbell, J.B., Finite difference techniques for ring capacitors, J. Engrg. Math. 9 (1975) 21-28.
5. Challis, N.V. and Burley, D.M., A numerical method for conformal mapping, IMA J. num. Analysis 2 (1982) 169-181.
6. Gaier, D., Konstrucktive Methoden der konformen Abbildung, Springer Verlag, Berlin, 1964.
7. Gaier, D., Determination of conformal modules of ring domains and quadrilaterals, Lecture Notes in Mathematics 399, Springer, New York, 1974.
8. Gaier, D., Capacitance and the conformal module of quadrilaterals, J. Math. Anal. Appl. 70 (1979) 236-239.
9. Gaier, D., Vorlesungen über Approximation im Komplexen, Birkhäuser, Basel, 1980.
10. Gutknecht, M.H., Numerical conformal mapping methods based on function conjugation, Preprint 1984 (To appear in J. Comp. Appl. Math.).
11. Kulikov, I.V., W_2^1, L_∞-convergence of Bieberbach polynomials in a Lipschitz domain. Uspekhi Mat. Nauk. 36 (1981) 177-178.
12. Levin, D., Papamichael, N. and Sideridis, A., The Bergman kernel method for the numerical conformal mapping of simply-connected domains, J. Inst. Maths. Appl. 22 (1978) 171-187.
13. Papamichael, N. and Kokkinos, C.A., Two methods for the conformal mapping of simply-connected domains, Comput. Meths. Appl. Mech. Engrg. 28 (1981) 285-307.
14. Papamichael, N. and Sideridis, A., The use of conformal transformations for the numerical solution of elliptic boundary value problems with boundary singularities, J. Inst. Math. Appl. 23 (1979a) 73-88.
15. Papamichael, N. and Sideridis, A., Subroutine CTM1, Tech. Rept. TR/89. Dept. of Maths., Brunel Univ., Uxbridge 1979b.
16. Papamichael, N. and Symm, G.T., Numerical techniques for two-dimensional Laplacian problems, Comput. Meths. Appl. Mech. Engrg. 6 (1975) 175-194.
17. Papamichael, N., Warby, M.K. and Hough, D.M., The determination of the poles of the mapping function and their use in numerical conformal mapping, J. Comp. Appl. Math. 9 (1983) 155-166.
18. Papamichael, N., Warby, M.K. and Hough, D.M., The treatment of corner and pole type singularities in numerical conformal mapping techniques. Preprint 1984 (To appear in J. Comp. Appl. Math.).
19. Papamichael, N. and Warby, M.K., Stability and convergence properties of Bergman kernel methods for conformal mapping. Tech. Rept. TR/15/84, Dept. of Maths., Brunel University, Uxbridge, 1984.

20. Simonenko, I.B., On the convergence of Bieberbach polynomials in the case of a Lipschitz domain, Math. USSR-IZV., 13, Pt1 (1978) 166-174.
21. Trefethen, L.N., Analysis and design of polygonal resistors by conformal mapping, ZAMP 35 (1984) 692-704.
22. Warby, M.K., Bergman kernel methods for the numerical conformal mapping of simply and doubly-connected domains. Ph.D. Thesis, Dept. of Maths. and Stats., Brunel University, Uxbridge, 1984.

Numerical Approximation of Partial Differential Equations
E.L. Ortiz (Editor)

LINEAR APPROXIMATION, ASYMPTOTIC EXPANSION AND MIXED PROBLEM FOR SOME SEMI-LINEAR WAVE EQUATION IN I DIMENSION

PHAM NGOC DINH ALAIN
Département de Mathématiques - Université d'Orléans
45046 ORLEANS CEDEX (FRANCE)

We study the nonlinear wave equation with a right-hand side of the form $f(t,u,u_t)$, $f \in C^1([0,\infty[\times \mathbb{R}^2)$. A theorem on local existence is proved and an asymptotic expansion of order 2 in ε is obtained with a function of the form εf, for ε sufficiently small. Then we consider the case where $f(u_t)=|u_t|^\alpha \operatorname{sgn}(u_t)$ ($0<\alpha<1$) with non homogeneous mixed conditions. We then prove the local existence of the solution under weaker initial conditions.

INTRODUCTION

The following initial and boundary value problem is considered

(I) $\quad u_{tt} - \Delta u = f(t,u,u_t)$

(2) $\quad u = 0$ on $\partial\Omega$ ($\Omega =]0,1[$)

(3) $\quad u(x,0) = \tilde{u}_0(x) \quad , \quad u_t(x,0) = \tilde{u}_1(x)$

In section I, it is proved that for $\tilde{u}_0$ in $H^2(\Omega)\cap H_0^1(\Omega)$, $\tilde{u}_1$ in $H_0^1(\Omega)$ and for $f \in C^1([0,\infty[\times \mathbb{R}^2)$, the problem admits a unique local weak solution. The existence of the solution is proved by associating to the initial equation linear recursive schemes which can serve as a starting point for computing algorithms based on best approximation pertubation techniques- the Tau method - If the function f belongs to $C^N([0,\infty[\times \mathbb{R}^2)$ then an asymptotic expansion of order N+1 in ε is obtained with a right hand side of the form $\varepsilon f(t,u,u_t)$, for ε sufficiently small, that extends to the partial differential equation of type (I) the results obtained in differential equations [I] or in partial differential equations with f taken equal to $\varepsilon f(t,u)$ [IO].

In section 2 we study the case $f(u_t) = |u_t|^{\alpha}\operatorname{sgn}(u_t)$ with non-homogeneous mixed conditions namely

$$u_x(0,t) = g(t) \quad , \quad u(1,0) = 0$$

We then prove the local existence and uniqueness of the solution with weaker initial conditions i.e. $\tilde{u}_0$ in $H_0^1(\Omega)$, $\tilde{u}_1$ in $L^2(\Omega)$ and $g(t)$ in $L^2(0,T)$ by using a Volterra nonlinear integral equation associated to a monotonic method generated by the non linear term. Here with f in the form $f(u_t) = |u_t|^{\alpha}\operatorname{sgn}(u_t)$ equation (I) governs the motion of a linear viscoelastic bar with nonlinear elastic contraints (nonlinear Voight model) .

SECTION I

We shall make the following assumptions

$$(4) \qquad \begin{aligned} &\text{i) } f \in C^1([0,\infty[\times \mathbb{R}^2) \\ &\text{ii) } f \in L^2(0,T;H_0^1(\Omega)) \\ &\text{iii) } \frac{df}{dt} \in L^2(0,T;L^2(\Omega)) \end{aligned}$$

Note that in this sequel we shall consider the equation (I) as an ordinary differential equation in a Banach space for $u(t)$, so that we shall write $\dot{u}(t)$ for $u_t(.,t)$.

Let $\|\cdot\|$ denote the $L^2(\Omega)$ norm. Then we have the following [4]:

Theorem 1

Suppose f satisfies (4) and let $\tilde{u}_0 \in H_0^1 \cap H^2, \tilde{u}_1 \in H_0^1$. Then there exists a $T>0$ such that the initial and boundary value problem (I)-(3) admits a unique solution $u(t) \in L^{\infty}(0,T;H_0^1 \cap H^2)$ and such that $\dot{u}(t) \in L^{\infty}(0,T;H_0^1)$ and $\ddot{u}(t) \in L^{\infty}(0,T;L^2)$. Furthermore, $u(t)$ is the limit of the sequence $\{u_m(t)\}$ of solutions of the following linear initial and boundary value problems

$$(5) \qquad \ddot{u}_m - \Delta u_m = f(t,u_{m-1},\dot{u}_{m-1}) \ , \quad m \geqslant 1 \quad ; \quad u_0 \equiv 0$$

(6) $\quad u_m = 0$ on $\partial\Omega$

(7) $\quad u_m(0) = \tilde{u}_0 \quad , \quad \dot{u}_m(0) = \tilde{u}_1$

The sequence $u_m(t)$ converges to $u(t)$ in the following sense:

$\|\nabla u_m(t) - \nabla u(t)\|^2 + \|\dot{u}_m(t) - \dot{u}(t)\|^2 \longrightarrow 0$ uniformly on $[0,T)$ for $m \longrightarrow \infty$

(we have set $L^2 = L^2(\Omega)$, $H_0^1 = H_0^1(\Omega)$, $H^2 = H^2(\Omega)$, $H_0^1(\Omega)$ and $H^2(\Omega)$ denoting the usual Sobolev spaces)

The proof of the theorem relies on a number of propositions.

Proposition I

Under the assumptions (4),there exists a $T > 0$ such that for each m the solution of the initial and boundary value problem (5)-(7) exists and lies in a bounded set of $L^\infty(0,T;H_0^1 \cap H^2)$. Furthermore the sequences $\{u_m\}$ and $\{\dot{u}_m\}$ lie in a bounded set of $L^\infty(0,T;H_0^1)$ and $L^\infty(0,T;L^2)$ respectively.

Proof

Suppose by induction that $u_{m-1}(t)$ is solution of the initial and boundary value problem (5)-(7) (with (m-1) instead of m) and satisfies the following condition:

$$\|u_{m-1}\|_{L^\infty(0,T;H_0^1 \cap H^2)} \leqslant M$$

$$\|\dot{u}_{m-1}\|_{L^\infty(0,T;H_0^1)} \leqslant M$$

$$\|\ddot{u}_{m-1}\|_{L^\infty(0,T;L^2)} \leqslant M$$

Consider a special basis of H_0^1 : $w_1,\dots,w_n,\dots$ formed by the eigenfunctions of the Laplacian Δ on H_0^1 .Let $(w_1,\dots w_k)$ be the linear space generated by $w_1,\dots w_k$. To show the existence of $u_m(t)$ let us introduce :

(8) $$u_m^{(k)}(t) = \sum_{j=1}^{k} \xi_{mj}^{(k)}(t) w_j$$

such that

$$(9)\quad \begin{cases} \langle \ddot{u}_m^{(k)}, w_j \rangle + a(u_m^{(k)}, w_j) = \langle f(t, u_{m-1}, \dot{u}_{m-1}), w_j \rangle, \quad 1 \leqslant j \leqslant k \\ u_m^{(k)}(0) = u_{ok} \quad , \quad \dot{u}_m^{(k)}(0) = u_{1k} \end{cases}$$

where

$$(10)\quad \begin{aligned} u_{ok} &\longrightarrow \tilde{u}_o \quad \text{in } H_o^1 \cap H^2 \text{ strong} \\ u_{1k} &\longrightarrow \tilde{u}_1 \quad \text{in } H_o^1 \text{ strong} \end{aligned}$$

and where

$$(11)\quad a(u,v) = \int_0^1 \frac{\partial u}{\partial x} \cdot \frac{\partial v}{\partial x}\, dx \quad , \quad \langle u, v \rangle = \int_0^1 uv\, dx$$

Let

$$(12)\quad \begin{aligned} K_o &= \sup_{(0 \leqslant t \leqslant T;\ |x|, |y| \leqslant MC)} (|f(t,x,y)|) \\ K_1 &= \sup_{(0 \leqslant t \leqslant T;\ |x|, |y| \leqslant MC)} \left(\left|\frac{\partial f}{\partial t}\right| + \left|\frac{\partial f}{\partial x}\right| + \left|\frac{\partial f}{\partial y}\right| \right) \end{aligned}$$

where

$$(13)\quad C \geqslant \sup_{u \in H^1} \left(\|u\|_{C^o(\bar{\Omega})} / \|u\|_{H^1} \right)$$

The constante M and the interval $[0,T]$ will be made precise later on.

Let us set

$$X_m^{(k)}(t) = \|\dot{u}_m^{(k)}(t)\|^2 + \|\nabla u_m^{(k)}(t)\|^2$$

$$Y_m^{(k)}(t) = \|\nabla u_m^{(k)}(t)\|^2 + \|\Delta u_m^{(k)}(t)\|^2$$

$$Z_m^{(k)}(t) = \|\ddot{u}_m^{(k)}(t)\|^2 + \|\nabla \dot{u}_m^{(k)}(t)\|^2$$

and

$$S_m^{(k)}(t) = X_m^{(k)}(t) + Y_m^{(k)}(t) + Z_m^{(k)}(t)$$

Therefore using (10) and hypothesises (4) we have the following inequality [4]:

$$(14)\quad S_m^{(k)}(t) \leqslant M^2/4 + (6K_1 + 2K_o + 20K_1 M) \int_0^t \sqrt{S_m^{(k)}(t)}\, dt$$

The function

(I5) $$S(t) = (M/2 + (3K_1 + K_o + 10K_1M)t)^2$$

is the maximal solution of the nonlinear Volterra equation

(I6) $$S(t) = M^2/4 + (6K_1 + 2K_o + 20K_1M)\int_o^t \sqrt{S(t)}\,dt$$

Hence according to the theory of Volterra inequations associated with a non-decreasing function [3], we can choose one $T > 0$ and one M such that

(I7) $$S_m^{(k)}(t) \leqslant S(t) \leqslant M^2 \qquad \text{a.e.} \quad t \in [0,T]$$

From (I7) and by using a compactness argument we can easily show the existence of $u_m(t)$ solution of the initial and boundary value problem (5)-(7).

Q.E.D.

<u>Proposition 2</u>

Let M and T defined as in proposition I and let furthermore $k_T^M = 16MK_1(C_o+1)T$. Then the sequences $\{u_m\}$ and $\{\dot{u}_m\}$ constructed in the proof of proposition I are Cauchy sequences in $L^\infty(0,T;H_o^1)$ and $L^\infty(0,T;L^2)$ respectively.

The proof of proposition 2 is immediate since $k_T^M < 1$ for a convenient T.

- We turn now to the

Proof of theorem I

It is immediate that there is at most one solution of (I)-(3).

Let

$$W = \left\{ u \in L^\infty(0,T;H_o^1) \;\middle|\; \dot{u} \in L^\infty(0,T;L^2) \right\}$$

W is a Banach space for the norm $\|u\|_{L^\infty(0,T;H_o^1)} + \|\dot{u}\|_{L^\infty(0,T;L^2)}$.

From proposition 2 we can deduce that there exists $u \in L^\infty(0,T;H_o^1)$ such that $u_m \longrightarrow u$ in W strong.

It follows from Riesz-Fischer theorem, Lebesgue's theorem on bounded convergence and hypothesis (4i) that there exists a subsequence $\{u_{mj-1}\}$ such that

(18) $\qquad f(t,u_{mj-1},\dot{u}_{mj-1}) \longrightarrow f(t,u,\dot{u})$ in $L^{\infty}(0,T;L^2)$ weak $*$

Then if we pass to the limit in the equation

(19) $\langle \ddot{u}_{mj}, w_i \rangle + a(u_{mj}, w_i) = \langle f(t,u_{mj-1}, \dot{u}_{mj-1}), w_i \rangle, \quad 1 \leqslant i \leqslant k$

we easily find from proposition I that u satisfies the equation

(20) $$\langle \ddot{u}, v \rangle + a(u,v) = \langle f(t,u,\dot{u}), v \rangle \text{ all } v \in H_0^1$$
$$u(0) = \tilde{u}_0 \quad , \quad \dot{u}(0) = \tilde{u}_1$$

Remark.- If the function f verifies beyond the hypothesis (4i) the following

(21) $\qquad f(t,0,0) = 0 \quad$ all $t \geqslant 0$

then $f(t,u_{m-1},\dot{u}_{m-1}) \in H_0^1$ a.e. $t \in [0,T]$ if u_{m-1} and $\dot{u}_{m-1}$ are in a bounded set of $L^{\infty}(0,T;H_0^1)$. Therefore we can prove as previou--sly the existence and uniqueness of the solution of the problem (1)-(3).

ASYMPTOTIC EXPANSION

In this part f is assumed fulfilling hypothesises

(22) $$f \in C^N([0,\infty[\times \mathbb{R}^2) \qquad (N \geqslant 1)$$
$$f(t,0,0) = 0 \quad \text{all } t \geqslant 0$$

Let $u_{\varepsilon}(x,t)$ be the unique solution of the problem

(23) $$\begin{cases} Lu = \varepsilon f(t,u,\dot{u}) \text{ with } L = \dfrac{\partial^2}{\partial t^2} - \dfrac{\partial^2}{\partial x^2} \; (0<x<1,\ 0<t<T,\ 0<\varepsilon\leqslant 1) \\ u(0,t) = u(1,t) = 0 \\ u(x,0) = \tilde{u}_0 \quad , \quad \dot{u}(x,0) = \tilde{u}_1 \end{cases}$$

Let us consider the sequence of functions $\{\hat{u}_p\}$ (p=o,1,...N) defined by

(24) $$\begin{cases} L\hat{u}_0 = 0 \qquad (0<x<1 \ , \ 0<t<T) \\ \hat{u}_0(0,t) = \hat{u}_0(1,t) = 0 \end{cases}$$

$$\hat{u}_o(x,0) = \tilde{u}_o \quad , \quad \dot{\hat{u}}_o(x,0) = \tilde{u}_1$$

$$(25) \quad \begin{cases} L\hat{u}_p = P_p(t, \hat{u}_o, \hat{u}_1, \dots, \hat{u}_{p-1}) \quad , \quad 1 \leqslant p \leqslant N \\ \hat{u}_p(0,t) = \hat{u}_p(1,t) = 0 \\ \hat{u}_p(x,0) = \dot{\hat{u}}_p(x,0) = 0 \end{cases}$$

where

$$P_p(t,\hat{u}_o,\dots,\hat{u}_{p-1}) = \sum_{m=1}^{p-1} \sum_{k=o}^{m} f^{(m)}_{u^{m-k}\dot{u}^k}(t,\hat{u}_o,\dot{\hat{u}}_o) . P_{u^{m-k}\dot{u}^k}$$

with

$$P_{u^{m-k}\dot{u}^k} = \sum_{\alpha_i,\beta_i} (u_1)^{\alpha_1} \dots (u_{p-m})^{\alpha_{p-m}} . (\dot{u}_1)^{\beta_1} \dots (\dot{u}_{p-m})^{\beta_{p-m}} . \frac{m!}{\alpha_1! \dots (\alpha_{p-m})! \, \beta_1! \dots (\beta_{p-m})!}$$

α_i, β_i integers $\geqslant 0$ satisfying

$$\sum_{i=1}^{p-m} (\alpha_i + \beta_i) = m \quad ; \quad \sum_{i=1}^{p-m} i(\alpha_i + \beta_i) = p-1$$

Then, we have the following

Theorem 2

Under the assumptions (21) with $u_o \in H^1_o \cap H^2$ and $u_1 \in H^1_o$ the unique solution $u_\varepsilon(x,t)$ of the initial and boundary value problem (23) has the asymptotic expansion :

$$(26) \quad \left\| u_\varepsilon - \sum_{r=o}^{N} \varepsilon^r \hat{u}_r \right\|_{L^\infty(0,T;H^1_o)} + \left\| \dot{u}_\varepsilon - \sum_{r=o}^{N} \varepsilon^r \dot{\hat{u}}_r \right\|_{L^\infty(0,T;L^2)} \leqslant C \varepsilon^{N+1}$$

the function $\hat{u}_r(x,t)$ being the solution of (24)-(25).

Proof

Let v be $v = u_\varepsilon - \hat{u}$ with $\hat{u}(x,t) = \sum_{r=o}^{N} \varepsilon^r \hat{u}_r(x,t)$ then we can show that v satisfies the equation

$$(27) \quad \begin{cases} Lv = (f(t,v+\hat{u},\dot{v}+\dot{\hat{u}}) - f(t,\hat{u},\dot{\hat{u}})) + \alpha(\varepsilon,x,t) \\ v(0,t) = v(1,t) = 0 \\ v(x,0) = \dot{v}(x,0) = 0 \end{cases}$$

with

$$(28) \quad |\alpha(\varepsilon,x,t)| \leqslant K\varepsilon^{N+1} \quad \text{for } \varepsilon \text{ sufficiently small}$$

Let us introduce the sequence of functions $\{v_p(t), p \geqslant o\}$ defined by

$$(29)\begin{cases} Lv_p = \varepsilon\left[f(t,v_{p-1}+\hat{u},\dot{v}_{p-1}+\dot{\hat{u}}) - f(t,\hat{u},\dot{\hat{u}})\right] + \alpha(\varepsilon,x,t) \ ; v_o \equiv 0 \\ v_p(0,t) = v_p(1,t) = 0 \\ v_p(x,0) = \dot{v}_p(x,0) = 0 \end{cases}$$

For $p=1$ we have $Lv_1 = \alpha(\varepsilon,x,t)$. Therefore we deduce from (28)

$$(30) \qquad \|v_1\|_{L^\infty(0,T;H_0^1)} + \|\dot{v}_1\|_{L^\infty(0,T;L^2)} \leqslant C\varepsilon^{N+1}$$

for a convenient T.

Then using a recurrence argument we show that v_p satisfies the inequality (30). Finally we obtain the asymptotic expansion through a limiting process.

NUMERICAL ANALYSIS

The linear recursive schemes developed in this paper enables us to use a perturbation technique based on the ideas of best uniform approximation by polynomials. Given a linear partial differential equation with polynomial coefficients

$$(31) \qquad Lu = f(t,x)$$

we attempt to solve a slightly pertubed form of the original problem, defined by the so-called Tau problem :

$$(32) \qquad Lu_{rs} = f(t,x) + \tau H_{rs}(t,x)$$

where $H_{rs}(t,x)$ is the product of best uniform approximations of zero,of degrees r and s respectively, on a given domain D. The parameter τ is chosen for $u_{rs}(t,x)$ to be a bivariate polynomial which satisfies the given or boundary conditions satisfied by u.

The theory of the Tau method , originally proposed by Lanczos in the late thirties , has been developped by Ortiz [5] and computational procedures for the numerical treatment of partial differential equations with polynomial coefficients have been discussed by Ortiz and Samara [9]. Ortiz and Pham Ngoc dinh [6], [7] have discussed aspects of the error analysis of the Tau method in connection with a class of singularly pertubed problems for differential equations [6] and with nonlinear ordinary differential equations of Riccati's type [7]. In [8] we have discussed the numerical solution of a semi-linear hyperbolic problem of the following type :

(33) $u_{tt} - \Delta u = f(t,u)$

and sufficient conditions for the quadratic convergence are given in this paper .

SECTION 2

Let us consider the initial and boundary value problem :

(34) $u_{tt} - \Delta u + |u_t|^{\alpha} \operatorname{sgn}(u_t) = 0 \quad , \quad (x,t) \in \,]0,1[\times]0,T[$

(35) $u_x(0,t) = g(t) \quad , \quad u(1,t) = 0$

(36) $u(x,0) = u_0(x) \quad , \quad u_t(x,0) = u_1(x)$

with

$0 < \alpha < 1$

We shall make the following assumptions :

(37) $u_0(x) \in H^1 \quad , \; u_1(x) \in L^2$

(38) $g(t), \; g'(t) \; L^2(0,T)$; $g(0)$ being defined

Let

$$V = \{ v \in H^1 \mid v(1) = 0 \} \quad , \quad f(u) = |u|^{\alpha} \operatorname{sgn}(u)$$

Then we have the following [2]:

Theorem 3

There exists $T > 0$ such that the initial and boundary value problem (34)-(36) under the assumptions (37), (38) has one and only one solution $u \in L^{\infty}(0,T;V)$ and such that $\dot{u} \in L^{\infty}(0,T;L^2)$.

Proof

Consider a special orthonormal basis of V

$$v_\mu(x) = \sqrt{2/(1+\lambda_\mu^2)}\cos\lambda_\mu x \quad , \lambda_\mu = \pi(2\mu+1)/2 \; , \mu\in N$$

formed by the eigenfunctions of the Laplacian Δ

Let the subspace $(v_1, v_2, \ldots, v_n)$ generated by the distinct basis elements $v_1, v_2, \ldots, v_n$ of V.

Consider the Galerkin problem associated to the initial and boundary value problem (34)-(36).

$$\text{(39)} \quad a(u_n(t), v_p) + \frac{d}{dt}\langle \dot{u}_n(t), v_p\rangle + \langle f(\dot{u}_n(t)), v_p\rangle + g(t)v_p(0) = 0$$

$$u_n(0) = u_{on} \longrightarrow u_o \quad \text{in } H^1 \quad \text{strong}$$

$$\dot{u}_n(0) = u_{1n} \longrightarrow u_1 \quad \text{in } L^2 \quad \text{strong}$$

where

$$\text{(40)} \qquad u_n(t) = \sum_{p=1}^{n} \xi_{pn}(t)v_p$$

In (39) $a(u,v)$ and $\langle\, ,\rangle$ have the same meaning (1I) as stated before in section I.

Then $u_n(0,t)$ can be written as

$$\text{(41)} \qquad u_n(0,t) = \gamma_n(t) - 2\int_0^t K_n(t-\theta)g(\theta)\,d\theta$$

with

$$\text{(42)} \qquad \gamma_n(t) = \sum_{p=1}^{n} v_p(0)(\alpha_{pn}\cos\lambda_p t + \beta_{pn}\sin\lambda_p t/\lambda_p) + \sum_{p=1}^{n} \frac{v_p(0)}{\|v_p\|} \cdot \int_0^t \frac{\sin\lambda_p(t-\theta)}{\lambda_p} \langle -f(\dot{u}_n(\theta)), \frac{v_p}{\|v_p\|}\rangle d\theta$$

and

$$\text{(43)} \qquad K_n(t) = \sum_{p=1}^{n} \frac{\sin\lambda_p t}{\lambda_p}$$

In connection with the function $\gamma_n(t)$ we have the lemmas :

<u>Lemma 1</u>

There exists two positive, continuous functions $D_1(t)$ and $D_2(t)$ independent of n such that

$$\int_0^t |\gamma_n'(\theta)|^2 d\theta \leqslant D_1(t) + D_2(t)\int_0^t \|f(\dot{u}_n(\theta))\|\, d\theta$$

for t sufficiently small.

The lemma I follows from the generalized Minkowski's inequality.

- Let

(44) $$\sigma_n = \|u_n(t)\|^2_V + \|\dot{u}_n(t)\|^2 + \int_0^t |\dot{u}_n(0,\theta)|^2 d\theta$$

Then we can show from the lemma I the following inequality

$$\sigma_n \leqslant D_3(t) + D_4(t)\int_0^t \sigma_n^\alpha(\theta)\,d\theta$$

where $D_3(t)$ and $D_4(t)$ are two positive, continuous functions.

The theory of nonlinear Volterra integral equations with non decreasing kernel applying to the inequality (44) involves that we can extract from the sequence $\{u_n\}$ a subsequence $\{u_\mu\}$ such that :

(45)
$$\begin{aligned}
&\text{i)} \quad u_\mu \to u \quad \text{in } L^\infty(0,T;V) \text{ weak } * \\
&\text{ii)} \quad \dot{u}_\mu \to \dot{u} \quad \text{in } L^\infty(0,T;L^2) \text{ weak } * \\
&\text{iii)} \quad \dot{u}_\mu(0,t) \to \dot{u}(0,t) \quad \text{in } L^2(0,T) \text{ weak} \\
&\text{iv)} \quad f(\dot{u}_\mu) \to \chi \quad \text{in } L^\infty(0,T;L^2) \text{ weak } *
\end{aligned}$$

- To pass to the limit we shall require the two following lemmas :

<u>Lemma 2</u>

Let u be the solution of the following problem :

$$\begin{cases} \ddot{u} - \Delta u + \chi = 0 \\ u_x(0,t) = g(t) \quad , \quad u(1,t) = 0 \\ u(0) = u_0 \quad , \quad \dot{u}(0) = u_1 \end{cases}$$

$u \in L^\infty(0,T;V)$ and $\dot{u} \in L^\infty(0,T;L^2)$,

then we have

$$\frac{1}{2}\, a(u_0,u_0) + \frac{1}{2}\|u_1\|^2 - \int_0^s \langle \chi,\dot{u}\rangle d\theta - \int_0^s g(\theta)\,\dot{u}(0,\theta)\,d\theta \leqslant \frac{1}{2}\, a(u(s),u(s)) + \frac{1}{2}\|\dot{u}(s)\|^2$$

a.e. $s \in]0,T[$

Lemma 3

The function $f(u) = |u|^{\alpha}\mathrm{sgn}(u)$ generates a monotonic and hemicontinuous operator from $L^2 \to L^2$.

- If we pass to the limit in the equation (39) with $n=\mu$ we find without difficulty from (45) that u satisfies the equation :

$$a(u,v) + \frac{d}{dt}\langle \dot{u},v\rangle + \langle \chi,v\rangle + g(t)v(0) = 0 \quad \text{all } v\in V$$

and the initial conditions (36).

By using the lemmas 2 and 3 we finally show that $\chi = f(\dot{u})$.

Uniqueness of the solution is demonstrated from a corollary of the lemma 2 (with $u_0 = u_1 = g(t) = 0$).

REFERENCES

I. Boujot J., Pham Ngoc Dinh A. and Veyrier J.P. - Oscillateurs harmoniques faiblement perturbés : l'algorithme des " pas de géant ". RAIRO Analyse numérique, vol.4(I980), pp.3-23

2. Dang Dinh Ang and Pham Ngoc Dinh A. - Mixed problem for some semilinear wave equation with a nonhomogeneous condition. To appear

3. Lakshmikantham V. and Leela S. - Differential and integral inequalities, vol.I, Academic press, New-York and London, I969

4. Nguyen Thanh Long and Pham Ngoc Dinh A. - Linear approximation and asymptotic expansion associated to the nonlinear wave equation in one dimension. To appear

5. Ortiz E.L. - The Tau method. SIAM J.Num.Anal., vol.6(I969), pp.480-492

6. Ortiz E.L. and Pham Ngoc Dinh A. - An error analysis of the Tau method for a class of singularly pertubed problems for differential equations. Math.Meth.in the Appl.Sci., vol.6 (I984), pp.457-466

7. Ortiz E.L. and Pham Ngoc Dinh A. - Convergence of the Tau method for nonlinear differential equations of Riccati's type. Nonlinear Anal.T.M.A., vol.9(I985), pp.53-60

8. Ortiz E.L. and Pham Ngoc Dinh A. - Linear recursive schemes associated with some nonlinear partial differential equation in one dimension. To appear

9. Ortiz E.L. and Samara H. - Numerical solution of partial differential equations with variable coefficients with an operational approach to the Tau method. Computers and Maths. with Appls., vol.IO(I984), pp.5-I3

IO. Pham Ngoc Dinh A. - Sur un problème hyperbolique faiblement non linéaire en dimension I. Demonstratio Mathematica, vol.I6 (I983), pp.269-289

Part V: APPLICATIONS

Differential Equations in Numerical Problems of Science and Engineering

Numerical Approximation of Partial Differential Equations
E.L. Ortiz (Editor)

APPLICATION OF CHARACTERISTICS METHOD WITH VARIABLE TIME-STEP TO STEADY-STATE CONVECTION-DIFFUSION PROBLEMS

Alfredo Bermúdez

José Durany

Departamento de Ecuaciones Funcionales. Universidad de Santiago.Spain.

Abstract.- This paper deals with the numerical simulation of partial differential equations of convection-diffusion type when the convection dominates the diffusion.

First, the steady-state equation is transformed into an evolution problem and then we use a mixture of characteristics method with variable time-step and finite element approximations.

1. Introduction.

In a previous paper (Bermúdez-Durany, 1985) the authors adapt a combination of characteristics method and finite element methods (Pironneau, 1982; Bercovier-Pironneau-Sastri, 1983; Benque-Ibler-Keraimsi-Labadie, 1980; Douglas-Russell, 1982) to solve steady-state convection-diffusion problems of the type

(1.1) $u\nabla y-\nu\Delta y= f.$

The idea is to transform (1.1) into an evolution equation by adding the term $\frac{\partial y}{\partial t}$ and then to take into account that $\frac{\partial y}{\partial t} + u\nabla y$ is the total derivative of y along characteristics.

Next, this total derivative is discretized with a classical two-step formulae and the problem wich arises approximated by finite element methods.

In the present paper an extension of these ideas is considered. It consists on using more general formulae dependend on the spatial variable x to approximate the total derivative.

We show that some well known numerical schemas to discretize the convective term, as those given by |Tabata, 1977| and |Bristeau-Pironneau-Glowinski-Periaux-Perrier, 1979| can be obtained by following this approach.

Finally, numerical results are given for some two-dimensional test examples.

2. The steady-state convection-diffusion problem.

Consider a domain Ω with boundary Γ in $\mathbb{R}^N$. Let y(x) be the solution of the partial differential equation

(2.1) $u.\nabla y-\nu\Delta y= f,$ for $y= g$ on Γ

where u denotes the velocity field and ν is the diffusion coefficient.

If $f \in L^{\infty}(\Omega)$ and $g \in H^{\frac{1}{2}}(\Gamma)$ then this equation has a unique solution in the Sobolev space $H^1(\Omega)$. Usually it governs several steady-state physical phenomena such as conductive-convective heat transfer, diffusion-advection of miscible pollutants, etc.

When convection dominates diffusion (high Peclet number), solving (2.1) is a difficult problem. It is well known that inestabilities can appear in the classical Galerkin finite element scheme. To avoid this problem some upwinding technique has to be introduced and refinements of the mesh have to be used.

Here, we shall study in more details an upwinding method based on the characteristics and althrough it is not essential, we shall assume for simplicity that

(2.2) $\nabla . u = 0$ in Ω , and

(2.3) $u.n = 0$ on Γ

where n is the unit normal vector to Γ.

3. Characteristics method

With the purpose of upwinding the convection term by approximating the total derivative, the stationary problem (2.1) is transformed into an evolution problem

(3.1) $\frac{\partial \bar{y}}{\partial t} + \bar{u}.\nabla\bar{y} - \nu\Delta\bar{y} = \bar{f}$ a.e. in $(0,T)$

(3.2) $\bar{y}|_{\Gamma} = \bar{g}$

(3.3) $\bar{y}(x,0) = y(x)$

where

(3.4) $\bar{y}(x,t) = y(x)$ for all $t \in (0,T)$

($\bar{u}$, $\bar{f}$ and $\bar{g}$ are defined in the same way).

Now, $\frac{\partial \bar{y}}{\partial t} + \bar{u}.\nabla\bar{y}$ can be written as $\frac{D\bar{y}}{Dt}$, the total derivative of $\bar{y}$ in the direction of the flow $\bar{u}$ and consequently the equation (3.1) yields:

(3.5) $\frac{D\bar{y}}{Dt} - \nu\Delta\bar{y} = \bar{f}$ in Ω.

On the other hand, if $X(x,t;\tau)$ is the unique solution of the autonomous equation

(3.6) $\frac{dX}{d\tau} = u(X)$; $X(x,t;t) = x$

then X denotes the trayectory of a flow particle (characteristic) which will be in x at time t and it is easy to see that

(3.7) $$\frac{D\bar{y}}{Dt}(x,t) = \frac{d}{d\tau}\bar{y}(X(x,t;\tau),\tau)\Big|_{\tau=t}.$$

Therefore it is reasonable to discretize the total derivative by using the following backward finite difference formula:

(3.9) $$\frac{D\bar{y}}{Dt}(x,t^{m+1}) \simeq \frac{\bar{y}(x,t^{m+1})-\bar{y}(X^m(x),t^m)}{k}$$

where $k=\Delta t$, $t^{m+1}=(m+1)k$ and $X^m(x)$ is an approximation of $X(x,t^{m+1};t^m)$ (notice that $X(x,t^{m+1};t^{m+1})=x$).

Remark 3.1. For evolution problems upwinding scheme (3.9) leads to a step by step algorithm in time and its main advantage is that at each iteration a symmetric and time-independent linear system has to be solved (a study of these schemes including error estimates is given in |Pironneau, 1982|).

For steady-state problems (3.9) can be rewritten

(3.10) $$\frac{D\bar{y}}{Dt}(x,t) \simeq \frac{y(x)-y(X^k(x))}{k}$$

where $X^k(x)=X(x,t;t-k)$ denotes the position of the particle x in k-time unities before.

If we consider the approximation scheme (3.10), an appropiate variational formulation of the equation (3.5) is given by

(3.12) $$\begin{cases} \frac{1}{k}(y(x),w(x)) - \frac{1}{k}(y(X(.)),w(x)) + \nu(\nabla y(x),\nabla w(x)) = (f(x),w(x)) \\ \qquad\qquad \forall w\in H^1_0(\Omega) \\ y\in H^1(\Omega),\ y|_\Gamma = g \end{cases}$$

where (.,.) denotes the usual $L^2(\Omega)$-scalar product.

This equation has a unique solution for $f\in L^\infty(\Omega)$ and $g\in H^{\frac{1}{2}}(\Gamma)$ (Bermúdez-Durany, 1985).

4. Finite element discretization

Let τ_h be a triangulation of the domain Ω in $\mathbb{R}^2$. The finite element spaces are taken to be

(4.1) $$V_h = \{v;\ v|_K \in P_0,\quad \forall K\in\tau_h\}$$

(4.2) $$W_h = \{w\in C^0(\Omega);\ w|_K \in P_1,\ \forall K\in\tau_h\}$$

(4.3) $$W_{h0} = W_h \cap H^1_0(\Omega)$$

with h being the size of the biggest side of triangles K, and P_0, P_1

the polynomial spaces of degree less or equal to zero and one respectively.

If u_h, g_h are any interpolates of u, g in V_h and W_h respectively, then the discretized problem is

$$(4.4)\quad \begin{cases} \frac{1}{k}(y_h(x),w_h(x)) - \frac{1}{k}(y_h(X_h^k(.)),w_h(x)) + \nu(\nabla y_h(x),\nabla w_h(x)) = (f,w_h), \\ \qquad\qquad \forall w_h \in W_{h0} \\ y_h \in W_h,\ y_h|_\Gamma = g_h \end{cases}$$

where $X_h^k(x) = X_h(x,t;t-k)$ and X_h is the solution of

$$(4.5)\quad \frac{dX_h}{d\tau} = u_h(X_h);\quad X_h(x,t;\ t) = x$$

It can be shown (see Bermúdez-Durany, 1985 for details) that (4.4), (4.5) has a unique solution with $O(h+\frac{h^2}{k}+k)$ accuracy.

5. Relationship to other upwind schemes: Characteristics method with variable time step.

If we consider a Galerkin's formulation of the initial problem (2.1) and a finite element approximation by using the space W_h (see (4.2)), the following linear system is obtained

$$(5.1)\quad (u.\nabla y_h,w_{hi}) + \nu(\nabla y_h,\nabla w_{hi}) = (f,w_{hi})$$

where $\{w_{hi},\ i=1,\ldots,N_h\}$ is a basis of W_h.

In |Tabata, 1977| the following discretization of the first term in (5.1) is proposed

$$(5.2)\quad (u.\nabla y_h,w_{hi}) = (\frac{\partial y_h}{\partial u},w_{hi}) \simeq \frac{S_i}{3}|u(b_i)|\frac{|y_h(b_i)-y_h(c_i)|}{|b_i-c_i|}$$

where S_i is the area of triangles surronding node b_i and c_i is the intersection of the straight line $\{p:\ p=b_i-\lambda u(b_i),\ \forall\lambda>0\}$ with the first edge of triangles met.

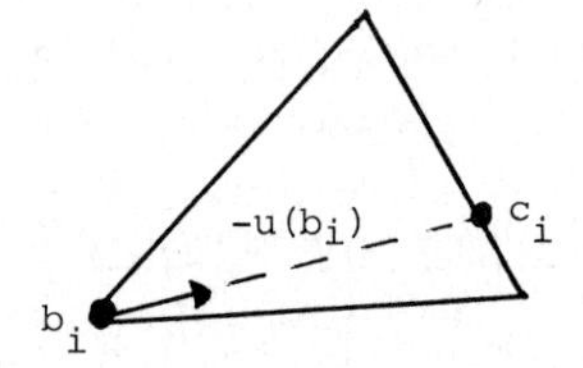

Figure 1.- Tabata's method.

On the other hand, we can modify the scheme (3.10) by employing a space-dependent time step k(x). Thus the discretization of the total derivative is given by

$$(5.3)\quad \frac{D\bar{y}}{Dt}(x,t) \simeq \frac{y(x)-y(X^k(x))}{k(x)}$$

where now $X^k(x)= X(x,t;t-k(x))$ and X is the solution of (3.6).

Therefore when we use a quadrature formula whose nodes are the vertices of each triangle, the finite element approximation of (5.3) becomes

$$(5.4)\qquad \left(\frac{1}{k(x)}\, y_h, w_{hi}\right) - \left(\frac{1}{k(x)}\, y_h(X_h^k(.)), w_{hi}\right) \simeq \frac{S_i}{3} \left| \frac{y_h(b_i)-y_h(X_h^k(b_i))}{k(b_i)} \right|$$

Now, if we take $u_h|_K = u(b_i)$ and

$$(5.5)\qquad k(b_i)= \frac{|b_i - c_i|}{|u(b_i)|} \qquad (\text{i.e. } X_h^k(b_i)= c_i)$$

then the Tabata's scheme (5.2) is obtained.

Moreover, this scheme is also obtained by taking k constant and equal to

$$(5.6)\qquad k= \min_i \{k(b_i)\}$$

because we are taking a finite element approximation of degree one. This fact is very interesting in computer numerical implementations.

In the next paragraph, we shall be interested in studying other methods with variable time step which can not be simplified to constant time step.

6. Schemes with three points.

In the previous paragraphs we have seen two-point formulas to discretize the total derivative with $O(k)$ accuracy. Now, we consider a backward scheme with three points of second order in the time step. This scheme is

$$(6.1)\qquad \frac{D\bar{y}}{Dt}(x,t) \simeq \frac{\frac{3}{2}\,\bar{y}(x,t) - 2\bar{y}(X^1(x),t-k) + \frac{1}{2}\,\bar{y}(X^2(x),t-2k)}{k}$$

where $X^1(x) \simeq X(x,t;t-k)$ and $X^2(x) \simeq X(x,t;t-2k)$.

This method can be considered as a generalization of the three points scheme given by |Bristeau and al., 1979| to discretize the convection term in (5.1).

In fact, the method of Bristeau and al. consists on approximating $\left(\frac{\partial y_h}{\partial u}, w_{hi}\right)$ by a parabolic interpolation with the three points b_i, c_i, d_i where c_i and d_i are on the line

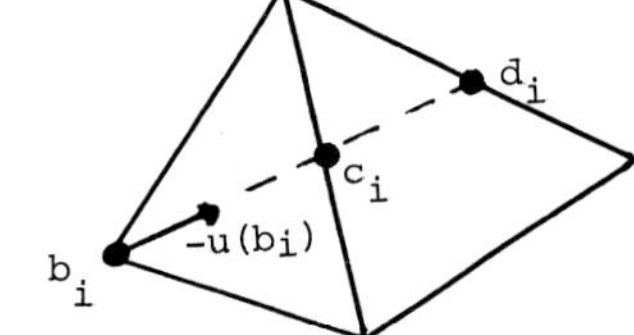

Fig. 2.-Bristeau and al.'s method

They obtain,

$$(6.2) \qquad (u.\nabla y_h, w_{hi}) = \left(\frac{\partial y_h}{\partial u}, w_{hi}\right) \simeq \frac{S_i}{3} . |u(b_i)| .$$

$$.\left[\frac{|b_i - c_i| + |b_i - d_i|}{|b_i - c_i| . |b_i - d_i|} y_h(b_i) - \frac{|b_i - d_i|}{|b_i - c_i| . |d_i - c_i|} y_h(c_i) + \right.$$

$$\left. + \frac{|b_i - c_i|}{|b_i - d_i| . |d_i - c_i|} y_h(d_i)\right]$$

Formula (6.1) can be generalized by taking time steps depending on x, as follows

$$(6.3) \qquad \frac{D\bar{y}}{Dt}(x,t) = a_1(x)\bar{y}(x,t) + a_2(x)\bar{y}(X^1(x), t-k_1(x)) + \\ + a_3(x)\bar{y}(X^2(x), t-k_1(x)-k_2(x))$$

where $X^1(x) = X(x,t;t-k(x))$, $X^2(x) = X(x,t;t-k_1(x)-k_2(x))$ and the functions a_1, a_2, a_3 are defined by

$$(6.4) \qquad \begin{cases} a_1(x) = \dfrac{2k_1(x)+k_2(x)}{k_1(x)(k_1(x)+k_2(x))} \\[2ex] a_2(x) = \dfrac{k_1(x)+k_2(x)}{k_1(x)k_2(x)} . (-1) \\[2ex] a_3(x) = \dfrac{k_1(x)}{k_2(x)(k_1(x)+k_2(x))} \end{cases}$$

In (6.4) $k_1(x)$, $k_2(x)$ are strictly positive real functions in Ω ($k_1(x)$, $k_2(x)$ denote the time steps from x to $X^1(x)$ and $X^1(x)$ to $X^2(x)$ respectively).

Now we use the same finite element approximation and quadrature formula as in the previous paragraph, so that (6.3), (6.4) becomes

$$(6.5) \qquad \left(\frac{Dy_h}{Dt}, w_{hi}\right) \simeq \frac{S_i}{3}\left[a_1(x)y_h(b_i) + a_2(x)y_h(X^1(b_i)) + a_3(x)y_h(X^2(b_i))\right]$$

with a_1, a_2, a_3 defined in (6.4) and

$$(6.6) \qquad \begin{cases} k_1(x) = \dfrac{|b_i - X^1(b_i)|}{|u(b_i)|}, \\[2ex] k_2(x) = \dfrac{|X^1(b_i) - X^2(b_i)|}{|u(X^1(b_i))|} \end{cases}$$

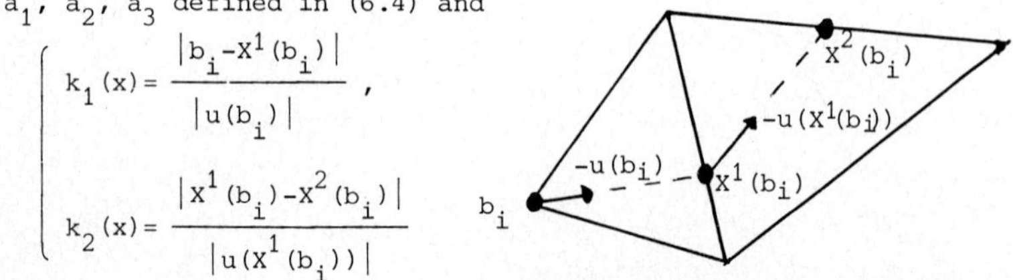

Figure 3.-Three point characteristics method. (Variable time step).

Clearly, if we take $u(X^1(b_i)) = u(b_i)$ the method of Bristeau and al.

(6.2) is deduced from (6.4), (6.5), (6.6).

7. Numerical implementation.

In this paragraph, we shall study some difficulties which appear in the numerical computations of the model problem (4.4), (4.5).

The second term in (4.4) can be written

$$(y_h(X_h^k(.)),w_h) = \sum_{i=1}^{N_h} \xi_{hi}(w_{hi}(X_h^k(.)),w_h). \tag{7.1}$$

Thus the contribution of this term to the matrix system is given by

$$a_{ij}^{(2)} = (w_{hj}(X_h^k(.)),w_{hi}) \tag{7.2}$$

and, consequently, a non-symmetric linear system is obtained.

The main computations to be made are the following:

a) Compute $X_h^k(x)$.

b) Compute (7.2) by numerical quadrature.

c) Solution of the non-symmetric linear system.

Remark 7.1. It is possible (see Remark 3.1) to move the second term in (4.4) into the right-hand-side of the equation and to look for the solution when $t\to\infty$ through a step by step algorithm in time. In this way a symmetric and time independent linear system has to be solved. However if $\nu\to 0$ (high Peclet number) or $k\to 0$ the convergence of this algorithm is very slow (see Bermúdez-Durany, 1985).

a) Compute $X_h^k(x)$: The numerical solution of (4.5) can be obtained by using different methods of approximation as Euler or Runge-Kutta - schemes (see Pironneau, 1982, Bercovier and al., 1983). But, if we want to use a quadrature formula to integrate (7.2) it is neccesary to find the triangle K such that $X_h^k(b_i)\in K$ (b_i quadrature node). For this, we can calculate the intersection of the line $(b_i,X_h^k(b_i))$ with all edges of triangles between b_i and $X_h^k(b_i)$. Moreover if u_h is constant per triangle then $X_h^k(b_i)$ can be computed exactly because it has a polygonal shape (see Figure 4).

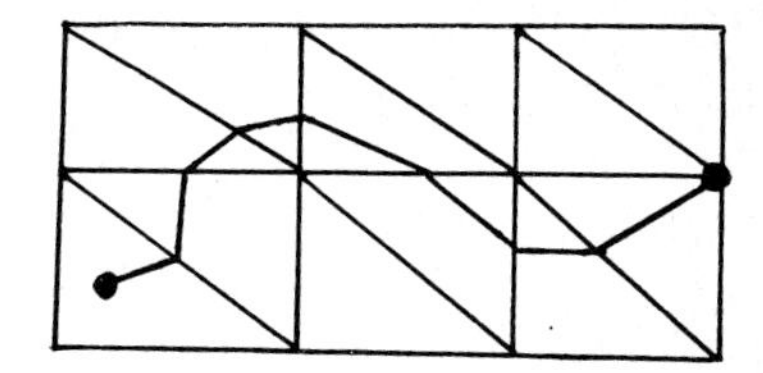

Figure 4.- Characteristic $X_h^k(b_i)$

b) We can use the quadrature formula

$$(w_{hj}(X_h^k(.)),w_{hi}) = \sum_{l=1}^{N_q} \alpha_l\, w_{hj}(X_h^k(b_l))(w_{hi}(b_l)) \tag{7.3}$$

with quadrature points at vertices of triangles or at the middle of the edges, etc. In practice, we take more quadrature points that nodal points per triangle in order to get small numerical dissipation.

c) The non-symmetric linear system (4.4) may be solved by preconditioned conjugate gradient algorithms. For more details see |Joly, 1984|.

8. Numerical examples.

The method presented in the paragraph 7 has been applied to solve the problem (4.4),(4.5) in $\Omega=(0,1)\times(0,1)$ by using a triangular mesh of 800 elements ($h=\frac{1}{20}$).

The velocity $u=(u_1,u_2)$ is given by

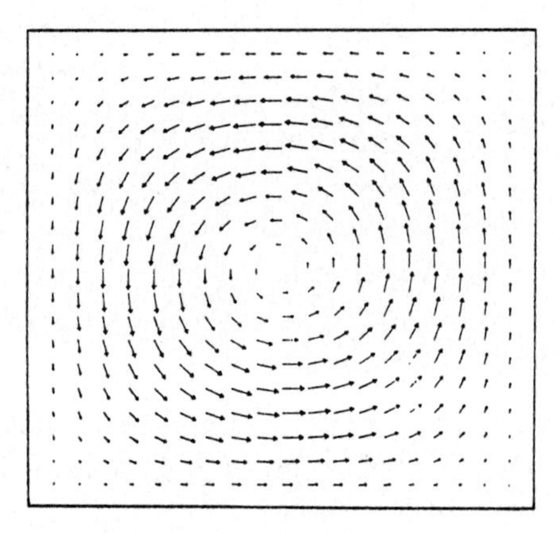

Figure 5.- Velocity.

$$(8.1)\quad \begin{cases} u_1(x_1,x_2)=20000\ x_1^2(1-x_1)^2x_2(1-x_2)(1-2x_2) \\ u_2(x_1,x_2)=20000\ x_2^2(1-x_2)^2x_1(1-x_1)(1-2x_1) \end{cases}$$

We take the functions f and g that the solution of the problem (2.1) is given by

$$(8.2)\quad y(x_1,x_2)=100\ x_1^2(1-x_1)^2+100\ x_2^2(1-x_2)^2$$

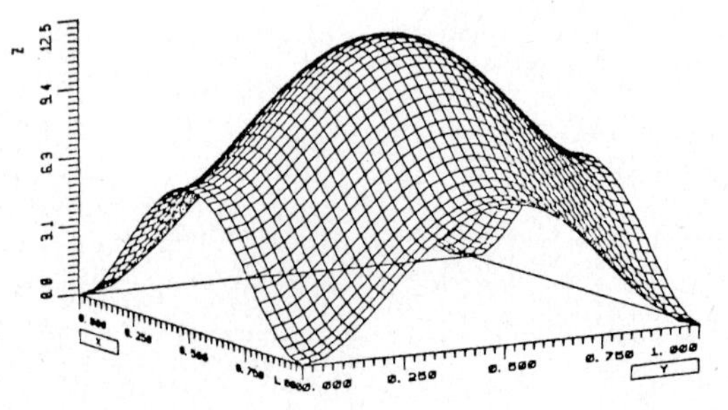

Figure 6.- Exact solution.

Figure 7 shows the approximated solution for small Peclet number ($Pe\simeq 10$). In this case this solution is independent of the parameter k within reasonable limits.

Remark 8.1. When there is not diffusion ($\nu=0$), the problem

(8.3) $\quad u.\nabla y= f$

has an infinity solutions for $\nabla.u=0$ and $u.n=0$.

The problem is well-possed if $\{X(x,t;\tau)\}$ intersects to Γ in $\tau(x)<t$ for all x. In this case the boundary condition has to be replaced by

(8.4) $\quad y|_{\Sigma}= g$

where

$\Sigma=\{x\in\Gamma;u(x).n(x)<0\}$.

When ν is very small in (4.4), some particular examples show that different values of k lead to different approximated solutions. These solutions are approximations of the solutions of the problem (8.3).

Figures 8, 9, show the approximated solutions for differents values of k and Peclet number equal to 400. Note the behaviour of the solutions that we have mentioned in the Remark 8.1.

However, in the following example with high Peclet numbers, we have obtained the same approximated solutions for different time-steps k and

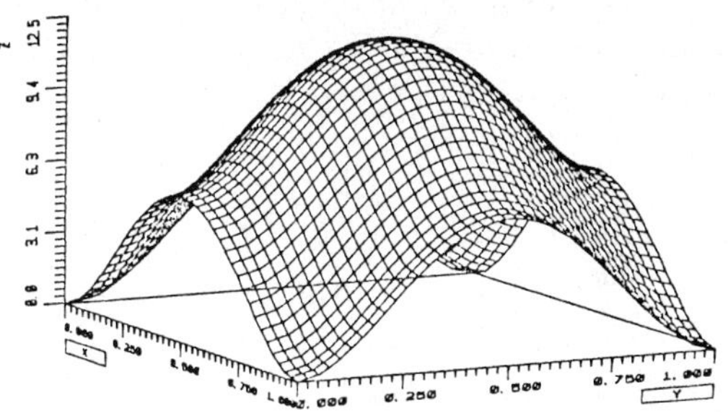

Figure 7.-Approximated solution $Pe\simeq 10$, k= 10^{-2}, 10^{-3}, 10^{-4}.

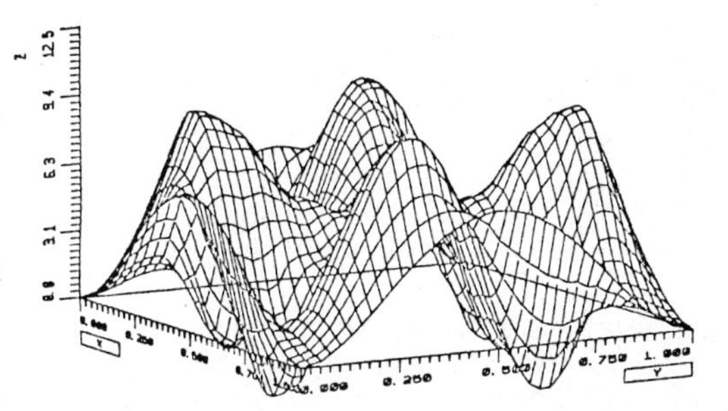

Figure 8.-Approximated solution $Pe\simeq 400$, k= 10^{-2}.

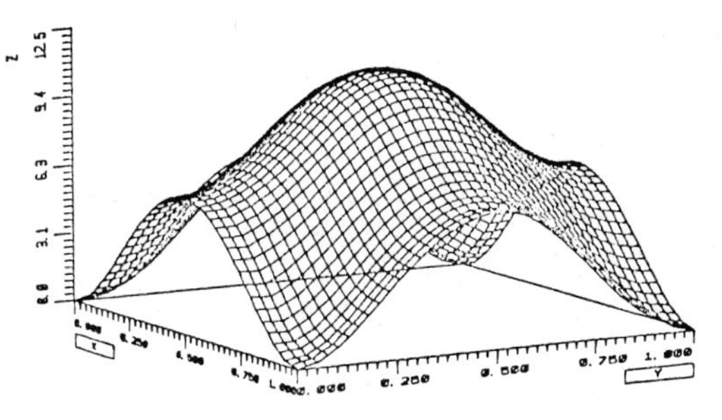

Figure 9.-Approximated solution $Pe\simeq 400$, k= 10^{-4}.

even though we do not know the exact solution it seems that the numerical method has a good behaviour.

In fact, we consider the previous velocity (8.1), f=0 and the boundary condition

$$\text{(8.5)} \quad \begin{cases} g(x_1,x_2)=1 & \text{if } x_1=0 \\ g(x_1,x_2)=0 & \text{if } x_1=1 \\ \dfrac{\partial g}{\partial n}=0 & \text{in other case} \end{cases}$$

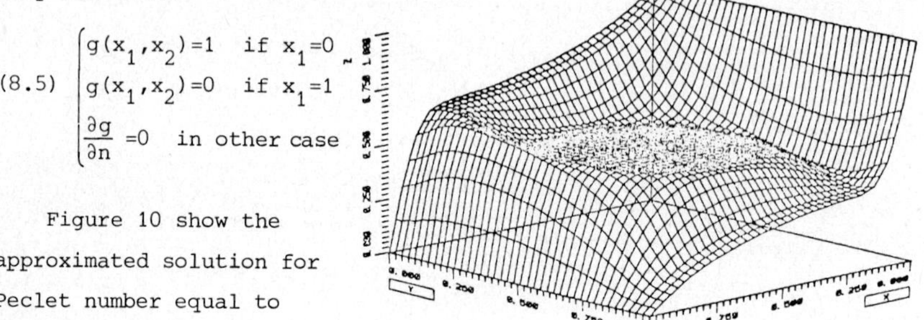

Figure 10 show the approximated solution for Peclet number equal to 1000.

Figure 10.- $\nu=0.12$, Pe=1000, $k=10^{-2}$, 10^{-3}, 10^{-4}.

REFERENCES

BENQUE, J., IBLER, B., KERAIMSI, A., LABADIE, G., A finite element method for Navier-Stokes equations, Proc. 3rd Int. Conf. finite elements in flow problems, Canada, 1980.

BERCOVIER, M., PIRONNEAU, O., SASTRI, V., Finite elements and characteristics for some parabolic-hiperbolic problems, App. Math. Modelling, Vol. 7, 1983.

BERMUDEZ, A., DURANY, J., La méthode des caractéristiques pour les problèmes de convection-diffusion stationnaires, Modélisation Mathématique et Analyse Numérique, Paris, 1985. (To appear).

BRISTEAU, M., PIRONNEAU, O., GLOWINSKI, R., PERIAUX, J., PERRIER, P., On the numerical solution of non linear problems in fluid dynamics by least squares. Compute Methods in Applied Mech. Vol. 17, 18, 1979.

DOUGLAS, J., RUSELL, T., Numerical methods for convection dominates diffusion problems, SIAM J. on Numerical Analysis. Vol. 19, nº 5, 1982.

JOLY, P., Méthodes de gradient conjugué, Publication of Laboratoire d'Analyse Numérique. Univ. Paris VI, 1984.

PIRONNEAU, O., On the transport-diffusion algorithm and its applications to the Navier-Stokes equations, Num. Math. Vol. 38, 1982.

TABATA M., A finite element approximation corresponding to the upwind differencing, Memoirs of Numerical Mathematics 1, 1977.

NOTE.- This work has been supported in part by C.A.I.C.Y.T., project n. 1800-82., and the Spanish-French "Acción Integrada" n. 123/85.

Numerical Approximation of Partial Differential Equations
E.L. Ortiz (Editor)

HOMOGENEIZATION OF SLIGHTLY COMPRESSIBLE INVISCID FLOWS

T. CHACON (Univ. SEVILLE & INRIA)
O. PIRONNEAU (Univ. PARIS NORD & INRIA)
INRIA BP.105 ROCQUENCOURT 78153 LE CHESNAY CEDEX

ABSTRACT :
This paper presents the homogeneization of a class of compressible ideal flows : We suppose that the flow evolves in two well separated scales, and the turbulent perturbation is incompressible about the small scale. By using homogeneization formal theory, we derive a system of deterministic equations verified by the mean quantities of the flow. We present finally some numerical results in 3D channel incompressible flow by comparison between our modelization and the direct simulation.

INTRODUCTION

The modelization of flows with highly oscillating data has found in the last years a very useful tool, named the homogeneization theory, which has opened a new way to simulate the turbulent flows, in addition to the direct simulation and to the classical simulations obtained from the physics of the problem.

The largest interest of this theory is that it allows a deterministic and systematic development of equations for the averaged quantities, instead of those obtained form Physics, which need heuristic reasoning, with the consequent loss of information.

Succesive approaches have been made to the homogeneization of the full Navier-Stokes equations. In some cases, it is possible to prove the convergence (see Tartar [3], Papanicolaou and Pironneau [4]), but usually the averaged equations remain purely formal.

In 1981, McLaughlin et al. (see [1]) obtained averaged equations for incompressible flow which evolves in two well separated scales. The model they obtained was tested later with good results (see [5]), and it gives some mathematical foundaticns of the known k-ε model (see [2]).

However, the application of homogeneization techniques to the modelization of compressible turbulent flows remains a difficult question, because of the double non-linearity appearing in the equations.

This paper presents the extension of the model of McLaughlin et al. [1] to a class of compressible flows : We consider a compressible perfect flow

with velocity fiel which can be decomposed in sum of a mean field plus a "microstructure" -two scaled- fluctuation. Under the hypothesis that this perturbation is incompressible about the "highly oscillating" variable, we derive a system of equations for the mean flow by using the techniques developed in [1].
We compare the equations thus obtained with those corresponding to incompressible flows, and finally we show some numerical test for incompressible 3D Channel flow.

1. TWO-SCALED TURBULENT FLOWS

Let u^ε and p^ε be the velocity and the pressure fields, respectively, corresponding to an ideal compressible flow with initial data in two scales, in the form

(1) $$u^\varepsilon(x,0) = u_o(x) + \varepsilon^{1/3}\, w_o(x/\varepsilon,x)$$

(2) $$\rho^\varepsilon(x,0) = \rho_o(x) + \varepsilon^{2/3}\, \sigma_o(x/\varepsilon,x)$$

Where u_o, ρ_o are the initial mean velocity and density, resp. w_o, σ_o are the initial perturbations of the velocity and density, resp., and ε is a small parameter : the ratio between the two considered scales.
Let us suppose that the initial flow is isentropic. It means that the corresponding pressure p^ε is given by

(3) $$p^\varepsilon = c\,(\rho^\varepsilon)^\gamma \quad ; \text{ with } c,\ \gamma \in \mathbb{R}.$$

These quantities $-u^\varepsilon$, ρ^ε and p^ε- are governed by the Navier-Stokes equations for ideal flows :

(4) $$u^\varepsilon{}_{,t} + u^\varepsilon.\nabla u^\varepsilon + 1/\rho^\varepsilon\ \nabla p^\varepsilon = 0,$$

(5) $$\rho^\varepsilon{}_{,t} + \nabla.(\rho^\varepsilon u^\varepsilon) = 0\ ;$$

in a domain $\Omega \subset \mathbb{R}^3$.

The initial conditions (1)-(2) correspond to the scales in ε obtained in [1], which accord to Kolmogorov theory.

We will suppose that the initial data are smooth enough to have regular solution -no shocks- at least during a finite time. The functions $w_o(y,x)$ and $\sigma_o(y,x)$ are periodic and have zero-mean in the y variable on the unit cube .

This is the way to express that w_0 and σ_0 are the initial perturbations of our velocity and density, resp.

We will see that w_0 and σ_0, which are independent in principle, must in fact verify a compatibility condition which expresses the incompressibility of the perturbation about the y variable (see(22)).

It is expected that the oscillations of these fluctuations when $\varepsilon \to 0$ will simulate the almost random behaviour of the turbulent perturbation in real flows.

2. AVERAGED EQUATIONS

To obtain averaged equations, we use homogeneization techniques : Let us consider the following formal expansions of u^ε, ρ^ε in powers of ε:

$$u^\varepsilon(x,t) = u(x,t) + \varepsilon^{1/3}\, w(a/\varepsilon\ ;\ x,t) + \varepsilon^{2/3}\, u^1(a/\varepsilon;\ x,t) + .. \tag{6}$$

$$\rho^\varepsilon(x,t) = \rho(x,t) + \varepsilon^{2/3}\, \sigma(a/\varepsilon\ ;\ x,t) + \varepsilon\, \rho^1(a/\varepsilon;\ x,t) + .. \tag{7}$$

Where $a = a(x,t)$ are the lagrangian co-ordinates of the problem (see [1]) :

$$a_t + u.\nabla a = 0,\ a(x,0) = x. \tag{8}$$

To find an expression derived from (7) for the pressure, it is better to consider first that

$$1/\rho^\varepsilon\ \nabla p^\varepsilon = p^\varepsilon_* \tag{9}$$

$$p_*^\varepsilon = \varphi_\gamma(\rho^\varepsilon),\quad \varphi_\gamma(\xi) = \begin{cases} c\,\gamma \log(\xi) & \text{if } \gamma = 1 \\ c\,\gamma/(\gamma-1)\xi^{\gamma-1} & \text{if } \gamma \neq 1 \end{cases} \tag{10}$$

Then the expression for the pressure is

$$p_*^\varepsilon = p_*(x,t) + \varepsilon^{2/3}\, \pi(a/\varepsilon,\ x,t) + \varepsilon p^1(a/\varepsilon\ ;\ x,t) + \ldots \tag{11}$$

Where

$$\begin{cases} p_* = \varphi_\gamma(\rho) \\ \pi = \varphi'_\gamma(\rho)\,\sigma \\ p^1 = \varphi'_\gamma(\rho)\,\rho^1 + 2/3\,\varphi''_\gamma(\rho)\,\sigma, \text{ etc.} \end{cases} \tag{12}$$

In (6)-(7), w, u^1 ... ; σ, ρ^1,... are zero-mean periodic functions on the unit cube. Thus, π, p^1, ... have zero mean because each p^k is alinear combination of σ, ρ^1, ..., ρ^k.

By following the technique developped in [1], [2], averaged equations are obtained :

$$u_{,t} + u.\nabla u + 1/\rho\ \nabla p + \varepsilon^{2/3}\, \nabla.(\rho R) = O(\varepsilon), \tag{13}$$

(14) $$\rho,_t + \nabla.(\rho u) = O(\varepsilon),$$

(15) $$p = c\rho \quad ,$$

(16) $$q,_t + u.\nabla q + R:\nabla u + 1/\rho\ \varepsilon^{1/3}\ \nabla.(\rho V) = O(\varepsilon^{2/3}),$$

(17) $$h,_t + u.\nabla h + S:\nabla u + \varepsilon^{1/3}\ \nabla.D = O(\varepsilon^{2/3})\ ;$$

where q is the kinetic turbulent energy and h is the global turbulent helicity, defined by

(18) $$q = 1/2 <|w|^2>, \quad h = <w.r>\ ;\ r = (\nabla a \nabla_y) \times w$$

and R, S, V, D are functions depending on the turbulent fluctuations :

(19) $$\begin{cases} R = <w \otimes w>, \quad S = <w \otimes r + r \otimes w>, \\ V = <(1/2|w|^2 + \pi)\ w>, \quad D = 2 <(1/2|w|^2 + \pi)\ r>, \end{cases}$$

This system generalizes the one found by McLaughlin et al. for incompressible flows. As has been said, the scales in ε accord with the Kolmogorov Theory. Notice that its generalization to viscous flows which viscosity is $O(\varepsilon^2)$ is immediate (see[6]). The first microstructure fluctuation -w- can be obtained through a canonical microstructure problem : Let $\tilde{w}$ be a solution of

(20) $$\begin{cases} \tilde{w}.\nabla_y \tilde{w} + C\nabla_y \tilde{\pi} = 0, \quad \nabla_y.\tilde{w} = 0, \\ <\tilde{w}> = 0,\ 1/2 <\tilde{w}.\ C^{-1}\tilde{w}> = q,\ <\tilde{w}.C^{-1}\tilde{r}> = h, \text{ where } \tilde{r} = (\nabla a)^T r \\ \tilde{w}, \tilde{\pi} \text{ periodic in the unit cube ;} \end{cases}$$

where C is the symetric matrix defined by $C = (\nabla a)^T(\nabla a)$. Then, w is defined by

(21) $$w = (\nabla a)^{-T}\ w.$$

From the condition $\nabla_y.\tilde{w} = 0$, it is clear that the perturbation $\tilde{w}$ is incompressible about the highly oscillating variable. To solve numerically the system (13)-(17), regular tabulation of the tensors R and S as functions of ∇a, q and h are needed. Then, we need the problem (20) to have locally isolated solutions which depend continuously on these parameters. In particular, this hypothesis implies that the initial fluctuating velocity $-w_o-$ and pressure $-\pi_o-$ verify

(22) $$w_o.\nabla_y w_o + \nabla_y \pi_o = 0\ ;\ \nabla_y.w_o = 0\ ;\ \text{with } \sigma_o = 1/c\ \rho_o^{2-\gamma}\pi_o.$$

This is the compatibility condition for the initial perturbation which in reality is defining σ_o as a function of w_o. From an other point of view, the problem (20) is the same obtained for the incompressible case. Then, the same code to solve (20) can be used to tabulate the closure functions in both the incompressible and the compressible cases.

3. NUMERICAL TESTS

The model which equations are (13)-(17) has been improved with the introduction of eddy viscosity terms, following the same technique as Begue et al. [6]. It has been solved numerically in the 3D channel for incompressible flows. The considered equations are :

$$(23)\quad \begin{cases} u,_t+u.\nabla u+\nabla p + \varepsilon^{2/3}\nabla.(qR) = \nu\Delta u + \varepsilon^{4/3}\, a\nabla.[\sqrt{q}(\nabla u + \nabla u^T)], \\ \nabla.u = 0, \\ q,_t+u.\nabla q + q\, R : \nabla u = \varepsilon b\Delta\, (\sqrt{q}^3) \\ a,_t + u.\nabla a = 0 \end{cases}$$

Where ν is the kinematic viscosity and a, b are positive parameters depending on some microstructure problems (see Chacon [8]).
The choice of suitable boundary conditions for the system (23) must be carefully made, because our model is not considering turbulence near the boundaries. The chosen boundary conditions are :

$$(24)\quad \begin{cases} u(x_1,x_2,x_3 = 0,2) = 0 \\ a(x_1,x_2,x_3 = 0,2) = (x_1,x_2,x_3) \\ q(x_1,x_2,x_3 = 0,2) = 0 \end{cases}$$

We look for (x_1-x_2)-periodic solutions of (23) on the cell $]0,2\pi/\alpha[\times]0,2\pi/\beta[$. The considered domain is the "Channel" given by

$$(25)\qquad \Omega =]0,2\pi/\alpha[\times]0,2\pi/\beta[\times]0,2[.$$

The problem (23) has been discretized with a spectral expansion, following Orszag and Patera [7] :

$$(26)\qquad u(x,t) = \sum_n \sum_m v_{nm}(x_3,t)\; e^{i(\alpha x_1 n+\beta x_2 m)}$$

The microstructure initial conditions are easy to introduce in this case :
We have taken in (1)

$$(27)\quad \begin{cases} u_o(x) = u_B(x) \\ w_o(y,x) = 2\,\mathrm{Re}\, v(x_3)\; e^{iN(\alpha y_1+\beta y_2)} \end{cases}$$

Where u_B is the poiseuille basic flow :

$$(28) \qquad u_B(x) = (1-(1-x_3)^2,0,0)^T.$$

To have a highly oscillating initial perturbation, the wavenumber $N(=1/\varepsilon)$ in (27) has to be taken as large as possible. In the experiments, we have taken a 32 x 32 x 33 grid which allows to have $\varepsilon=.1$. Our main experiment has been the comparison between the solution of (23) and the direct solution of the Navier-Stokes equations, obtained by using the same spectral discretization (see (26)). We obtain good agreement between both simulations. The same time scale, and similar behaviour of velocity and microstructure energy are found, as seen on the figures.

CONCLUSION

As has been said, the considered turbulent perturbation has to be incompressible about the "fast" variable. This means, in particular, that our case is not considering fast sound waves, the existence of which is a particular characteristic of tubulence in compressible flows. However, the averaged equations correspond to a compressible flow : we are thus considering a kind of slightly compressible flows.

The application of Homogeneization theory to slighlty compressible flows shows its power and possibilities, and suggests a possible future application to full-compressible flows.

Captions to the figures

Figures 1,2 : Velocity field and microstructure kinetic energy corresponding to the direct simulation with a parabolic function v in (27).

Figures 3,4 : The same corresponding to the modelization. In this case, the equations (23) can be reduced to a 1D system.

Notice the good agreement between both cases.

Notations :

Let A, B be two matrix, and a, b be two vectors. Then,

$$A : B = A_{ij}B_{ij}, \quad \langle w\rangle = \int_Y w(y,x)dy, \quad Y =]0,1[^3; \quad \langle a\otimes b\rangle_{ij} = \langle a_i b_j\rangle$$

REFERENCES

[1] D. McLAUGHLIN, G. PAPANICOLAOU, O. PIRONNEAU : Convection of microstructures and related problems. (To appear in SIAM).

[2] T. CHACON, O. PIRONNEAU : On the mathematical foundations of the k-ε turbulent model (To appear).

[3] L. TATAR : Remarks on oscillations and Stoke's equations. Proceedings Workshop on Numerical Simulation of Turbulence. Nice, Dec 1984.

[4] G. PAPANICOLAOU, O. PIRONNEAU : On the asymptotic behavior of motions in random flow. Stockastic Nonlinear Systems. Arnold-Lefever eds., Springer, Berlin, 1981, pp 36-41.

[5] T. CHACON, O. PIRONNEAU : Convection on microstructures by incompressible and slightly compressible flows. Proceedings Congress on Oscillations Theory in P.D.E. Minneapolis, March 1985. H. Weinberger ed (To appear).

[6] C. BEGUE, T. CHACON, D. McLAUGHLIN, G. PAPNICOLAOU, O. PIRONNEAU : Convection of microstructures II. Proceedings VI Congress on Computing Methods in Applied Sciences and Engineering. Glowinski-Lions ed. North-Holland, 1983.

[7] S.A. ORZAG, A.T. PATERA : Secondary instability of wall bounded shear flows. J. Fluid. Mech. (1983), vol 128, pp 347-285.

[8] T. CHACON : Contribucion al estudio del modelo M.P.P. de turbulencia. Thesis Univ. of Seville. Sept. 1984.

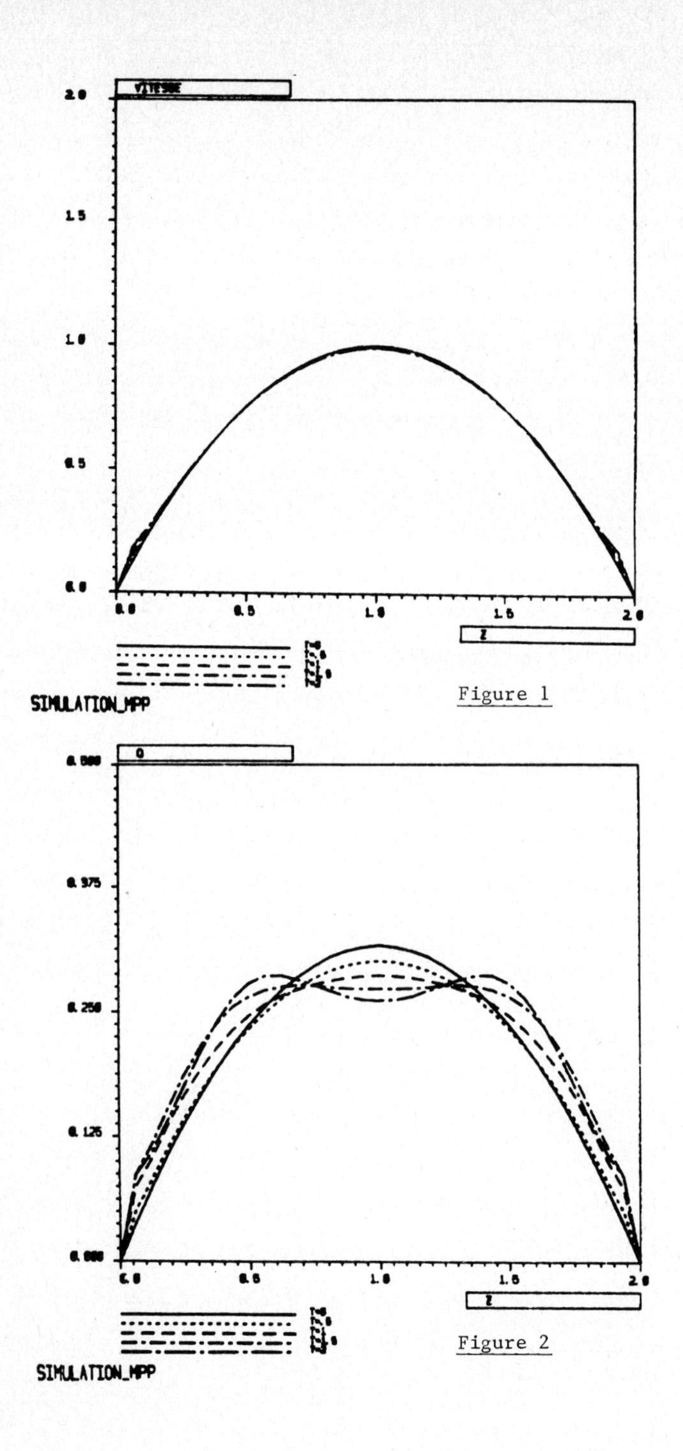

Figure 1

Figure 2

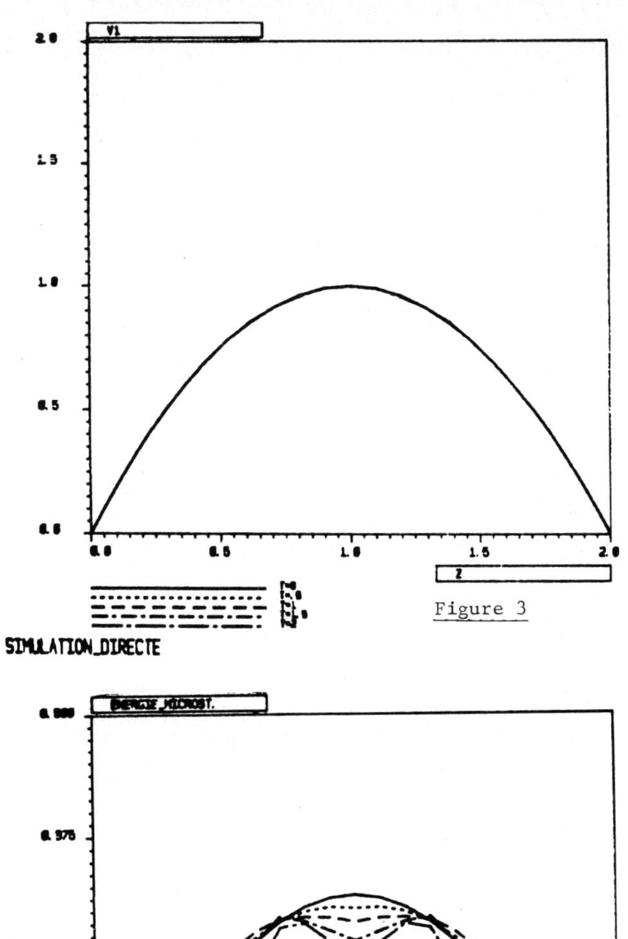

Figure 3

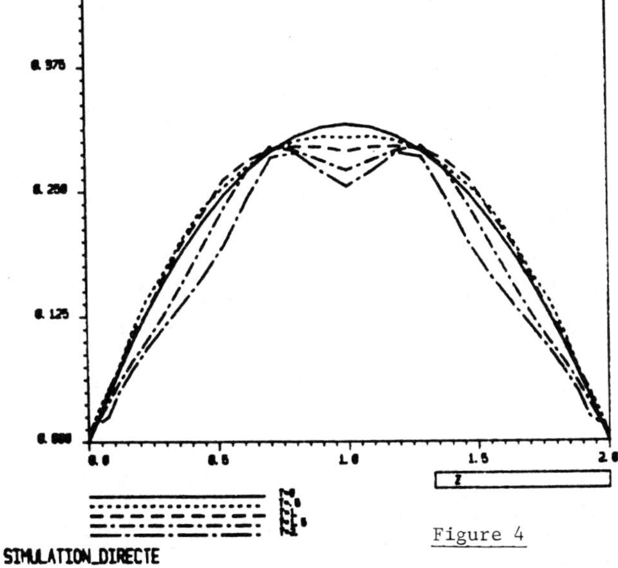

Figure 4

Numerical Approximation of Partial Differential Equations
E.L. Ortiz (Editor)

CONTINUATION OF PERIODIC SOLUTIONS IN ORDINARY DIFFERENTIAL EQUATIONS WITH APPLICATION TO THE HODGKIN-HUXLEY MODEL

Martin Holodniok and Milan Kubíček

Prague Institute of Chemical Technology

Prague, Czechoslovakia

The DERPER algorithm was constructed for the continuation of periodic solutions (Holodniok and Kubíček, 1984). The algorithm is based on the shooting method and on the arc-length continuation algorithm DERPAR (Kubíček, 1976; Kubíček and Marek, 1983).

The use a simple shooting method is ineffective in cases, where one or more eigenvalues of the monodromy matrix are large; in such cases the necessary computing time sharply increases. A multiple shooting method (Stoer and Bulirsch, 1980) was therefore used instead of a simple shooting method for the computation of periodic solutions in such cases.

New algorithm is tested on the Hodgkin-Huxley model of the conduction of the nervous impulse and is compared with the simple shooting algorithm. The increased effectiveness of the new algorithm is demonstrated.

1. Introduction

Let us consider mathematical models in the form of systems of nonlinear ordinary differential equations depending on a physical parameter. Several different approaches can be designed for computation and continuation of periodic solutions. The easiest one consists of a dynamic simulation of the studied system; only stable periodic orbits are computed in this way. The finite difference methods can be used to obtain both stable and unstable orbits (Rinzel and Miller, 1980). A simple shooting method coupled with continuation along the arc-length of solution locus (Kubíček, 1976) forms a basic of the algorithm DERPER (Holodniok and Kubíček, 1984). The algorithm was successfully applied to a number of practical problems. The complete picture of the dependences of periodic solutions on the parameter with many examples of different types of periodic solutions for the Lorenz model was published in (Holodniok, Kubíček and Marek, 1982). Extensive study of periodic solutions was also perfomed for the problem of two coupled well mixed cells with the Brusselator chemical reaction schema (Schreiber, Holodniok, Kubíček and Marek, 1985).

However, the DERPER algorithm failed in specific cases. In the study of periodic solutions for the Hodgkin-Huxley model (Hodgkin and Huxley, 1952) the continuation algorithm with a simple shooting technique did not lead to required results; the method failed for the larger part of the dependence of periodic solutions (e.g. Hassard, 1978). The complete dependence of periodic solutions was obtained by means of the finite difference method (Rinzel and Miller, 1980); the task of the stability verification and of the detecting of bifurcation points, however, is here complicated.

The use of the multiple shooting method was proposed in earlier paper (Holodniok and Kubíček, 1984), and studied also by other authors (Deuflhard,1984; Seydel, 1985). However, applications for continuation of periodic solutions were not reported.

2. Development of an algorithm

Consider an autonomous system of ordinary differential equations

$$\frac{dy_i}{dt} = f_i(y_1, \ldots, y_n, \alpha) \quad i=1,2, \ldots, n. \tag{1}$$

A periodic solution with the period T satisfies

$$y_i(t+T) = y_i(t), \quad i=1,2,\ldots, n. \tag{2}$$

The transformation $t = T z$ produces

$$\frac{dy_i}{dz} = T f_i(y_1,\ldots, y_n, \alpha), \quad i=1,2,\ldots, n \tag{3}$$

and boundary conditions (2) (conditions of periodicity) appear in the form

$$y_i(1) - y_i(0) = 0, \quad i=1,2,\ldots, n. \tag{4}$$

In the multiple shooting technique (see Stoer and Bulirsch, 1980) the interval $[0,1]$ is divided into m-1 subintervals by means of grid points

$$0=z_1 < z_2<\ldots<z_m=1, \quad \Delta z_j=z_{j+1}-z_j, \quad j=1,2,\ldots, m-1. \tag{5}$$

Let us denote $y=(y_1,\ldots,y_n)^T$. Then for $j=1,\ldots,m-1$ the initial conditions are chosen,

$$y(z_j) = x_j = (x_{j1}, \ldots, x_{jn})^T. \tag{6}$$

For fixed value of α we choose also the value of the period T and after integration of (3) on the interval $[z_j, z_{j+1}]$ with the initial condition (6) we obtain solution at the point z_{j+1}; let us denote it $y(z_{j+1}|x_j, T)$. This solution has to satisfy

$$y(z_{j+1}| x_j, T) - x_{j+1} = 0, \quad j=1,2,\ldots, m-1; \tag{7}$$

from the periodicity it follows $x_1=x_m$. The conditions (7) form a system of $n(m-1)$ nonlinear equations for $n(m-1)+1$ unknowns $(x_1, x_2, \ldots, x_{m-1}, T)$. Therefore, let us choose a fixed value for one unknown at the point z_1, x_{1k} (also called the phase condition). The chosen value of x_{1k} must actually exist on the trajectory of the k-th component of the wanted periodic solution $y_k(z)$, $z \in [0,1)$, i.e., $x_{1k} = y_k(\bar{z})$ for certain $\bar{z} \in [0,1)$. Denote $\tilde{x}_1 = (x_{11}, \ldots, x_{1,k-1}, x_{1,k+1}, \ldots, x_{1n})$. Then in (7) we have only $n(m-1)$ unknowns $(\tilde{x}_1, x_2, \ldots, x_{m-1}, T)$.

The system (7) is solved by the Newton's method; values of x_{1k} and α remain fixed during iteration.

For continuation of the solution of (7) with respect to the parameter α we rewrite Eq. (7) into the form

$$F_j \equiv y(z_{j+1} \mid x_j, t, \alpha) - x_{j+1} = 0, \quad j=1,2,\ldots, m-1, \qquad (7')$$

$$x_m = x_1 .$$

Now we need to continue the branch of solutions of $n(m-1)$ equations for $n(m-1)+1$ unknowns $(\tilde{x}_1, x_2, \ldots, x_{m-1}, T, \alpha)$. We use the predictor-corrector algorithm DERPAR (Kubíček, 1976) for the continuation along an arc-length of the solution locus. For more details and an adaptive control of x_{1k} see (Holodniok and Kubíček, 1984). The evaluation of the Jacobi matrix of Eqs. (7') is necessary in the DERPAR-continuation. The Jacobi matrix can be written in the following form :

$$\begin{bmatrix} \tilde{G}_1 & -I & 0 & \cdots & & \partial F_1/\partial T & \partial F_1/\partial \alpha \\ 0 & G_2 & -I & \cdots & & \partial F_2/\partial T & \partial F_2/\partial \alpha \\ \cdots & & & & & \vdots & \vdots \\ & & 0 & G_{m-2} & -I & \cdot & \cdot \\ -\tilde{I} & 0 & \cdots & 0 & G_{m-1} & \partial F_{m-1}/\partial T & \partial F_{m-1}/\partial \alpha \end{bmatrix} \qquad (8)$$

where n by n matrix G_j is

$$G_j = \left\{ \frac{\partial y_i(z_{j+1} \mid x_j, T, \alpha)}{\partial x_{j\ell}} \right\}, \quad i, \ell = 1, \ldots, n. \tag{9}$$

$\widetilde{G}_1$ is formed from G_1 by deleting the column with the index k; analogously for $\widetilde{I}$.
The partial derivatives are obtained on the basis of variational equations

$$\frac{dU}{dz} = T \frac{\partial f(y(z), \alpha)}{\partial y} U, \tag{10}$$

where U is n by n matrix. If $U(z_j) = I$ is chosen as initial condition the integration of (10) on the interval $[z_j, z_{j+1}]$ gives $G_j = U(z_{j+1})$. For $\partial F_j / \partial T$ it holds

$$\partial F_j / \partial T = \Delta z_j \, f(y(z_{j+1} \mid x_j, T, \alpha)) \tag{11}$$

and similar variational equations as (10) can be set up for $\partial F_j / \partial \alpha$.

It is very easy to prove, that the monodromy matrix has a form

$$M = G_{m-1} G_{m-2} \ldots G_2 G_1. \tag{12}$$

Then it follows from the theory of periodic solutions that for an autonomous system is one eigenvalue of M always equal to unity; let $\lambda_1 = 1$, and then the other ones determine stability of solutions. Denote $\lambda_2 : |\lambda_2| \geq |\lambda_s|$, $s=3,\ldots,n$.

3. Application

To test the effectivity of the above algorithm we have chosen the Hodgkin-Huxley model of the conduction of the nervous impulse (Hodgkin and Huxley, 1952) in the following form (Hassard, 1978):

$$
\begin{aligned}
dy_1/dt &= -\alpha - G \\
dy_2/dt &= [(1-y_2)\beta_2(y_1) - y_2\gamma_2(y_1)]\phi \\
dy_3/dt &= [(1-y_3)\beta_3(y_1) - y_3\gamma_3(y_1)]\phi \\
dy_4/dt &= [(1-y_4)\beta_4(y_1) - y_4\gamma_4(y_1)]\phi
\end{aligned} \tag{13}
$$

with $G(y_1,\ldots, y_4) = 36\, y_3^4\,(y_1-12)+120 y_2^3\, y_4(y_1+115) + 0.3(y_1+ 10.599)$

$\beta_2(y_1) = 1/\text{expc}(v_{25})$, $\quad\gamma_2(y_1) = 4\,\exp(y_1/18)$,

$\beta_3(y_1) = 0.1/\text{expc}(v_{10})$, $\quad\gamma_3(y_1) = \exp(y_1/80)/8$,

$\beta_4(y_1) = 0.07\,\exp(y_1/20)$, $\quad\gamma_4(y_1) = 1/(1+\exp(v_{30}))$,

where $v_{10} = (y_1+10)/10$, $v_{25} = (y_1+25)/10$, $v_{30} = (y_1+30)/10$ and expc is the function $\text{expc}(x) = (e^x-1)/x$, $x \neq 0$, $\text{expc}(0)=1$, and $\phi = 3^{(\tau-6.3)/10}$. The value of the parameter $\tau = 6.3$ was chosen, as the model exhibits here an interesting switch-back of periodic solutions in dependence on the current stimulus α.

The use of the simple shooting algorithm (Hassard, 1978) did not bring satisfactory results. Also our computation using the algorithm DERPER did not give better results, cf. Fig. 1. The technique worked only in regions of parameters, where $|\lambda_2| \lesssim 10^5$.

The dependence of periodic solutions on the parameter α, obtained by means of the continuation algorithm DERPER using the above multiple shooting method is shown in Fig.1 (Holcová, 1985). The value of $m=11$ was chosen, i.e., the interval $[0,1]$ was divided uniformly; in all computations $k=1$. The computed dependence agrees with the one given in the paper (Rinzel and Miller, 1980). Part of the dependence given in Fig. 1 is shown in more detail in Fig. 2

together with corresponding values of λ_2. The crosses (x) denote parts of the curves obtained by the simple shooting method (the starting point was chosen either close to the Hopf bifurcation point or at the branch of stable periodic solutions).

4. Effective choice of mesh points

Let us consider fixed value of α . In the case of stable periodic solutions is the choice m=11 too conservative; a simple shooting method would be sufficient. To study in more detail the behaviour of the multiple shooting algorithm, we have chosen five testing solutions (cf. Fig. 1, Tab. 1), denoted as A,B,C,D, and E, for different values of λ_2, and computed convergence properties for different equidistant and non-equidistant choice of the mesh points.
The solutions obtained for m=6 are given in Table 2 at the mesh points (to five significant figures).

A comparison of the relative consumption of the computer time (number of iterations, NI, resp.) for different choices of mesh points z_j are for solutions A,B,C, and E shown in Tables 3,4,5. The accurate solutions truncated to 2,3, 5, and 6 valid figures at the mesh points were chosen as starting estimates in each test. The relative error with respect to the number of mesh points was computed as
$ER = (\sum_{j=1}^{m-1} F_j^2)^{1/2}/(m-1)$. The iteration process ended, when $ER < 10^{-10}$. The cases of overflow, convergence to different solution (than the estimated one) was considered as divergence and denoted as D. RT denotes relative time of computations (related to the best obtained time).

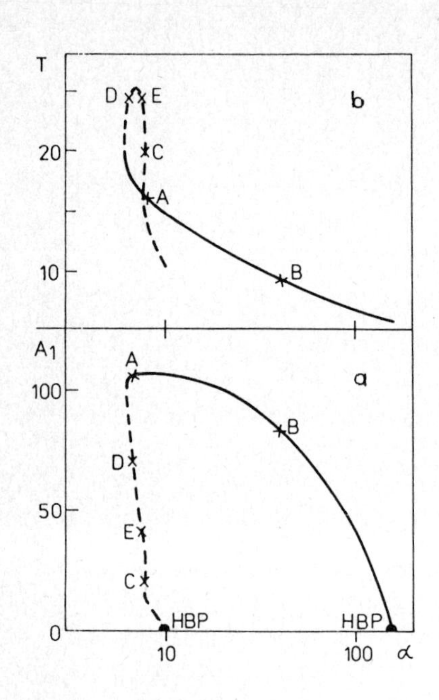

Fig.1: Dependence of
a) amplitude of y_1
b) period
on the parameter α .
HBP - Hopf bif.point
—— - stable per.solution
---- - unstable per.solution

Table 1: Testing solutions

Sol.	α	T	λ_2
A	7.87	16.131	7.1E-2
B	41.112	9.124	1.8E-1
C	7.922	20.680	-3.5E3
D	6.775	24.583	1.1E6
E	7.562	24.733	6.3E8

It follows from the Table 3 that the simple shooting method is the best method to compute stable periodic solutions. The solution C in the Table 4 has relatively worse convergence properties in comparison with the solution E (table 5); the solution C is located close to the limit point and the period-doubling bifurcation point. Rich structure of various solutions exists in the neighbourhood of C and this explains the difficulties with convergence.

A non-equidistant mesh points were chosen for m=4 and

Table 2: Periodic solutions A,B,C and E for m=6

Solution A

z_i	y_1	y_2	y_3	y_4
0.0	-40.832	0.92468	0.75107	0.079787
0.2	8.1803	0.018897	0.54464	0.28875
0.4	2.3949	0.037919	0.41411	0.44029
0.6	-3.2668	0.073609	0.37688	0.48115
0.8	-8.7817	0.12952	0.39441	0.44635

Solution B

z_i	y_1	y_2	y_3	y_4
0.0	-40.830	0.92108	0.72618	0.065104
0.2	4.7273	0.030436	0.63657	0.14252
0.4	- 1.3241	0.056120	0.54034	0.24076
0.6	- 7.8515	0.11309	0.49664	0.27982
0.8	-19.524	0.25977	0.50306	0.25660

Solution C

z_i	y_1	y_2	y_3	y_4
0.0	-5.27	0.19046	0.53315	0.19083
0.5	-7.8048	0.12174	0.39430	0.41230
0.7	-12.301	0.18933	0.44494	0.33492
0.85	-16.270	0.26506	0.49438	0.25412
0.95	-18.452	0.32907	0.53263	0.19529

Solution E

z_i	y_1	y_2	y_3	y_4
0.0	-5.27	0.32297	0.62123	0.091008
0.5	-8.4600	0.13036	0.39961	0.41128
0.7	-13.937	0.21895	0.46576	0.30338
0.85	-19.940	0.34307	0.53293	0.19622
0.95	-30.561	0.58306	0.60806	0.10416

Table 3: Convergence for different nets

		Solution A				Solution B			
		2 s.f.		3 s.f.		3 s.f.		5 s.f.	
m	net	RT	NI	RT	NI	RT	NI	RT	NI
2		3.1	4	2.6	3	1.9	3	1.4	2
6	equi.	3.8	5	2.5	3	1.9	3	1.4	2
21	equi.	3.7	4	3	3	2.4	3	1.7	2

s.f. - significant figures

RT - relative time

NI - number of iterations

equi. - equidistant mesh points

— - not computed

Table 4: Convergence for different nets (solution C)

		3s.f.		5 s.f.		6 s.f.	
m	net	RT	NI	RT	NI	RT	NI
2	simple shooting	D		D		D	
5	equi.	D		D		D	
5	0,0.5, 0.75, 0.9	D		2.2	6	1.3	3
6	equi.	D		D		D	
6	0,0.5, 0.7, 0.85, 0.95	D		1.6	4	1.3	3
11	equi.	D		1.4	3	1.0	2
21	equi.	D		1.7	3	1.3	2

D - divergence

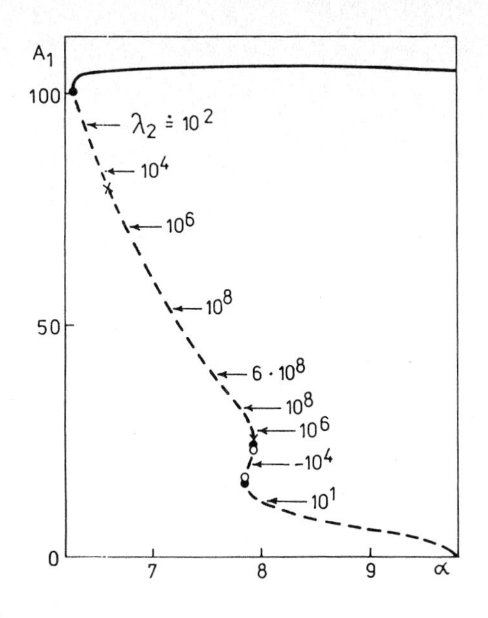

Fig. 2:

Dependence of amplitude of y_1 on parameter α

- • - limit point
- o - period doubling bif. point
- × - end of continuation with simple shooting method

cf. also legende to Fig. 1

and m=5 in the Table 5. Our choice was influenced by an apriori knowledge of the fundamental matrix U(z), (the elements of the matrices G_j are in the absolute value less than 10^5). This choice, as it follows from the Table 5, gives comparable results with the choice where m=11 and m=21. Similar apriori information can be in the case of continuation obtained on the basis of the computed periodic solutions and the chosen net of mesh-points can be varied adaptively in the course of continuation.

5. Conclusion

The multiple shooting method is an effective technique for the continuation of periodic solutions in cases, where some of the multipliers are larger than 10^5. It follows from the detailed study (cf. Tab.3-5) that good results can be obtained for low number of mesh points when these are located properly. An adaptive algorithm with variable number of

Table 5: Convergence for different nets (solution E)

		3 s.f.		5 s.f.		6 s.f.	
m	net	RT	NI	RT	NI	RT	NI
2	simple shooting	D		D		D	
4	equi.	-		-		D	
4	0,0.7, 0.85	-		2.5	4	-	
5	equi.	D		D		2.9	5
5	0,0.5, 0.75, 0.9	D		D		2.4	4
5	0,0.7, 0.8, 0.9	-		2.0	3	-	
6	equi.	D		4.5	8	2.9	5
6	0,0.5, 0.7, 0.85, 0.95	D		2.5	4	1.9	3
8	0,0.2, 0.4, 0.6, 0.7,0.8, 0.9	3.5	6	2.0	3	1.5	2
11	equi.	3.7	6	2.1	3	1.5	2
21	equi.	3.0	4	1.8	2	1.8	2

mesh points and an adaptive non-equidistant step size may be developed on the basis of accumulated experience.

Acknowledgements

Authors appreciate discussion on multiple shooting technique with Prof. Roland Bulirsch. They would like also to thank to Prof. Miloš Marek for very useful discussion and help with the preparation of the manuscript.

References

1. Deuflhard P., Computation of periodic solutions of nonlinear ODEs, BIT 24, 1984, 456-466.

2. Hassard B., Bifurcation of periodic solutions of the Hodgkin-Huxley model for the squid axon, J. theor.Biol. 71, 1978, 401-420.

3. Hodgkin A.L., Huxley A.F., A quantitative description of membrane current and its application to conduction and excitation in nerve, J. Physiol. (Lond.) 117, 1952, 500-544.

4. Holcová H., MSc Thesis, Faculty of Mathematics and Physics, Charles University Prague, 1985.

5. Holodniok M., Kubíček M., DERPER-An algorithm for the continuation of periodic solutions in ODEs, J.of Comp. Physics 55, 1984, 254-267.

6. Holodniok M., Kubíček M., Marek M., Stable and unstable periodic solutions in the Lorenz model, Tech.Univ. München, preprint TUM - M8217, 1982.

7. Kubíček M., Algorithm 502: Dependence of solution of nonlinear systems on a parameter, ACM TOMS, 2, 1976, 98-107.

8. Kubíček M., Marek M., Computational methods in bifurcation theory and dissipative structures, Springer Verlag, New York, 1983.

9. Rinzel J., Miller R.N., Numerical calculation of stable and unstable periodic solutions to the Hodgkin-Huxley model equations, Math. Biosciences, 49, 1980, 27-59.

10. Schreiber I., Holodniok M., Kubíček M., Marek M., Periodic and aperiodic regimes in coupled dissipative chemical oscillators, submitted, 1985.

11. Seydel R., New methods for calculating the stability of periodic solutions, personal communication.

12. Stoer J., Bulirsch R., Introduction to numerical analysis, Springer Verlag, New York, 1980.

Numerical Approximation of Partial Differential Equations
E.L. Ortiz (Editor)

A NEWTON/BI-CONJUGATE GRADIENT CONTINUATION PROCEDURE FOR BUOYANCY FLOWS

S. Sivaloganathan & J.S. Rollett
Oxford University Computing Laboratory
Oxford, UK

A solution procedure based on Newton Raphson linearisation and a preconditioned bi-conjugate gradient algorithm is presented and applied to a finite element discretisation of the buoyancy driven cavity problem. We present solutions for Rayleigh numbers up to 10^8, and also discuss cheaper preconditioning strategies with a view to tackling three dimensional problems.

1. Introduction

The buoyancy driven cavity problem provides a suitable problem for evaluating the performance of numerical methods for viscous fluid flow problems. This model problem has many practical applications ranging from nuclear-reactor insultation, double-glazing in houses to solar energy collection.

We describe a computational procedure for the solution of the non-linear system of equations that arise from a finite element discretisation of the Boussinesq equations over the whole region of flow. The results presented are for cavities of aspect ratio 1 and Prandtl number of .71, however the method can be used for cavities of different aspect ratio as well as for different Prandtl numbers. We omit intentionally any comparison with other work (see Jones et al 1983) since our computations have been performed on very crude meshes and since our objective has been to obtain solutions which reflect the general flow features and which could be used as initial guesses for finer grid solutions, as opposed to obtaining highly accurate solutions.

2. Governing equations

The conservation equations for momentum mass and energy, in the Boussinesq approximations, are given by

$$u\frac{\partial u}{\partial x} + v\frac{\partial u}{\partial y} + \frac{\partial p}{\partial x} - Pr\nabla^2 u = 0 \qquad (2.1)$$

$$u\frac{\partial v}{\partial x} + v\frac{\partial v}{\partial y} + \frac{\partial p}{\partial y} - Pr\nabla^2 v - (Ra)(Pr)T = 0 \qquad (2.2)$$

$$\frac{\partial u}{\partial x} + \frac{\partial v}{\partial y} = 0 \qquad (2.3)$$

$$u\frac{\partial T}{\partial x} + v\frac{\partial T}{\partial y} - \nabla^2 T = 0 \qquad (2.4)$$

where x and y are the horizontal and vertical axes, and u and v the corresponding non-dimensional horizontal and vertical velocities, p denotes the (non-dimensional) pressure and T the non-dimensional temperature (here the non-dimensionalisation of Mallinson and de Vahl David (1977) has been adopted). $Pr(= \nu/\kappa)$ is the Prandtl number and $Ra(= g\alpha D^3(T_h^* - T_c^*)/\kappa\nu)$ is the Rayleigh number, where ν is the kinematic viscosity. κ is the diffusivity, α the coefficient of volumetric expansion and g the gravitational acceleration. D is a characteristic length and T_h^*, T_c^* are characteristic hot and cold temperatures. The boundary conditions for (2.1) - (2.4) are then the no slip condition on the boundary

i.e. $u = v = 0$ on $\partial\Omega$

T = +.5 and -.5 on the left and right vertical walls and the horizontal walls are adiabatic (i.e. $\frac{\partial T}{\partial y} = 0$). The region Ω is subdivided into a number of finite elements. For our purposes, the elements will be eight-noded rectangles. Two sets of basis functions are then defined. one set W_i which take the value 1 at node i and zero at all other nodes of the element, and the second set χ_s (defined only at corner nodes) which interpolates 1 at the Sth corner node and is zero at all other corner nodes. The Galerkin-finite element equations are then constructed in the now familiar manner see (Hutton, 1976) and (Mitchell & Wait, 1977). We then have a discrete system of equations non-linear in the nodal values of velocity and temperature see (Hutton, 1976) and (Rollett & Sivaloganathan, 1985).

3. Solution of Discrete Equations

Thus the differential equations have been reduced to a set of non-linear discrete equations and we solve this by a combination of Newton-Raphson linearisation and a pre-conditioned biconjugate gradient method. Applying Newton-Raphson linearisation, we then have to solve a linear system:

$$J\underline{\delta x} = \underline{f} \qquad (3.1)$$

where $\underline{\delta x}$ is the vector of shifts in velocity, pressure and temperature to be calculated, and J is a sparse, indefinite asymmetric matrix (see Rollett & Sivaloganathan, 1985). Hence we cannot use the conjugate gradient method. We use instead, the bi-conjugate gradient method of Fletcher (Fletcher, 1975). The computational procedure is as follows:

Choose initial guess $\underline{\delta x}_0$.

Calculate initial residual and bi-residual $\underline{r}_0 = \tilde{r}_0 = \underline{f} - J\underline{\delta x}_0$. Set search direction and bi-direction equal to $\underline{r}_0$, $\underline{p}_0 = \tilde{\underline{p}}_0 = \underline{r}_0$. The successive approximations $\underline{\delta x}_k (k \geq 1)$ to the solution of (3.1) are determined recursively by:

$$\underline{\delta x}_{k+1} = \underline{\delta x}_k + \alpha_k \underline{p}_k \tag{3.2}$$

$$\alpha_k = \begin{cases} \underline{\text{either}} & \langle \tilde{\underline{r}}_k, \underline{r}_k \rangle / \langle \tilde{\underline{p}}_k, J\underline{p}_k \rangle \\ \underline{\text{or}} & \langle \tilde{\underline{r}}_k, \underline{p}_k \rangle / \langle \tilde{\underline{p}}_k, J\underline{p}_k \rangle \\ \underline{\text{or}} & \langle \underline{r}_k, \tilde{\underline{p}}_k / \langle \tilde{\underline{p}}_k, J\underline{p}_k \rangle \end{cases} \tag{3.3}$$

The new residual is given by

$$\underline{r}_{k+1} = \begin{cases} \underline{\text{either}} & \underline{f} - J\underline{\delta x}_{k+1} \\ \underline{\text{or}} & \underline{r}_k - \alpha_k J \underline{p}_k \end{cases} \tag{3.4}$$

and the bi-residual is given by

$$\tilde{\underline{r}}_{k+1} = \tilde{\underline{r}}_k - a_k J^T \tilde{\underline{p}}_k \tag{3.5}$$

The new search direction and bi-directions are derived from

$$\underline{p}_{k+1} = \underline{r}_{k+1} + \beta_k \underline{p}_k \tag{3.6}$$

$$\tilde{\underline{p}}_{k+1} = \tilde{\underline{r}}_{k+1} + \beta_k \underline{p}_k \tag{3.7}$$

where

$$\beta_k = \begin{cases} \underline{\text{either}}\ \langle \underline{r}_{k+1}, \underline{r}_{k+1}\rangle / \langle \underline{r}_k, \underline{r}_k\rangle \\ \underline{\text{or}}\quad - \langle \tilde{\underline{r}}_{k+1}, J\underline{p}_k\rangle / \langle \tilde{\underline{p}}_k, J\underline{p}_k\rangle \\ \underline{\text{or}}\quad -\langle \underline{r}_{k+1}, J^T\tilde{\underline{p}}_k\rangle / \langle \tilde{\underline{p}}_k, J\underline{p}_k\rangle \end{cases} \tag{3.8}$$

The process is repeated until some convergence criteria on the residual is satisfied.

The solution procedure proposed is to solve (3.1) on the first Newton-Raphson linearisation by the frontal procedure to obtain the LU factors of $J^{(1)}$ and then on subsequent linearisations to solve the system by the bi-C.G. method using these factors as pre-conditioners so that at Newton-Raphson linearisation $r(r\geq 2)$ we solve:

$$B^{(r)}\ \underline{\delta x} = d^{(r)} \tag{3.9}$$

where $B^{(r)} = u^{-1}L^{-1}J^{(r)}$, $\underline{d}^{(r)} = U^{-1}L^{-1}\underline{f}^{(r)}$ and $J^{(r)}\underline{\delta x} = \underline{f}^{(r)}$ is the linear system produced on the r-th Newton-Raphson iteration. The pre-conditioned system (3.9) is then solved by the bi-CG method. At a Rayleigh number of about 10^5 (or higher), the procedure fails to converge since Newton-Raphson fails to converge (starting from a creeping flow approximation). However, when coupled with a continuation strategy, the solution procedure proves to be efficient.

4. Continuation

The continuation procedure we have developed to solve at high Rayleigh numbers is summarised in Fig 1. A more detailed description is given in (Rollett & Sivaloganathan, 1985). It is based on the following simple observation of the Newton method on non-linear problems in general:

(a) When the starting iterate is far from the true

solution, there are commonly two or three cycles in which the shifts are of similar size. The error at the end of a cycle is hardly smaller than at the start.

(b) There is then a cycle in which the shifts are smaller than those that went before, by a factor of between say, 0.5 and 0.1.

(c) Finally quadratic convergence sets in, with shifts which are more than one order of magnitude smaller than preceding shifts.

At stage (a), we are close to the size of error at which the process may not converge. We must avoid this. At stage (c) we are changing the parameters so little that we are not making progress efficiently. We wish to operate, so far as possible, at or near stage (b). We can use the ratio of the second Newton shift to first Newton shift as a good indicator of whether we are in the region of "safe convergence".

5. CPU Time Estimates

Under certain assumptions, we make some time estimates that indicate that for reasonable three dimensional problems even one LU decomposition becomes infeasible. We assume

1. Variables are velocity components u,v,w, pressures p and temperatures T.
2. Each finite element is a rectangular box with eight corner nodes, each shared between eight elements, 12 mid-edge nodes, each shared between four elements - so an average of 1 corner node and 3 mid-side nodes per element.
3. u, v, w, T are calculated at all nodes, p is

calculated only at corner nodes - so, on average, each element has 5+4+4+4 = 17 unknowns.

4. For one full LU decomposition, the operation count is (no. of unknowns) x $(\text{bandwidth})^2$.
5. For an n x n x n element mesh, no. of unknowns = $17n^3$ and the bandwidth = $17n^2$. Hence for one full LU operation, the count is $17^3 \times n^7 = 5000 \times n^7$.
6. For one step of ILU-preconditioned bi-CG iteration the operation count is 4 x (no. of unknowns) x (no. of non zeros/row), i.e. the work is dominated by 4 matrix by vector products. The extra work for vector operations might multiply this by a factor of 2.
7. The number of non-zeros per row is ~ 50 for a cuboid mesh.
8. For 1 ILU/bi-CG step, the operation count is ~ $3400n^3$ ($= 50 \times 4 \times 17n^3$). That ILU/bi-CG can be made to converge in 100 steps (most users of CG, with a good pre-conditioner, report satisfactory convergence after a small number of steps which is almost independent of n, see (Parlett, B.N. & B. Nour Imid, 1982).
9. That full LU-preconditioning would converge so fast (say O(10) steps) that one full LU decomposition would precondition 2 linearisations and would dominate the work for them. That 10 linearisations would solve a reasonably difficult problem, such as the buoyancy driven cavity problem at $Ra = 10^6$.
10. That the number of millions of floating point operations/second on the five machines are

(IBM)3081	FPS	DAP	CRAY 1	CYBER 205
0.2	10	20	100	200

Time estimates for use of LU decomposition to pre-condition bi-CG

We consider a mesh of n x n x n elements, the full LU operation count is $5000n^7$ so if n = 15, $5000n^7 \doteq 8 \times 10^{11}$; if n = 30, $5000n^7 \doteq 1 \times 10^{14}$. If 5 such operations dominate the time for a problem, then

Time for 1 problem

n \ MACHINE	3081	FPS	DAP	CRAY 1	CYBER 205
15	200 days	110 hrs	55 hrs	11 hrs	5.5 hrs
30	25000 days	500 days	250 days	50 days	25 days

Table 5.1

Time estimates for ILU-preconditioned bi-CG

We consider, again, a mesh of n x n x n elements. The operation count for 1 iteration step is $3400n^3$. If n = 15, $3400n^3 \doteq 1 \times 10^7$. If n = 30, $3400n^3 \doteq 9 \times 10^7$. we assume 10 linearisations are required to solve the problem and 100 bi-CG steps to solve each linearisation, then

Time for 1 problem

n \ MACHINE	3081	FPS	DAP	CRAY 1	CYBER 205
15	14 hrs	1000 sec	500 sec	100 sec	50 sec
30	125 hrs	2.5 hrs	1.25 hrs	900 sec	450 sec

Table 5.2

Thus from Table 5.1, it is quite clear that using a full LU decomposition, the problem for n = 30 is more or less out of reach, but n = 15 would be just about within reach on a "dedicated FPS" and fairly accessible on a CYBER 205. However, using incomplete LU, the problem for n = 30 is well within reach on a "dedicated FPS" and n = 15 is feasible on a 3081.

6. ILU Preconditioning

The heuristic used to ensure that the pre-conditioning is inexpensive to implement, is to restrict the factors to have the same sparsity structure as the original Jacobian. We have implemented the incomplete LU factorisation popularised by Meijerink and Van der Vorst (Meijerink & van der Vorst, 1977) and the modified incomplete LU factorisation of Gustafsson (Gustafsson, 1978) as preconditioners, but have met with very little success (i.e. bi-CG has not converged even after a number of iterations well past the order of the matrix). However, a shifted incomplete LU factorisation has proved to be much more effective as a preconditioner. We chose the shift factor to be mid-way between the smallest and largest eigenvalue. For the problem under consideration at a Rayleigh no. of 10^3, the smallest eigenvalue is located near 0 and the largest is O(200). Thus we perform an incomplete LU factorisation on J + 100I (where I is identity matrix) and use these factors to precondition the original system. Using a pressure correction equation, we require six Newton Raphson linearisations and on average, about 90 bi-CG cycles per linearisation to obtain the solution to the buoyancy driven cavity problem on a 5 x 5 uniform mesh (contrasted with a total of 3LU decompositions to solve the problem using a frontal procedure). However, if the temperature field is fixed at the correct temperature solution for this Rayleigh number, then we require only 3 Newton-Raphson linearisations and on average 35 bi-CG cycles/linearisation to solve the problem. The indications are that it is the coupling between the energy equation (2.4) and the momentum equation (2.2) that causes problems in the numerical solution. Thus, for non-buoyancy flows, the shifted ILU provides a good pre-conditioner for the bi-CG method and coupled with the continuation procedure of Section 4 would provide an efficient method of solution at high Reynolds numbers. Further investigation is necessary to obtain an efficient sparse pre-conditioner in the case of buoyancy flows.

7. Results

The solutions presented in this section were obtained using the continuation strategy with the LU preconditioned bi-CG method. Figure 2 is the velocity distribution at a Rayleigh no. of 10^5, the crude 5 x 5 uniform mesh still manages to pick out the secondary recirculating flows; at a Rayleigh of 10^6 (Fig 3) we require a uniform 9 x 9 mesh to pick out the flow features, and by a Rayleigh no. of 10^7 (Fig 4), a uniform 9 x 9 mesh does not resolve the boundary layers well enough to avoid the spurious recirculations near the vertical walls. Fig 5 is the solution obtained at $Ra = 10^8$. A more detailed and critical discussion of the results is given in (Rollett & Sivaloganathan, 1985).

One of us (S.S.) wishes to thank the SERC for a research studentship and the CEGB for support of this studentship and for use of computing facilities.

REFERENCES

1. Fletcher, R., Conjugate gradient methods for indefinite systems, Proc. Dundee Conf. on Numerical Analysis, ed Watson, G.A. (1975).

2. Gustafsson, I., A class of first order factorisations, BIT 18, 142-156 (1978)

3. Hutton, A.G., A survey of the theory and application of the finite element method in the analysis of viscous incompressible Newtonian flow, CEGB Report No. RD/B/N3049 (1976).

4. Jones, I.P. & de Vahl Davis, G., Natural Convection in a square cavity: a comparison exercise, Int J. for Numerical Methods in Fluids, Vol 3, 227-248 (1983).

5. Mallinson, G.D. & de Vahl Davis, G., Three-dimensional natural convection in a box: a numerical study, J. of Fluid Mechanics, 83, 1-31 (1977).

6. Meijerink, J.A. & van der Vorst, H.A., An iterative method for linear systems of which the coefficient matrix is a symmetric M-matrix, Math. of Comp. 31, 148-162 (1977).

7. Mitchell, A.R. & Wait, R., The finite element method in partial differential equations, Wiley (1977).

8. Nour-Omid, B. & Parlett, B.N., Element Preconditioning, PAM-103, Centre for Pure & Applied Maths, Berkeley (1982).

9. Rollett, J.S. & Sivaloganathan, S., High Rayleigh number solutions of the buoyancy driven cavity problem, Proc. Int. Conf. on Comp. Fluid Dynamics, eds Baines, M.J. & Morton, K.W., (1985).

10 Wong, Y.S., Iterative methods for problems in numerical analysis ,DPhil Thesis,Oxford University (1978).

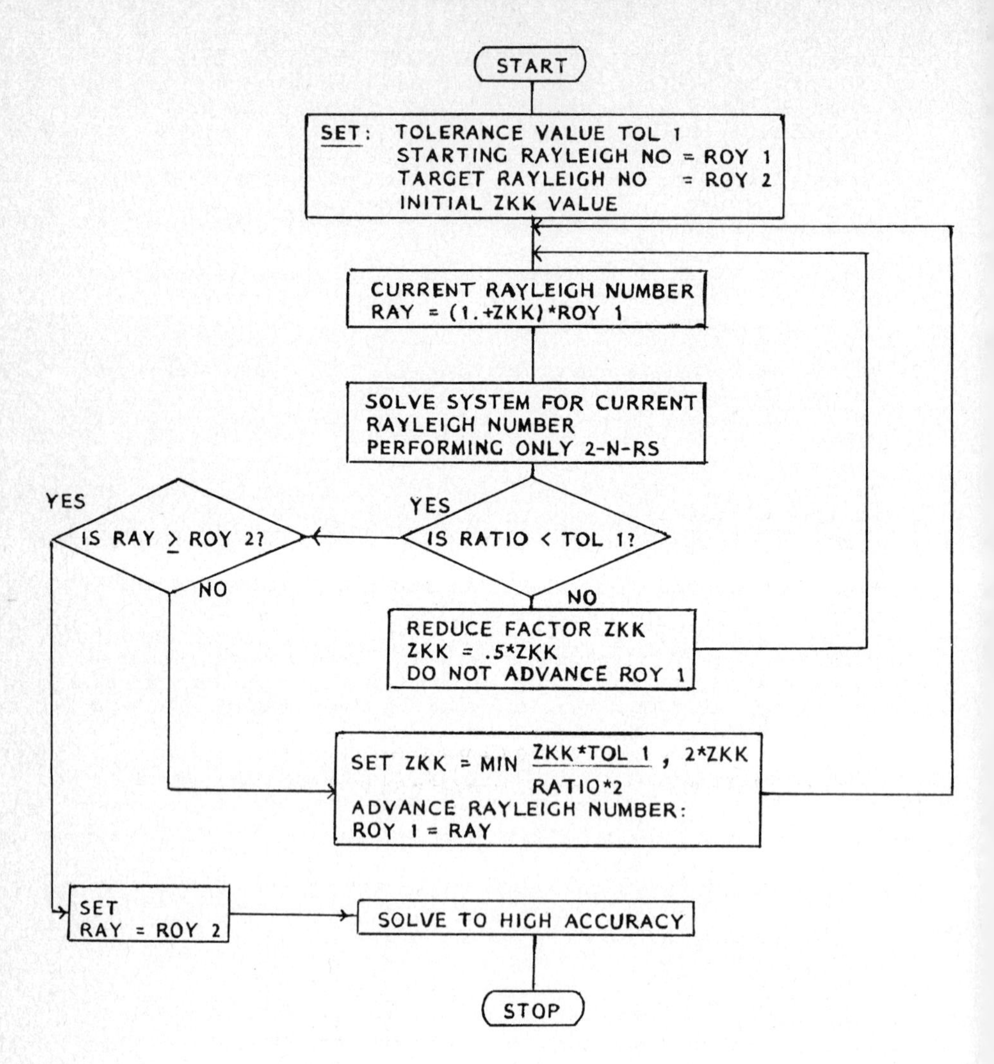

FLOW DIAGRAM OF CONTINUATION STRATEGY

FIGURE 1

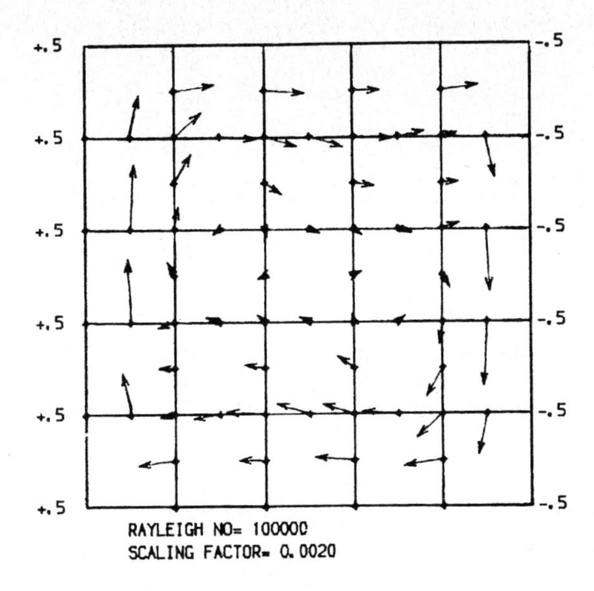

FIGURE 2

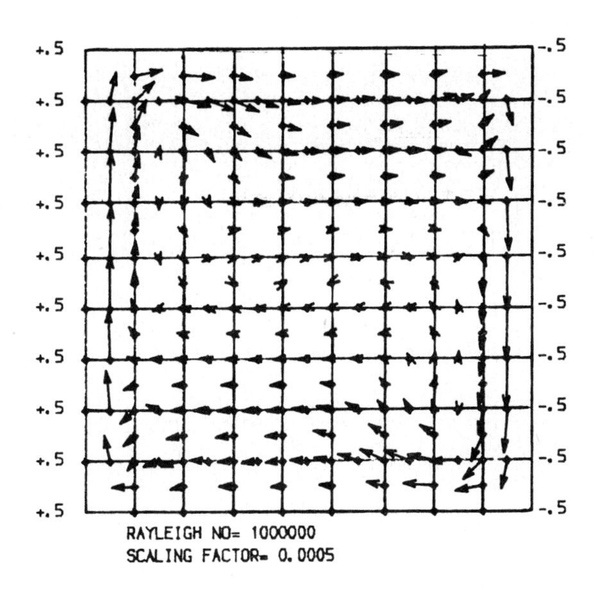

FIGURE 3
VELOCITY FIELDS AT VARIOUS RAYLEIGH NUMBERS

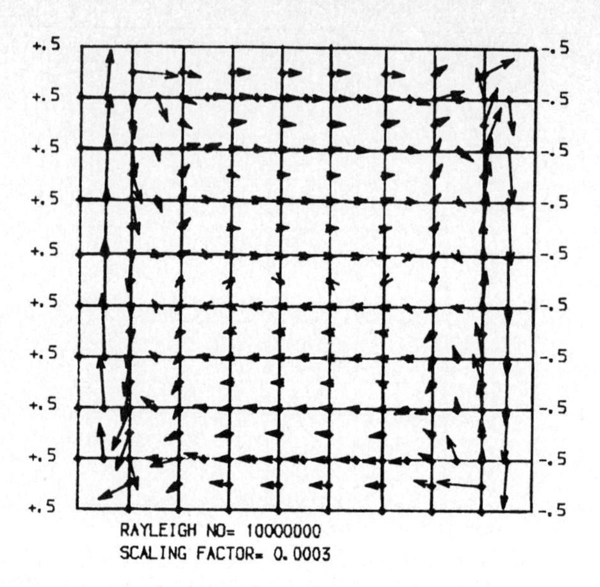

FIGURE 4

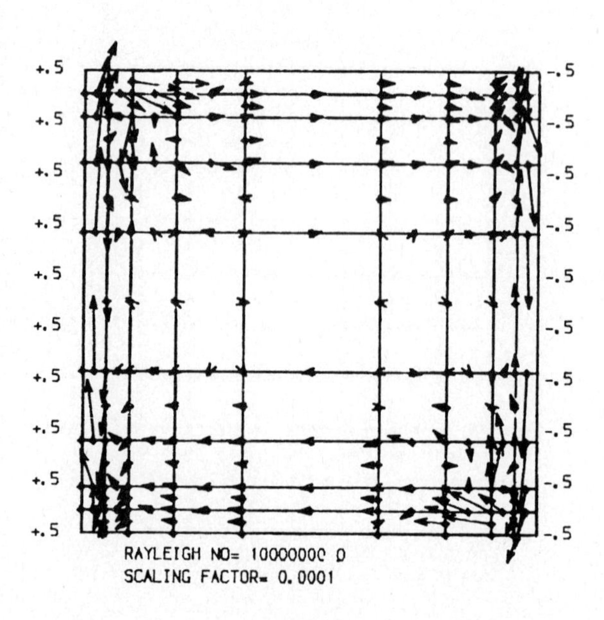

FIGURE 5
VELOCITY FIELDS AT VARIOUS RAYLEIGH NUMBERS

Numerical Approximation of Partial Differential Equations
E.L. Ortiz (Editor)

IST NUMERICAL SCHEMES FOR NONLINEAR EVOLUTION EQUATIONS OF PHYSICAL INTEREST

Thiab R. Taha
Mark J. Ablowitz*
The University of Georgia
Athens, GA U.S.A.

*Clarkson University
Potsdam, NY U.S.A.

This paper deals with numerical solutions of nonlinear evolution equations solvable by the inverse scattering transform (IST), namely the nonlinear Schrödinger (NLS), the Korteweg-de Vries (KdV) and the modified Korteweg-de Vries (MKdV) equations. These equations describe a wide class of physical phenomena (Ablowitz and Segur, 1981). Comparisons between schemes constructed by methods related to the IST and certain other known numerical methods (a) finite difference and (b) finite Fourier (pseudo-spectral) methods are obtained. Experiments have shown that the IST schemes compare very favorably with known numerical methods. In this paper a summary of the results of the performance of the IST schemes for the NLS and KdV equations will be discussed, and a new results of the performance of the IST scheme for the MKdV equation will be presented.

1. Introduction

In (1977) Ablowitz and Ladik found partial difference equations, based on IST, which can be used as a numerical scheme for the NLS equation. This numerical scheme maintains many of the important properties of the original equation and has many advantageous properties (Ablowitz and Ladik, 1975, 1977). This work motivated us (Taha and Ablowitz, 1984) to implement this scheme and compare it to other known numerical schemes such as: (i) The classical explicit method, (ii) Hopscotch method, (iii) Implicit-Explicit method, (iv) Crank-Nicolson implicit scheme, (v) The split step Fourier method by Tappert, and (vi) Pseudospectral (Fourier) method by Fornberg and Whitham for the NLS equation

$$iU_t = U_{xx} + 2|U|^2U \tag{1.1}$$

Our approach for comparison is to (a) fix the accuracy (L_∞) for computations beginning at t = 0 and ending at t = T; and (b) leave other parameters free (e.g., Δt, Δx) and compare the computing time required to attain such accuracy for various choices of parameters. Similarly we (Taha and Ablowitz, 1984) have found a numerical scheme, based on IST, for the KdV equation

$$U_t + 6U\,U_x + U_{xxx} = 0 \qquad (1.2)$$

This scheme is compared to other known schemes such as: (i) Zabusky and Kruskal scheme, (ii) Hopscotch method, (iii) Goda Scheme, (iv) A scheme suggested by M. Kruskal, (v) Split step Fourier method by Tappert, and (vi) Pseudospectral method by Fornberg and Whitham. Our approach for comparison was the same as in the case of the NLS equation. In the above equations one and two soliton solutions are used as initial conditions. Periodic boundary conditions are employed. Moreover, the numerical solution is compared to the exact solution. In addition, two of the conserved quantities are computed, namely, the second and the third.

Our numerical experiments indicate (for the range of amplitudes we considered) that the Ablowitz-Ladik global scheme is faster than all the finite difference methods, but somewhat slower than the finite Fourier (Pseudospectral) methods. We note that the local version did not perform as well as the global one in the case of high amplitudes. On the other hand the local version of the IST scheme for the KdV equation proved to be faster than both the finite difference and the finite Fourier methods we considered. We note that the global version of the IST scheme for the KdV equation did not perform as well as its local one. Recently we have implemented the IST scheme (Taha and Ablowitz, 1984) for the MKdV equation

$$U_t + 6U^2U_x + U_{xxx} = 0 \qquad (1.3)$$

and we have compared it to other known numerical methods such as: (i) An implicit scheme, (ii) Split step Fourier method, and the (iii) Pseudospectral method by Fornberg and Whitham. Our approach for comparison is the same as in the case of the NLS and KdV equations. One soliton solution is used as initial condition. Periodic boundary conditions are employed.

2. The Representation of the MKdV Equation Using Numerical Methods

(i) The proposed global scheme which is based on the IST is (Taha and Ablowitz, 1984)

$$\begin{aligned}\Delta^m R_n^m = {} & R_{n+2}^m \underline{A}^{(4)} - R_{n+2}^{m+1}\gamma_{n+1}\underline{D}^{(4)} + R_{n+1}^m S_{n+1} \\ & - R_{n+1}^{m+1}P_n - [R_{n-2}^{m+1}\underline{A}^{(4)} - R_{n-2}^m\gamma_{n-2}\underline{D}^{(4)} + R_{n-1}^{m+1}S_{n-2} \\ & - R_{n-1}^m P_{n-1}] + R_n^m\Big\{\underline{D}^{(0)} - \sum_{l=-\infty}^{n}[R_l^{m+1}\{R_{l-2}^{m+1}\underline{A}^{(4)} \\ & - R_{l-2}^m\gamma_{l-2}\underline{D}^{(4)} + R_{l-1}^{m+1}S_{l-2} - R_{l-1}^m P_{l-1}\} \\ & - R_l^m\{R_{l+2}^m\underline{A}^{(4)} - R_{l+2}^{m+1}\gamma_{l+1}\underline{D}^{(4)} + R_{l+1}^m S_{l+1} - R_{l+1}^{m+1}P_l\}]\Big\} \\ & - R_n^{m+1}\Big\{\underline{A}^{(0)} + \sum_{l=-\infty}^{n-1}[R_l^{m+1}\{R_{l-2}^{m+1}\underline{A}^{(4)} - R_{l-2}^m\gamma_{l-2}\underline{D}^{(4)} \\ & + R_{l-1}^{m+1}S_{l-2} - R_{l-1}^m P_{l-1}\} - R_l^m\{R_{l+2}^m\underline{A}^{(4)} - R_{l+2}^{m+1}\gamma_{l+1}\underline{D}^{(4)} \\ & + R_{l+1}^m S_{l+1} - R_{l+1}^{m+1}P_l\}]\Big\},\end{aligned} \tag{2.1}$$

where

$$S_n = \underline{A}^{(2)} + \underline{A}^{(4)}F_n + \underline{D}^{(4)}\sum_{j=-\infty}^{n} H_j, P_n = (\underline{D}^{(2)} + \sum_{j=-\infty}^{n}[\underline{A}^{(4)}E_j + \underline{D}^{(4)}G_j]\eta_j)\gamma_n,$$

$$\gamma_n = \prod_{=-\infty}^{n}(\delta_i^{m+1}/\delta_i^m),\ \delta_i^m = 1 + R_i^m,$$

$$\eta_n = \gamma_n^{-1}/\delta_n^m, H_n = -\{R_n^m R_{n+1}^{m+1}\delta_n^{m+1} - R_{n-1}^m R_n^{m+1}\delta_n^m\}\beta_{n-1},$$

$$\beta_n = \gamma_n/\delta^m_{n+1}, F_n = [R^{m+1}_{n+1}R^{m+1}_n - \sum_{j=-\infty}^{n} \Delta^m(R^m_j R^m_{j+1})],$$

$$G_n = (R^{m+1}_n R^{m+1}_{n+1} - R^m_n R^m_{n-1})\gamma_{n-1}\delta^{m+1}_n,$$

$$E_n = (R^m_n R^{m+1}_{n-1}\delta^{m+1}_n - R^m_{n+1}R^{m+1}_n \delta^m_n),$$

$$\underline{A}^{(2)} = -\frac{2}{3}\underline{A}^{(0)} + \frac{1}{2}\alpha, \quad \underline{D}^{(2)} = -\frac{2}{3}\underline{A}^{(0)} - \frac{1}{2}\alpha,$$

$$\underline{A}^{(4)} = \frac{1}{6}\underline{A}^{(0)} - \frac{1}{4}\alpha, \quad \underline{D}^{(4)} = \frac{1}{6}\underline{A}^{(0)} + \frac{1}{4}\alpha,$$

$$\alpha = \frac{\Delta t}{(\Delta x)^3}, \quad \underline{A}^{(0)} = \text{arbitrary constant},$$

$R = \Delta x U$, and $|n| < p$ (half the length of the interval of interest, and $m > 0$). This scheme is implemented with the value of $\underline{A}^{(0)} = \frac{3}{2}\alpha$.

(ii) The proposed local scheme which is derived from equation (2.1) with $\underline{A}^{(0)} = \frac{3}{2}\alpha$ is

$$\begin{aligned}
\frac{U^{m+1}_n - U^m_n}{\Delta t} = {} & \frac{U^{m+1}_{n-1} - 3U^{m+1}_n + 3U^{m+1}_{n+1} - U^{m+1}_{n+2}}{2(\Delta x)^3} + \frac{U^m_{n-2} - 3U^m_{n-1} + 3U^m_n - U^m_{n+1}}{2(\Delta x)^3} \\
& - \frac{1}{2\Delta x}\Big[U^{m+1}_{n+2}\{(U^m_{n+1})^2 + (U^m_n)^2\} - U^m_{n-2}\{(U^{m+1}_{n-1})^2 + (U^{m+1}_n)^2\} \\
& + \frac{U^{m+1}_{n+1}}{2}[U^m_n U^m_{n+1} + U^{m+1}_n U^{m+1}_{n+1} \\
& + 2U^m_{n-1}U^m_n] - \frac{U^m_{n-1}}{2}[U^m_n U^m_{n-1} + U^{m+1}_n U^{m+1}_{n-1} + 2U^{m+1}_n U^{m+1}_{n+1}] \\
& + \frac{U^m_n}{2}(U^{m+1}_n U^{m+1}_{n+1} + U^m_n U^m_{n+1}) - \frac{U^{m+1}_n}{2}(U^{m+1}_{n-1}U^{m+1}_n + U^m_{n-1}U^m_n) \\
& - 3[(U^m_n)^2 U^{m+1}_{n+1} - (U^{m+1}_n)^2 U^m_{n-1}]\Big]
\end{aligned} \tag{2.2}$$

(iii) An implicit scheme (Kruskal, 1981):

$$\frac{U_n^{m+1} - U_n^m}{\Delta t} = \frac{U_{n-1}^{m+1} - 3U_n^{m+1} + 3U_{n+1}^{m+1} - U_{n+2}^{m+1}}{2(\Delta x)^3} + \frac{U_{n-2}^m - 3U_{n-1}^m + 3U_n^m - U_{n+1}^m}{2(\Delta x)^3}$$

$$- \frac{1}{2(\Delta x)^3}\Big\{\theta[(U^3)_{n+1}^{m+1} - (U^3)_{n-1}^{m+1} + (U^3)_{n+1}^m - (U^3)_{n-1}^m]$$

$$+ 3(1-\theta)[(U^2)_n^{m+1}(U_{n+1}^{m+1} - U_{n-1}^{m+1}) + (U^2)_n^m (U_{n+1}^m - U_{n-1}^m)]\Big\} \quad (2.3)$$

Numerical experiments indicate that $\theta = \frac{2}{3}$ gives the best results.

(iv) Split step Fourier Method (Tappert, 1974):

For convenience the spatial period is normalized to $[0,2\pi]$, then Eg. (1.3) becomes

$$U_t + 6\frac{\pi}{p} U^2 U_X + \frac{\pi^3}{p^3} U_{XXX} = 0 \quad (2.4)$$

where p is half the length of the interval of interest, and $X = (x+p)\frac{\pi}{p}$. In order to apply the split step Fourier method for Eg. (2.4) we (a) advance the solution using only the nonlinear part

$$U_t + 6\frac{\pi}{p} U^2 U_X = 0 \quad (2.5)$$

This can be approximated by using an implicit method such as

$$\tilde{U}_n^{m+1} = U_n^m - \frac{\Delta t}{12\Delta x}\frac{\pi}{p}\Big\{[8(\tilde{U}^3)_{n+1}^{m+1} - 8(\tilde{U}^3)_{n-1}^{m+1} - (\tilde{U}^3)_{n+2}^{m+1} + (\tilde{U}^3)_{n-2}^{m+1}]$$

$$+ [8(U^3)_{n+1}^m - 8(U^3)_{n-1}^m - (U^3)_{n+2}^m + (U^3)_{n-2}^m]\Big\} \quad (2.6)$$

where $\tilde{U}$ is a solution of Eg. (2.5); (b) advance the solution according to

$$U_t + \frac{\pi^3}{p^3} U_{XXX} = 0 \quad (2.7)$$

by means of the discrete Fourier transform

$$U(X_j, t + \Delta t) = F^{-1}(e^{ik^3 \frac{\pi^3}{p^3} \Delta t} F(\tilde{U}(X_j,t))) \tag{2.8}$$

(v) Pseudospectral method (Fornberg and Whitham, 1978):

The pseudospectral method for Eg. (2.4) is

$$U(X, t + \Delta t) - U(X, t - \Delta t) + 12i\frac{\pi}{p} U^2(X,t)F^{-1}(k(F(u)))$$
$$- 2iF^{-1}\{\mathrm{Sin}(\frac{\pi^3 k^3}{p^3} \Delta t) F(u)\} = 0 \tag{2.9}$$

3. Conclusions

Our numerical results indicate that the IST schemes compare very favorably with the other numerical schemes. A summary of the results for the NLS and KdV equations are given in the text of this paper, (for more details see Taha and Ablowitz, 1984). For the MKdV equation, the proposed global schemes, based on IST, proved to be faster than all the methods we considered, followed by its local version (see Figures 1 and 2). It is worth noting that the proposed global scheme behaves much better than the other utilized schemes either when better accuracy is required or for large amplitudes. However we remark that the pseudospectral method becomes competitive with the IST global scheme when both high accuracy and large amplitudes are involved. The results support our belief that the IST schemes will perform better than the "standard" numerical schemes for nonlinear evolution equations solvable by IST.

It is worth mentioning that these IST schemes can also be used in combination of other numerical schemes to study a wider class of nonlinear evolution (physically important) equations. For example they can be used to study the perturbed forms of the KdV, MKdV, and NLS equations (Kodama and Ablowitz, 1981).

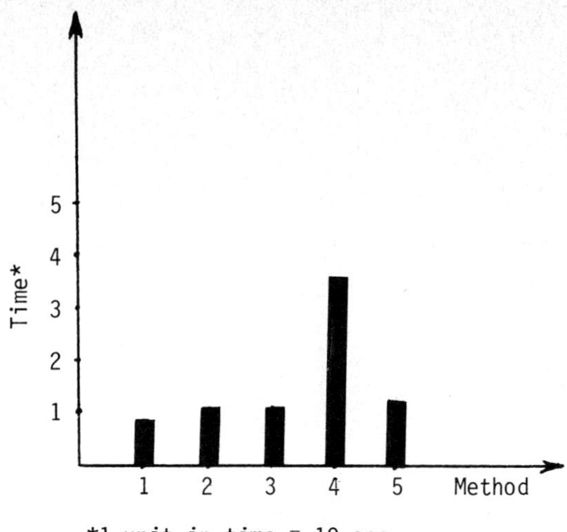

Fig. 1. Displays the computing time which is required to attain an accuracy ($L\infty$) < 0.005 for computations beginning at t = 0 and ending at t = 1.0, for the competitive methods:

1. The proposed global scheme
2. The proposed local scheme
3. An implicit scheme
4. Tappert scheme
5. Fornberg and Whitham scheme

utilized in solving the MKdV equation. 1-soliton as an initial condition with amplitude = 1 on the interval [-20,20].

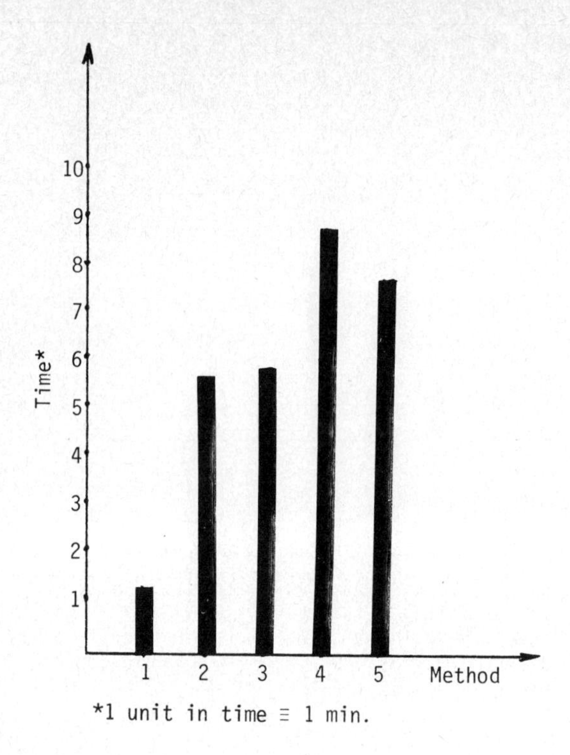

Fig. 2. Displays the computing time which is required to attain an accuracy ($L\infty$) < 0.02 for computations beginning at t = 0 and ending at t = 1.0, for competitive methods:

1. The proposed global scheme
2. The proposed local scheme
3. An implicit scheme
4. Tappert scheme
5. Fornberg and Whitham scheme

utilized in solving the MKdV equation. 1-soliton as an initial condition with amplitude = 2 on the interval [-20,20].

Acknowledgements

This research was partially supported by the Research Foundation of the University of Georgia (by way of a Faculty Research Grant and a Michael Award), the Office of Naval Research under grant N00014-76-C-0867, the Air Force Office of Scientific Research under grant 78-3674-D and the National Science Foundation under grant MCS-82-02117.

REFERENCES

1. Ablowitz M. and Segur H., Solitons and the Inverse Scattering Transform, SIAM, Philadelphia, 1981.
2. Ablowitz M. and Ladik J., Nonlinear differential-difference equations, Jour. Math. Phys. 16, 1975, 598-603.
3. Ablowitz M. and Ladik J., On the solution of a class of nonlinear partial difference equations, Stud. Appl. Math. 57, 1977, 1-12.
4. Fornberg B. and Whitham G.B., A numerical and theoretical study of certain nonlinear wave phenomena, Philos. Trans. Roy. Soc. London, Ser. A, 289, 1978, 373-404.
5. Kodama Y. and Ablowitz, M.J., Perturbations of solitons and solitary waves, Stud. Appl. Math. 64, 1981, 225-245.
6. Kruskal M., Private communications.
7. Taha T. and Ablowitz M., Analytical and numerical aspects of certain nonlinear evolution equations. I. Analytical, Jour. Comp. Phys. 55, 1984, 192-202.
8. Taha T. and Ablowitz M., Analytical and numerical aspects of certain nonlinear evolution equations. II. Numerical, Nonlinear Schrödinger equation, Jour. Comp. Phys. 55, 1984, 203-230.
9. Taha T. and Ablowitz M., Analytical and numerical aspects of certain nonlinear evolution equations. III. Numerical, Korteweg-de Vries equations, Jour. Comp. Phys. 55, 1984, 231-253.
10. Tappert F., Numerical solutions of the KdV equation and its generalization by the split-step Fourier method, in Nonlinear Wave Motion (ed. A.C. Newell), Lectures in Appl. Math. 15, AMS, Providence, RI, 1974, 215-216.